W9-AJM-189

TECHNOLOGY IN WAR

The Impact of Science on Weapon Development and Modern Battle

TECHNOLOGY IN WAR

KENNETH MACKSEY

An Arco Military Book
published by
PRENTICE HALL PRESS
New York

Technology in War *was produced by Grub Street, 4 Kingly Street, London W1R 5LF*

The idea for the book was conceived by Martin Bronkhorst

An Arco Military Book
Published by Prentice Hall Press
A Division of Simon & Schuster, Inc.
Simon & Schuster Building
Rockefeller Center
1230 Avenue of the Americas
New York, New York 10020

Published in Great Britain by Arms and Armour Press Limited.
This edition published by arrangement with Grub Street

PRENTICE HALL PRESS is a trademark of Simon & Schuster, Inc.

Designed by Grub Street, London
Manufactured in Italy
Printed and bound by Sagdos

Library of Congress Cataloging-in-Publication Data

Macksey, Kenneth.
Technology in war.

1. Weapons systems – History. 2. Military history, Modern. 3. Military art and science. 4. Technology.
I. Title.
UF530.M23 1986 355.8'2'09 85-24387
ISBN 0-671-61954-3

Contents

$= \frac{S \, . \, 10334 \, . \, p \, . \, n \, . \, h}{60}$ Kilogrammmeter oder gleich $\frac{\ldots}{60 \, . \, 75}$ Pferdekräfte.

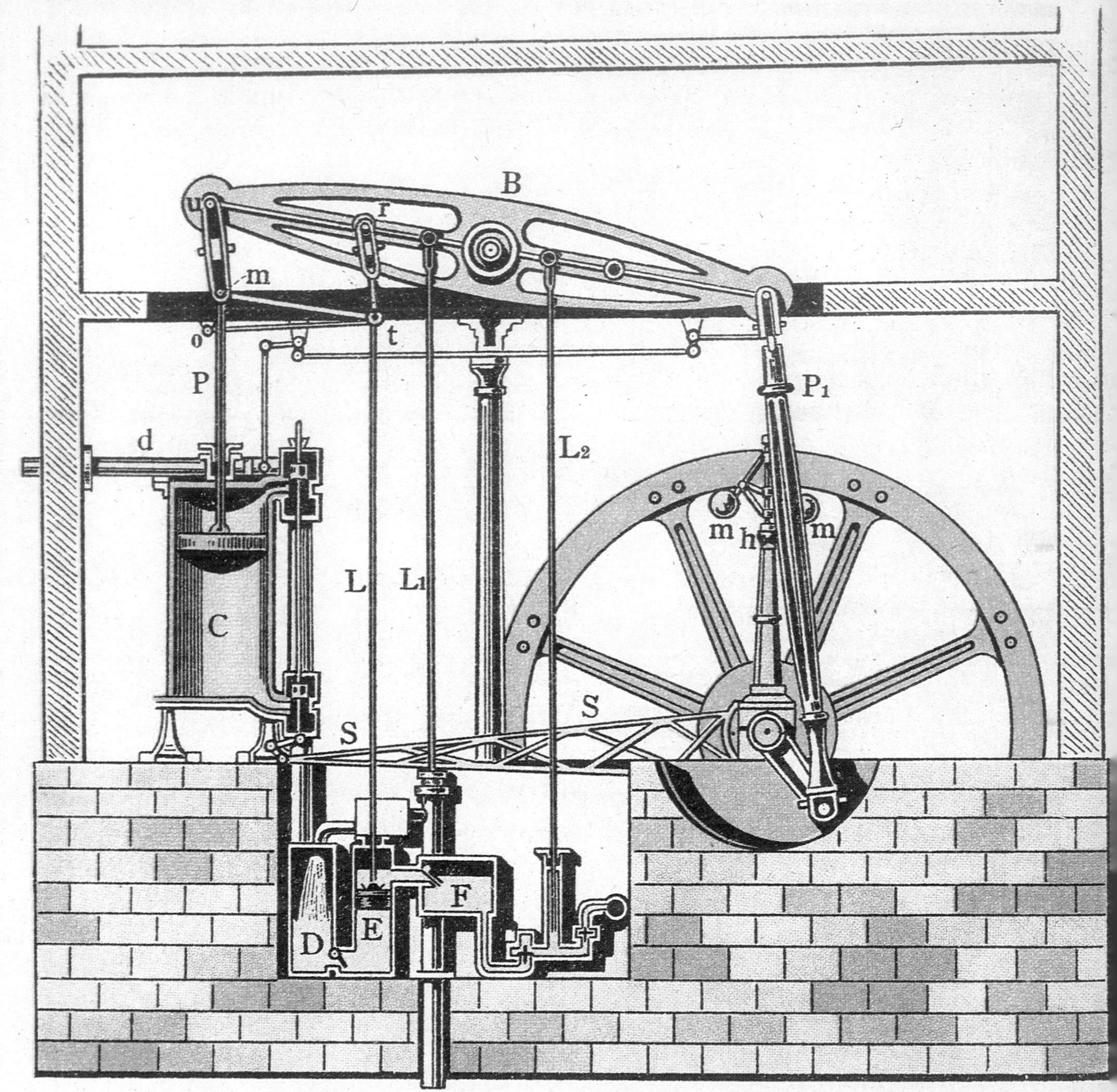

Fig. 46. Wattsche Dampfmaschine.

C Dampfcylinder. d Dampfzuflußrohr. B Balancier. R Schwungrad. D Kondensator. LE Luftpumpe. L_1F Kaltwasserluftpumpe des Kondensators. L_2 Warmwasserluftpumpe des Dampfkessels. S Exzenterstange der Steuerung. P_1 Pleuelstange mit Kurbel.

Beispiel. $p = 2{,}25$ Atmosphären, $S = 500 \text{ cm}^2 = \frac{1}{20} \text{ m}^2$, $h = 0{,}4$ m, $n = 120$; der Effekt $\mathfrak{E} = \frac{1}{20} \, . \, 10\,334 \, . \, 2{,}25 \, . \, \frac{120}{60} \, . \, 0{,}4 = 930{,}06$ kgm $= 12{,}4$

Chapter 1

After James Watt

1763-1849

Historians compiling the story of a Third World War at the end of the 20th century, were they lucky enough to have survived, would be compelled to recognize the connecting threads of technology which were identifiable in the 18th century, formally established by the mid-19th century and dominant by the 1950s — at the starting point of such a Third World War. The first Industrial Revolution struck its roots in war and continued to be motivated by recurrent themes devised by inventors, promoted by entrepreneurs and governments seeking riches, fulfilment, aggrandizement, security and power. The advancement of knowledge moved from one development to another; sudden technical breakthroughs, so beloved of journalists when announcing new discoveries, were rare in the extreme. As a result, unbroken links of causation can be seen to connect 18th-century flag signals, 19th-century telephones and 20th-century radio; guns and rockets; balloons, zeppelins, aeroplanes and spacecraft; wooden, steel, aluminium and plastic ships; black powder, gas and atomic bombs. Such a succession of inventions and improvements James Watt, at the peak of his achievements, cannot have foreseen.

***Left:** The transmission of information; Watt's steam engine as depicted in a German textbook.*

Freedom For Technology

Until the year 1763, the development of the art and technology of war had been orthodox and gradual, the child of minds imprisoned by a tiny élite of educated men who held tight to the reins of power and tended to repress anything which appeared to challenge their power and their monopoly of narrow innovation. Not even the 14th-century use of gunpowder as a propellant for missiles from cannon had produced a rapid and profound change. Indeed, in terms of accuracy and rate of fire, the long-bow that dominated battlefields in the 14th and 15th centuries remained superior to the 17th-century musket still in front-line service at the Battle of Waterloo in 1815. Technological evolution, dependent upon freedom of thought to lead to the implementation of ideas, lay in the hands of individuals, and often the self-interested kind.

It is impossible to credit one person or to specify

any one moment as initiating a new trend, although permissible to suggest that commercial and humanitarian motives lie at the root of much improvement. For example, it can be argued that the movement against the slave trade that began in Britain in 1673, when Robert Baxter claimed it was inefficient (without condemning it), initiated the strong current of liberal thought expressed by philosophers and philanthropists in the century to come, leading to the demand for political freedom which found major expression in 1775 when the American War of Independence began. Be that as it may, it is no coincidence that the invention in 1763 by James Watt of the first economically practical steam engine sparked off a truly gigantic explosion of inventiveness in all fields which continues apace to the present day.

Greater ease of written communication by means of more efficient and competitive publishing methods naturally hastened the process of change. By the mid-18th century far more people could read, and so the incentive to write and print burgeoned. The economic improvements that James Watt made to existing steam engines in 1763 by adding, among other items, a condenser, were promoted by his knowledge of a machine invented by John Wilkinson which was capable of accurately boring out a cylinder. Rapid dissemination of the news of Watt's engine, along with a further stream of modifications to its design and application, led to discussion of its enormous potential in substituting machine for muscle power, both to multiply production in factories and provide steam traction for vehicles. The day in France in 1769 when Nicolas Cugnot's steam tricycle carried four passengers for 20 minutes at a speed of 2.25 mph marked the beginning of the end for four-legged animals as instruments of mobility. And the translation of ideas by profound thinkers into widely-read pamphlets and books accelerated the pace of change. Adam Smith's *Wealth of Nations*, published in 1776, not only originated the liberal, political, economic philosophy of *laissez faire* that was to stimulate the embryo Industrial Revolution, but also described in precise form the division of labour which would create the mass production system at the heart of every expanding industrial enterprise.

Stimulation of industrial growth created a demand for labour that was met by the doubling of world population — to 950 million — between 1650 and 1798, a rise created by fundamental changes of attitude to public health and diet. Daniel Defoe, for example, had tackled the subject in 1698; John Pringle wrote a study on the army in 1752; while James Lind's essay of 1757 on *The most effectual Means of Preserving the Health of Seamen in the Royal Navy* would lead to sweeping reforms. First put into practice by Captain James Cook in his three long voyages of the 1760s and 1770s, Lind's recommendations resulted in immense saving of life as scurvy was eliminated by a balanced diet and disease controlled by strict personal hygiene. Progressively, as enlightened consciences came to place a higher value on individual dignity, efforts to conserve life became widespread and were recognized also as being of economic advantage.

Impact On Weapons And Warfare

Paradoxically interwoven with advances in humanitarianism, economics and industrial awareness and invention, burgeoned the study and practice of deadlier ways of destruction. The 18th century was graced by several great works on the military art, but none were more formative of the future than Marshal Maurice de Saxe's *Reveries upon the Art of War*, King Frederick of Prussia's *Military Instructions for the Generals* and the Comte de Guibert's *Essai générale de tactique* — all three seeking to formalize doctrine and methods for reducing the weight of weapons while increasing their killing power. It was Guibert who tackled the subject on the broadest sociological front by proposing in 1772 that enlarged populations should be exploited by conscripting men into popular citizen's forces; and too late when in 1779, in his *Defense du système de guerre moderne*, he retracted the proposal for a citizen army. For by then the idea had taken hold in America, as soon it would in France, amidst another popular revolution. Having publicized a number of unorthodox ideas, Guibert had achieved distinction in the eyes of many influential followers who faithfully executed his concepts; it was an example which inventors and industrialists, besides generals, were quick to copy.

The introduction of inventions, weapons among them, to everyday use was inevitably slow and confined almost exclusively to Britain. Quite apart from the problem of convincing people, the mass of whom the world over had their minds closed to new ideas, there was the problem of manufacture when factories were few and far between. Not until 1796 was the first flow-line engineering works laid down by the English firm of Boulton and Watt in what was called the Soho Factory, and it was another two years before Eli Whitney in the USA manufactured the first mass-production muskets. Nearly everything was laboriously hand-made by craftsmen, of whom there was a shortage, working to standards of their own setting. Only after Henry Maudsley, an employee of the

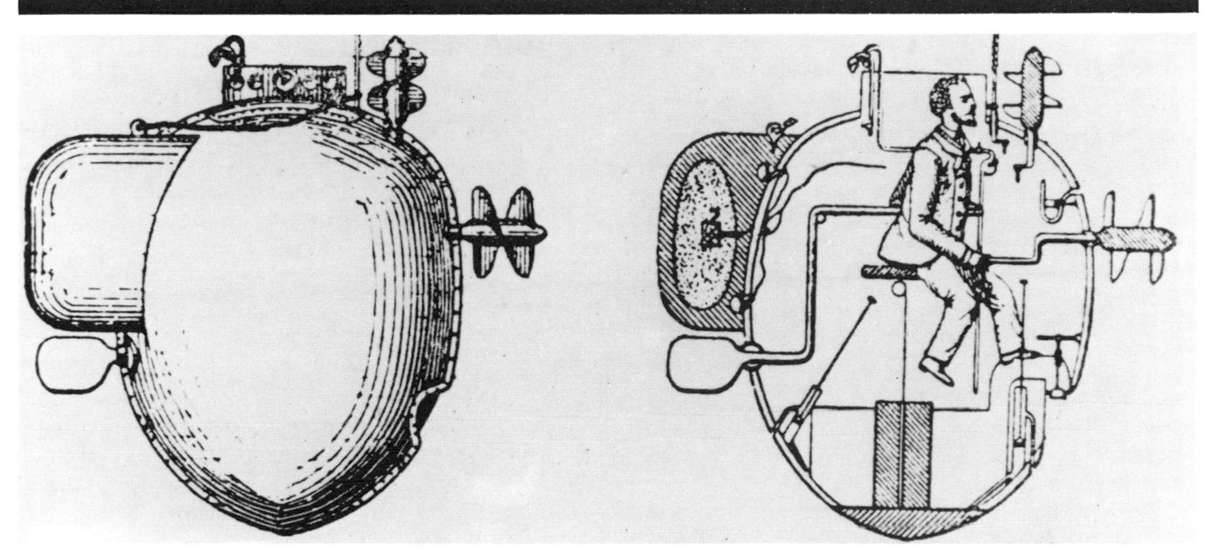

Above: **Bushnell's man-powered spherical submarine *Turtle* of 1776 attacked Lord Howe's flagship *Eagle* but was let down by weapon failure.**

Woolwich Arsenal in England, invented the precision, screw-cutting lathe at the turn of the century was it possible for semi-skilled workers to apply exact standards to the shop floor on a large scale. And only when these new tools and methods were combined in common use was it possible to make the vast number of weapons and equipment needed by citizens' forces.

One invention of the American Revolutionary War almost instantly put to use was the *Turtle*, a wooden-hulled, hand-propelled, spherical submarine designed in 1776 by a Yale student, David Bushnell. Crewed by an American revolutionary, Sergeant Ezra Lee, it floated down the Hudson River in an abortive attempt to screw a time-fused charge to the flagship of the British blockading fleet. The effect of this attack was convulsive in that, not only was the blockade relaxed when the target was compelled to shift berth, but henceforward every surface vessel was immeasurably more vulnerable to attack at its weakest points below the water line, against which no previous defence had been necessary. Yet this too was a method slow in development. When the American Robert Fulton tried to interest first his own country, and then France and Britain, in his wood and iron-clad *Nautilus* submarine in 1800, all were united in rejection — and equally unenthusiastic in 1805 by his so-called moored 'torpedo', a weapon later known as the mine. Underwater attack, it seems, was frowned on as underhand.

Hesitant also was the introduction of steam power to replace sail at sea. Although a steam-powered, water-jet propulsion boat was tried on the River Potomac by John Rumsey in 1775, it was another eight years before the Marquis d'Abban's steam-paddle steamer chugged along the River Saone in France; 1813 before Fulton's steam-propelled, paddle-wheel armoured warship *Demologos* came off the stocks; and 1837 before the Swede John Ericsson's screw-driven *Surveyor* demonstrated its superiority on the River Thames — without overcoming the assertions of the admirals who witnessed it that screws were impracticable for warships. Sail reigned supreme until midway through the 19th century and the substitution of iron for wooden hulls was generally resisted until timber supplies began to run out.

When it came to acquiring improved weapons, however, sailors could be quite venturesome. At the short ranges of engagement enforced by existing artillery pieces, gunners had a choice of trying to cripple the enemy by firing at extreme range against rigging (practised most successfully by the French) or endeavouring (like the British) to smash hulls, guns and men by firing at close range with heavy carronades, the invention of General Robert Melville and first tried in action in 1779. All manner of new weapon systems were on offer. In 1780, Hyder Ali of Mysore had fired rockets at the British in India, an event witnessed by William Congreve, who later produced rockets which were used from small boats in 1806 against the French at Boulogne. And in 1784 Lieutenant Henry Shrapnel, Royal Artillery, designed a shell which could be burst in the air to spray bullets among closely-massed ranks of enemy armies. This weapon was employed successfully in 1788 from Russian longboats at the mouth of the Liman River, in the Sea of Azov, to destroy a superior squadron of

Developing Firearms Technology And Its Effect On Artillery

The Prussian breech-loading 'needle' gun (right) shows the elements of all modern firearms to come: sealed breech, cartridge case, elongated bullet, fulminate cap and firing needle. The same technology had an immediate effect on artillery (far right) which, in the 1840s and 1850s, reached the point of change. Here a muzzle-loading cannon is being fired in the flat, lower register, while, as representative of the old system, a low-velocity mortar firing in the upper register lobs projectiles from behind cover over protective enemy works.

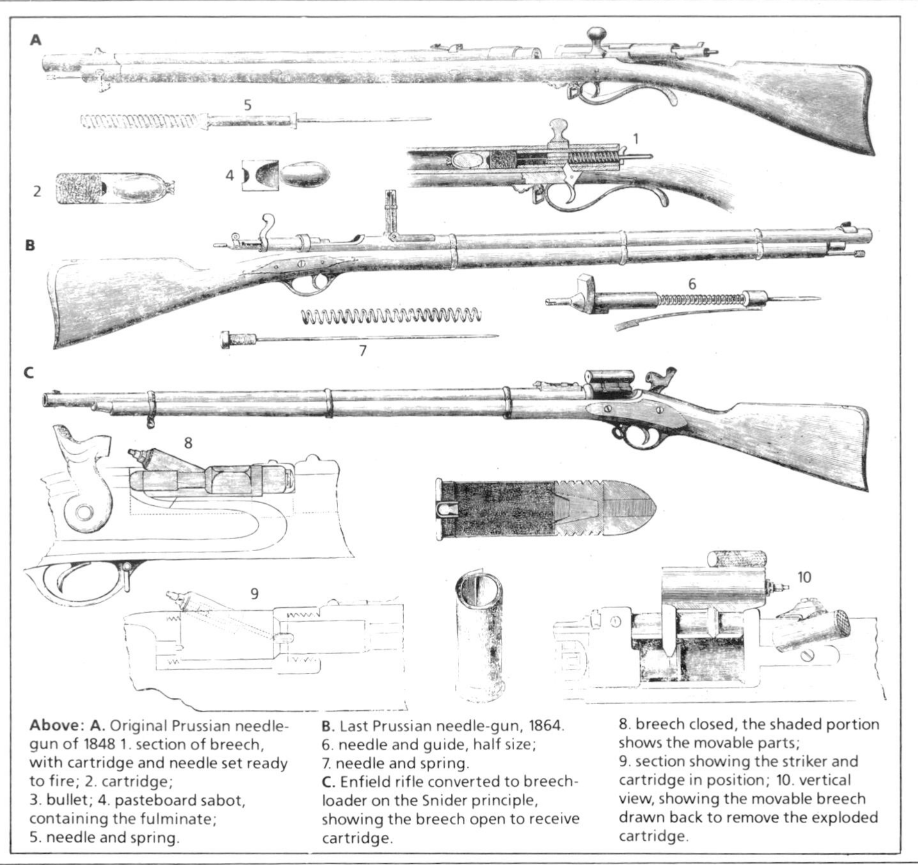

Above: A. Original Prussian needle-gun of 1848 1. section of breech, with cartridge and needle set ready to fire; 2. cartridge; 3. bullet; 4. pasteboard sabot, containing the fulminate; 5. needle and spring. **B.** Last Prussian needle-gun, 1864. 6. needle and guide, half size; 7. needle and spring. **C.** Enfield rifle converted to breech-loader on the Snider principle, showing the breech open to receive cartridge. 8. breech closed, the shaded portion shows the movable parts; 9. section showing the striker and cartridge in position; 10. vertical view, showing the movable breech drawn back to remove the exploded cartridge.

Turkish ships. But the device with perhaps most revolutionary potential was invented in 1805 by an English parson, Alexander Forsyth, who produced the tiny fulminate of mercury percussion cap. This was to be exploited in 1816 by the American sea captain J E Shaw and lead the way to a completely new and devastating range of breech-loading, cartridge-type firearms and artillery, as well as the development of hollowed-out shells, filled with high explosive and detonated by elaborate percussion fuses.

Easily the most important innovation of the late 18th century was manned flight, brought to fulfilment after ages of endeavour by François de Rozier in a Montgolfier-designed hot-air balloon in France in 1783 — an event which led to visions by 'futurologists' who competed to prophesy the shape of mechanized air warfare to come. Benjamin Franklin came close to the mark in 1784, when he postulated the descent of 10 000 men from the clouds doing 'an infinite deal of mischief before a force could be brought together to repel them'. A new mobility was foreseen for airborne armies, but it was as a scout to look over the hill at the opposition's preparations that a tethered balloon, with Captain Coutelle in its basket, first saw action at the Battle of Fleurus in June 1794. The development of lighter-than-air craft would proceed at a more leisurely pace than even that of steam-powered ships. The isolation of hydrogen in 1766 by Henry Cavendish in England contributed to lift an unmanned balloon three months before de Rozier went aloft, but powered steerable flight eluded the inventors because they lacked a reliable engine which did not threaten also to ignite the inflammable hydrogen.

The Anglo-French wars of the 18th and early 19th centuries only flirted with new technology. Post-Revolutionary France under the hand of Napoleon Bonaparte expanded vastly as large citizen armies, meagrely trained but daringly led, overcame smaller, less well-motivated, supinely-controlled professional forces. The weapons with which the opposing forces began the campaigns of 1793 were, in the main, those with which they ended; and the tactics steering superior force against the enemy's weaker links were those of antiquity — with defeat the penalty for those

ignoring basic rules. Numbers applied with verve counted enormously, and could only be supplied with weapons by manufacturers who concentrated in wartime upon what they best knew how to make in quantity. But the direction of vast forces was immensely improved by more efficient communications systems. Of these, the installation in 1792 by Claude Chappé of chains of semaphore relay stations throughout the French Empire was of fundamental importance. Whereas messages might previously have taken days to traverse a long distance by relays of horsemen, a semaphore message could travel at 600 miles an hour. Control could thus be centralized, enabling a far more efficient passage of intelligence and flexible deployment of resources. Similarly, the introduction by the Royal Navy in 1790 of a flag signalling system between ships enabled an admiral at sea to control the movements of individual vessels within his sight — a vastly superior method of control to the existing Fighting Instructions which lacked flexibility and could not hope to cope with unexpected and changing situations.

The reward for the Royal Navy was victory at the Battle of Trafalgar on 21st October 1805, when 27 British ships of the line under Admiral Lord Nelson eliminated 18 out of 33 French and Spanish opponents. The British achieved a 3 to 1 concentration of gunfire using a tactical deployment which enabled them to attack the enemy successively and hit him piecemeal. The secret of success lay in Nelson's plan, enhanced by flag signals which made possible last-minute adjustments in manoeuvre as well as the opportunity for the famous exhortation cheering each man to do his duty.

Just as Trafalgar was the last great battle at sea in which the old wooden-hulled ships slugged it out at close quarters, so Waterloo, on 18th June 1815, was the last great European battle employing the tactics and weapons of the pre-Industrial Revolution era. Yet Waterloo, like Trafalgar, also demonstrated the decisive importance of concentrated gun power, plus the need to mitigate its worst effects. Napoleon failed in his initial massed assault upon Wellington's emplaced, numerically inferior army; first, because he was unable to project sufficient artillery fire against the Anglo-Dutch army to shake, let alone destroy it; secondly, because Wellington lined up his men under the protection of a reverse slope and had them lie down to minimize the threat of the guns.

Transport And Communications

For forty years in the aftermath of Waterloo, the future impact of technology upon warfare was masked by the absence of any major campaign, most wars in that period being of a limited, colonial nature between ill-matched opponents. After N von Dreyse designed his needle gun in 1827, it was another 20 years before the Prussian Army adopted it, and it took even longer to establish in battle the vast supremacy over the musket of this breech-loading rifle with its faster rate of fire, enhanced by use of a paper cartridge detonated by a fulminate cap. Indeed, the adaptation to practical military usage of the steam-driven railway train and the telegraph was quicker because they also improved communications and expanded trade. The practical application in England of passenger and freight railway transport between Stockton and Darlington in 1825 was rapidly copied throughout the rest of the world. A crucial point in military employment was reached in 1848, when the Prussians transported by rail to Cracow a complete corps of 12 000 men, horses and guns. And, in that same year of extensive political revolution throughout Europe, the forces of law and order in France, Germany, Austria, Italy and Russia were moved to trouble-spots by steam train along a newly-built network of lines.

Faster yet in adaptation after 1829 was a network of cable message communication by means of the electric telegraph. The product of generations of research, no one man can lay unchallenged claim to the invention of the telegraph, any more than one country can be positive it introduced the first system in commercial use. It is sufficient to record that, once the principle had been demonstrated early in the 19th century, the stimulus to design various practical methods was unbounded. Certainly, much news of revolution in 1848 was carried by telegraph, in parallel with the use of this vast improvement upon semaphore to expedite counter-measures by ordering troop movements and controlling railway trains. Indeed, the sophistication of railway control would have been hampered without the introduction of the telegraph and the creation in 1850 by Samuel Morse of the universally-adopted code of dashes and dots which simplified and hastened the transmission and receipt of messages.

Weapons In Waiting

In the mid-19th century, when peoples' thoughts were overlain with the fear and instigation of political revolution, the seeds of military revolution stirring beneath the surface were slow to emerge due to limited opportunity and lack of an inspiring philosopher catalyst. Although Karl Marx was already expounding his liberal thoughts in the 1840s and wrote *The Communist Manifesto* in 1848, his influence was at that time minimal. Likewise, Karl von Clausewitz's celebrated work *On War*, posthumously published in 1832, would only slowly impose upon nations and military leaders his considerably misunderstood concept of war as an expression of total national effort, through the harnessing of entire populations and industrial capacity to the struggle.

Although several more wars of a relatively orthodox nature were to be fought before von Clausewitz's and Marx's writings became text books, it has to be noted that the re-equipment of sea and land forces with the latest weapons was tightly restricted by lack of incentive and the allocation of sizeable funds. The weapons of Waterloo served the purposes of combating primitive opponents well enough. In the absence of any pronounced threat to national security between major powers, rearmament was pointless — until, that is, internal pressures caused by liberal elements seeking further extensions to political liberty and expression sparked off revolutions that compelled governments to take physical counter-measures. As so often has been the case, struggles by the under-privileged for greater democratic freedom led to tighter self-protective control by the authorities and the emergence of new dictatorships reinforced by stronger armed forces. Inevitably, in the nature of history, dictators began to turn predatory eyes on neighbours in order to deflect attention from their own internal troubles, thus instigating further instability and the upsurge in rearmament and new technology which feature in the 1850s and every decade to follow.

Henceforward, as machines began to assume a dominant influence upon the well-being of individuals, nations and communities, it was no longer possible safely to defer the adoption of new and revolutionary weapons when they were designed or even suggested. The demands of sheer survival only precluded the making of arms when those concerned with security were prepared to risk or forfeit their existence and that of their followers.

The 'Arms Race' had become an essential fact of life which would from now onwards accelerate at an ever-increasing and uncontrollable rate.

Chapter 2

The Basis of Change

1850-1870

The revolutionary movements and uprisings of 1848, with their fading echoes of violence in 1849, spurred the first major European war since Waterloo. The Russian Emperor's designs upon the 'sick' Ottoman Empire led to his attempt in 1853 to dominate Turkey and attain possession of the Bosporous with access to the Mediterranean Sea. France's ambitions under the dictatorial Napoleon III, along with her rivalry and fears that the Russians would threaten her prestige in the Near East, became allied to British anxiety of a disturbance of the 'Balance of Power' and prompted these two nations to side with Turkey in sending naval forces to Constantinople. The Turks declared war on Russia in October, and at once a naval battle within Sinope harbour on 30th November revealed the changes wrought by new technology. In six hours, nine warships of a Russian squadron, armed with the latest percussion-fused bursting shells, had ripped apart and set fire to an equal-sized flotilla of Turkish ships (including two steamers). The day of the 'wooden walls' was over.

The Russian victory at Sinope and her invasion of Bulgaria served to convince the British and French that Russian ambitions had to be curbed. Eliminating the Russian fleet through the capture of its main Black Sea base at Sevastopol in the Crimea was seen as the way to achieve this limited aim, even after Russia had abandoned her invasion of Bulgaria in face of an Austrian and Prussian alliance.

The old men in command drew their experience from the campaigns of four decades ago, and projected battles with the tactics of 1812 using modern weapons. They frequently failed to take into account the immensely increased firepower made possible by the introduction of the Russian shell, as well as the breech-loading Stutze rifle used by the Russians and the Minié rifle with which both the French and British were equipped. While it had been hazardous enough in the past to advance against muzzle-loaders fired by massed ranks of musketeers, it was now suicidal to do so, just as it was suicidal for men and beasts to stand with shells bursting among them. If the Russians under Prince Alexander Menshikov had attempted to resist the initial Allied landing across open beaches on 13th September 1854, they might well have halted the invasion there and then against troops weakened by a voyage in terrible weather which prolonged their disembarkation over a period of five days. And if the Allies had then pressed hard their advance to

Above: ***Charge of the Heavy Brigade at Balaclava — a massed-formation, horse-heavy tactic on the verge of obsolescence owing to the intensity of modern fire-power.***
Right: ***Naval bombardment at the Crimea. Note the man with a telescope spotting fall-of-shot and the officers calculating data.***

Sevastopol after the Battle of the Alma on 20th September (when 5709 Russians out of 36 400 and 3000 Allied soldiers out of 51 000 fell), they might possibly have overrun fortifications which were by no means complete. But both sides were operating on a logistic shoestring at the end of extended and tenuous lines of communication and therefore as much intent upon making themselves secure as seeking a quick decision. The Allies shifted to new base positions before reaching the outskirts of Sevastopol on 8th October and it was the 17th before they felt able to begin an assault.

Superhuman efforts by a Russian engineer, Col Frants Todleben, inspired the defenders to stiff resistance against an improvised Allied siege train. Driving his people to work frantically to throw up earthworks, Todleben removed the warships' guns and incorporated them into the outer and coastal fortifications before sinking the ships in the harbour mouth to prevent entry by the enemy fleet. Not only did his naval and army gunners manage to contain the inadequate Allied effort on land, they also shot with devastating effect against the Allied fleet when it joined in the bombardment. Shell and heated shot tore among the ships, setting alight the wooden hulls and rigging. Over 500 casualties were inflicted on the British crews alone before the attempt was abandoned and the fleet withdrew.

The campaign now reverted to a pattern that was to become all too familiar in wars to come — a struggle to exhaustion in which men were overwhelmed by weapons while each side groped to extract itself from an impasse by intensifying physical destruction. With restricted room to manoeuvre, and with their lines of communication channelled through single, vulnerable avenues of approach, each of two Russian attempts to cut off the besiegers from their newly-established bases at Balaclava and Kamiesch were doomed to become direct assaults along predictable axes of advance. On 25th October came the Battle of Balaclava — an encounter engagement best remembered by posterity for a misguided and ineffectual charge by 673 horsemen of the British Light Cavalry Brigade into a torrent of fire from Russian artillery emplaced at the end and on both sides of a valley, with the loss of 247 men and 497 horses. In fact the battle was much more notable in terms of action by the Heavy Brigade and the 93rd Highlanders. For had not the 500 infantry of the 93rd stood firm and the Heavy Brigade charged home at the crucial moment against 3000 enemy cavalry, the base at Balaclava would have been overrun and the entire Allied force placed in jeopardy.

As it was, the volume of volleys generated by the 'thin red line' of Scottish infantry was such that the routine drill of forming a compact square to meet enemy cavalry could be dispensed with, particularly since the ground was broken. Opening fire at a range in excess of 100 yards (beyond which the accuracy of ball shot from the smooth-bore Minié rifle fell off badly) the ability to reload much more quickly through the breech instead of ramming from the muzzle as in the past enabled each man to fire rapidly enough to bring the enemy to a halt well short of the infantry's shoulder-to-shoulder ranks.

The lesson was obvious to behold. Henceforward cavalry stood little chance against unshaken and trained infantry in position — let alone artillery — as would be confirmed on a foggy 5th November when two columns of about 42 000 Russians, supported by artillery, blundered into a dispersed British infantry division at Inkerman. Caught by surprise, a small fraction of the total force of 8500 British managed to hold a vital ridge against the relatively low proportion of Russians at first engaged. Piecemeal and inadequately supported, the Russians threw in more men to the assault as, company by company, the British reinforced the threatened front without their commanders ever really being aware of the exact situation. But when the Russians drew off in the afternoon the casualty figures spoke for themselves: some 12 000 Russians to 3300 British and French. Again, it was rapid shooting at close range which had brought the decision during a battle in which the professionalism of regular troops had been demonstrated as a match-winning factor. Neither side had given way in face of carnage, though their efforts were blunted by the inability of the high command to read the battle because information from a confused front-line had not come quickly enough for them to communicate their orders and impose their will on the fight.

The stalemate which spread along the lines of circumvallation about Sevastopol was the product of exhaustion created by a hard winter and exacerbated by administrative chaos. A severe storm on 14th November wrecked 30 transports in their unprotected anchorage off Balaclava and destroyed much material and supplies. Across sodden ground churned into mud, horses and men struggled to haul waggons carrying stores to the front — compelled to go that way because Russian guns commanded the only hard-core road available. A deathly malaise of malnutrition and disease (chiefly cholera) took their toll and lowered the morale of all the contestants. In campaigns of old, this sort of creeping disaster through incompetence would have passed unnoticed at home until fatal damage had been inflicted. But this time the communications revolution speeded up news-relaying and so fuelled another development. Telegraph reports from the theatre of war, above all those from William Russell, correspondent of *The Times*, went beyond the reporting of battles and indulged in investigative journalism of such compulsion and immediacy that the British government was brought down and Lord Palmerston became Prime Minister. Urgent measures were taken to replace losses, create a siege train and improve general administration. The building of a new road, out of sight of enemy guns, to link Balaclava with the lines before Sevastapol was but one major project put in hand hastily.

Health And Welfare

With equal vigour (and enforced by the sheer insistence of political survival in a democracy) reforms demanded by public compassion and indignation were directed at improving the welfare of the troops and, above all, the treatment of the wounded. Up to this moment military administration was sometimes a matter of chance, and hospital staffing mediocre. Nursing was carried out by women whose status was that of superior domestic servants, whose training was rudimentary and whose addiction to alcohol was more prevalent than desirable. Surgery was still frequently carried out without anaesthetics and never with sterilized instruments (antiseptics were not yet in general use); hygiene was primitive; administration and layout of wards were disorderly; and treatment haphazard. Two women tackled this problem in the

Crimea and at the base hospital at Scutari.

As Inspector-General Surgeon of the British Army, Miranda Stuart (who spent her entire medical career disguised as Dr James Barry because women were forbidden to qualify as doctors) supervised the restoration of better order in the system at a time, incidentally, when chloroform was first being used in operations upon the wounded. And at the request of the Secretary of State for War, Florence Nightingale led a team of nurses to reorganize the care of the wounded by introducing revolutionary training and administrative reforms in the hospital. Her subsequent *Notes on Matters affecting the Health, Efficiency and Hospital Administration of the British Army* and *Notes on Nursing* became bibles of the worldwide reform of nursing, along with the design of hospitals and their wards. These strong-minded women were supported, of course, by a swell of public concern which had risen in the Britain of 1840, calling for overall improvements in public health. They were helped, too, by a general wish to alleviate pain, allied to a willingness to reject religious objections to the use of anaesthetics, knowledge of which had been published by Humphry Davy in 1800.

Blockade And Siege

With the defenders of Sevastopol penned behind their fortifications (although never entirely isolated from their field army inland) and their fleet in both the Black and Baltic seas either destroyed or blockaded, the final year of Russia's war against Turkey and her Anglo-

Logistics In The Crimea

Above: *A mass of shipping unloads at a newly-improvised port. Note the sail/steam-driven ships and, in the foreground, port-handling facilities including a small steam-driven crane. The railway appears to be man and horse powered, while, in the immediate foreground (left) bullock-drawn carts carry stores to the front.*
Right: *Florence Nightingale in a ward at Scutari. Its space and cleanliness greatly improved the survival chances of the wounded, although her main contribution was in raising nursing standards.*

French allies resolved itself into a diplomatic move aimed at a settlement in which none of the contestants would lose face. Meanwhile, Allied attempts were made by land and sea to subjugate Sevastopol and to raid or destroy by naval bombardment other Russian coastal facilities.

The conditions of siege warfare which had settled upon Sevastopol were, in the main, conducted by the ritual methods of old. That is, the besiegers drew close to the defences by driving forward entrenchments and constructing emplacements from which smooth-bore, muzzle-loaded artillery, firing at ranges rarely in excess of 1500 yards (and often much shorter for maximum effect), attempted to smash a breach in the enemy earthworks prior to launching a desperate infantry assault. The defenders meanwhile endeavoured to destroy the attackers by counter-bombardment, forays and attempts at relief from their army outside. The intensity of bombardment by modern shells nevertheless imposed new forms upon the old ritual. A ten-day bombardment between 8th and 18th April wrecked large sectors of the Russian defences and caused over 6000 casualties among men who were compelled to stand-to, awaiting an assault but lacking adequate protection. It was ironic that the assault failed to materialise fully owing to another manifestation of technological improvement. The telegraph brought interference from apprehensive governments in London and Paris endeavouring, for the first time in history, to dictate operations hour by hour in order to save losses and political credibility.

During successive attempts to overcome the Russians in the weeks to follow, one aspect of the revolution to come in warfare was clearly exposed. A network of trenches, dug deeper than usual and therefore requiring a longer period of bombardment preparatory to their destruction, was never wholly obliterated. In taking the outer defences on 7th June there were just under 7000 Allied casualties to 8500 Russian. Ten days later, ineffectually planned assaults on the key Malakoff and Redan works failed with losses of 4000 men under furious fire from over 100 Russian guns that had survived the preliminary Allied bombardment. For the next three months there took place what, in due course, would come to bear the label 'Attrition' — each side endeavouring to wear down the other by persistent shelling. The Allies, their lines of communication providing a substantial supply of ammunition, had the edge and inflicted several hundred losses a day upon the Russians. Nevertheless, defence proved superior to offence. When the Russian Army attempted the relief of Sevastopol from outside on 16th August, their attacking troops lost 8000 men to 2000 of their opponents in trying to advance uphill against the Traktir Ridge. Finally, on 8th September, the British were again repulsed at the Redan. But by meticulous planning and with senior commanders controlling and leading from the front (aided for the first time in a scheduled assault by the use of synchronized watches to achieve a co-ordinated start) the French overran the Malakoff. It was noteworthy that in what counted as a victory compelling the Russians to withdraw from the city, the losses of the assailants were 10 000 compared to 13 000 among the defenders. Though hardly commented upon, the age-old belief that attack was by far the most expensive type of operation was no longer necessarily valid.

Innovation At Sea

It was from sailors that many innovations emanated — even cutting across the prerogative of soldiers, as, for instance, when Admiral Lord Dundonald proposed the use of sulphur fumes to 'gas out' the defenders of Sevastopol, only to be rebuked by Palmerston on the grounds that 'no honourable combatant would use such means'. At sea, however, the long-abhorred form of underwater attack was at last put to a full practical test by the Russians, who tried out cone-shaped, zinc-canister 'torpedoes' filled with explosive which fired on impact by the breaking of a tubular glass detonator filled with acid. This was but a chemical modification of an electrically-fired static 'Torpedo' demonstrated by Samuel Colt in the USA in 1843 when he blew up a 500-ton brig for the edification of members of Congress. It was the first practical sea mine — come to fruition a mere forty years after Robert Fulton had invented it.

The gun versus armour competition was, however, of overriding importance at sea. The French led the way by constructing five special floating batteries driven by steam engines. These clumsy craft, 64 ft long by 42 ft in beam and 18 ft in draught, were protected by 45 ins of iron plate laid over thick timber and armed with sixteen 56-pounder, shell-firing cannon. When they floated in to trade shots with Russian shore batteries at Sevastopol, and against the Kinburn Forts at the mouth of the River Bug on 16th October, they managed to destroy their targets despite sustaining over 130 hits. Indeed they retired intact with only two killed and 24 wounded — a considerable improvement on the Royal Navy's results a year previously.

The ending by diplomacy of the Crimean War in February 1856 left the military members of the contestants in thoughtful mood. The Russian Emperor Alexander II appreciated that defeat was to no small

The Armoured Warship

***La Gloire*, and its British rival, *Warrior*, were two ships that represented not only the change from the old sail to the sail and steam layout, but also stimulated competition between Britain and France to achieve naval supremacy and led to the major revolution in warship design over the next 60 years.**

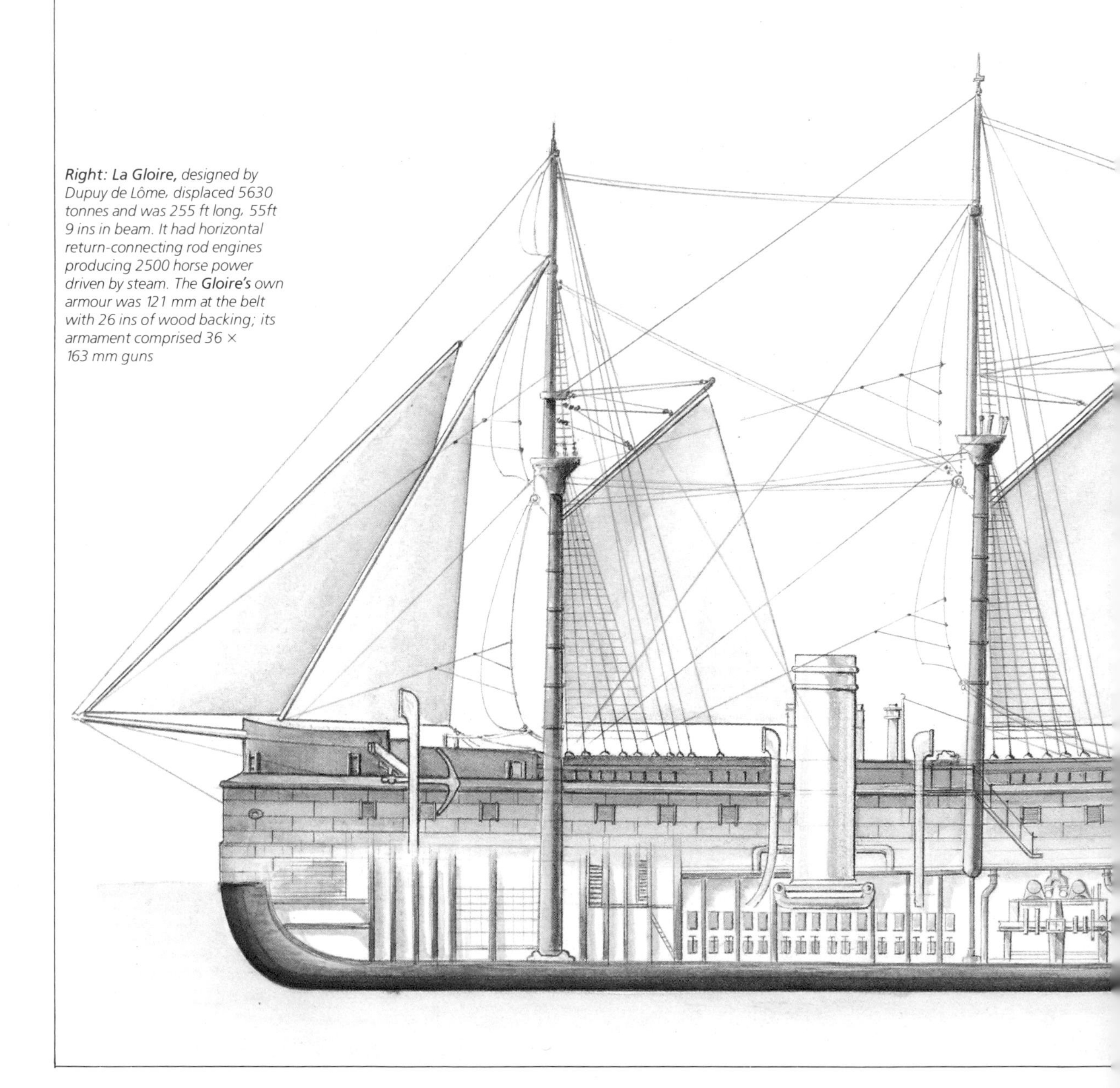

Right: ***La Gloire,*** *designed by Dupuy de Lôme, displaced 5630 tonnes and was 255 ft long, 55ft 9 ins in beam. It had horizontal return-connecting rod engines producing 2500 horse power driven by steam. The* ***Gloire's*** *own armour was 121 mm at the belt with 26 ins of wood backing; its armament comprised 36 × 163 mm guns*

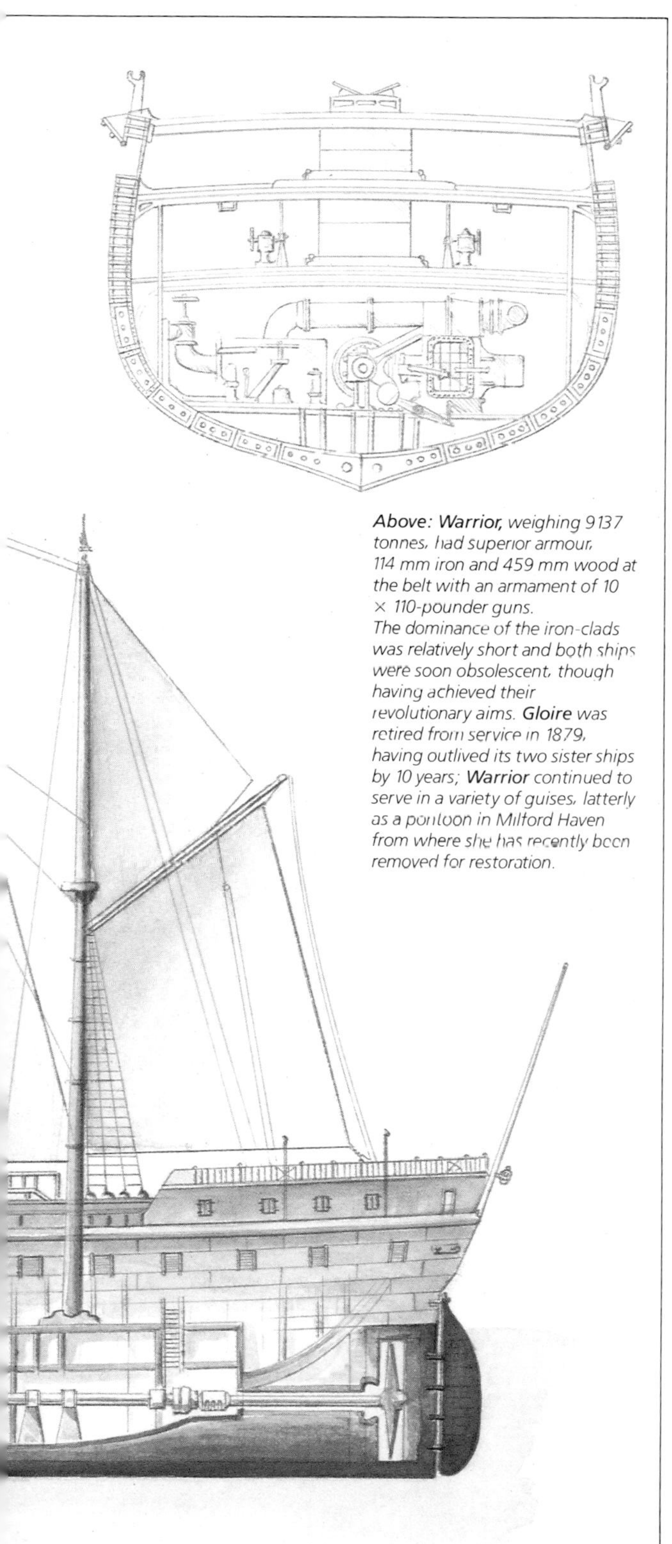

Above: ***Warrior,*** *weighing 9137 tonnes, had superior armour, 114 mm iron and 459 mm wood at the belt with an armament of 10 × 110-pounder guns. The dominance of the iron-clads was relatively short and both ships were soon obsolescent, though having achieved their revolutionary aims.* ***Gloire*** *was retired from service in 1879, having outlived its two sister ships by 10 years;* ***Warrior*** *continued to serve in a variety of guises, latterly as a pontoon in Milford Haven from where she has recently been removed for restoration.*

extent due to technical backwardness and lack of a strong manufacturing base. The victors and other European nations realized that they would have to bow to the irresistible power of new weapons and extensively re-equip their navies and armies.

The Solferino Campaign

Just over three years later, in June 1859, the lessons of the Crimea were underlined when Austria invaded Piedmont. France, anxious under its military dictator Napoleon III to recapture the glory of Napoleon I, came to the latter's assistance. The occasion provided the first opportunity for the extensive strategic and tactical movement of troops and supplies by rail, with deliveries made close to the front and, during the Battle of Magenta on 4th June, almost into the front line. A bloodbath of appalling dimensions can be laid at the door of incompetent generals on both sides, but in mitigation it has to be stated that, as in the Crimea, the changed circumstances of modern warfare made it extremely difficult for leaders to read the battle and communicate across the front with sufficient speed.

Cavalry scouts probing the enemy positions for information were finding it harder to penetrate defensive screens held by opponents strong in long-range firepower. Having at last obtained information, the method of transmitting it remained that of the fastest horse, which often was slower than the speed that forces could move by rail. Unexpected encounters led to extemporized battle plans in which men and horses repeatedly found themselves flung at each other with scant attempt at preparation. Plans were based on incorrect information or inspired guesses, resulting in attacks launched on false premises against the enemy's main strength instead of weakness. At Solferino on 24th June, throughout the decisive engagement between an Austrian army of 120 000 men with 451 guns, and a Franco-Piedmontese army of 118 600 with 320 guns, attempts at out-flanking went awry and were converted into headlong assaults at dreadful cost. In a battle lasting under 12 hours, in which spectacular artillery duels ending in favour of the French were predominant, the casualties inflicted were enormous — some 22 500 defending Austrians against just over 17 000 of their victorious attackers, who finished so exhausted that they were quite incapable of pursuit. Not that pursuit was needed. A decision had been reached and a fresh step made in the creation of modern Italy.

Once more public revulsion, focused by vivid descriptions of the carnage in a book published shortly after the battle by a Swiss observer, Henri Dunant, led to further outcry and efforts by politicians to ameliorate the plight of soldiers through an international organization. As a result the Red Cross Society was set up in 1864 to care for wounded men and prisoners of war. But in the meantime the means to wage war with yet greater ferocity had been magnified and multiplied by the latest weapons and the major campaigns which followed the wars of the 1850s.

The Technical Revolution Gathers Motion

Among so many weapons of enhanced power, none were more formidable than artillery, which had undergone a transformation since 1840. The old method of casting solid barrels from iron and then boring them out was superseded either by the invention of Daniel Treadwell, in which the barrel was built up by iron or steel hoops shrunk round a central cylinder; or by Thomas Rodman's way of casting the barrel hollow around a removable water-cooled core. These changes not only facilitated mass production but also made feasible the manufacture of barrels which could withstand far higher pressures than those of old and thus permit substantial additions to range and weight of projectile fired. Of equal, if not greater, importance in the race to lengthen range was the introduction of rifled barrels and the use of elongated shells in place of the old spherical shot. A spun, streamlined projectile had much truer ballistic properties for improved accuracy and range, but the combination did not achieve its full potential until Rodman invented a powder that burned more efficiently, giving increased velocity. This explosive nevertheless still possessed the defect of all black powder in that it generated clouds of smoke, obscuring observation of the fall of shot to the gunners and advertising the guns' position to the enemy. For all their uncured defects, the new guns made obsolete at a stroke all existing artillery because they could out-range smooth-bore cannon by some 8700 to 2400 yards, and at a range of 1000 yards had a 7 to 1 better chance of hitting their target. This the Austrians had discovered at Solferino, when the latest French rifled guns stood off and smashed smooth-bore guns that could not reach, let alone hit, their tormentors.

Above: ***The Whitworth homogenous steel breech-loading rifled cannon which, at twice the price of older pieces, possessed almost double their range with far greater accuracy and destructive effect.***

It was but a beginning of the artillery revolution. Improved rates of fire allied to simplified loading procedures had become possible when, in the early 1840s, a practical breech-loading gun was designed by a Swede, Baron Wahrendorf. Its development was essential if only because increased weights of projectiles and propellants, along with lengthened barrels, were making manual ramming through the muzzle over-laborious. The rifled, breech-loading guns produced in the late 1850s by William Armstrong and W G Whitworth in England also incorporated the latest construction by shrinking metal rings round the inner barrel lining. They led to yet another re-thinking by competitors in both the manufacturing and political arenas. For not only did the Armstrong Whitworth guns appear easier, cheaper and more flexible than the gun drilled out of a solid steel block by Alfred Krupp in Germany in 1850; they also dictated that existing fortifications and redoubts would be out-matched.

So seriously did Lord Palmerston, the British Prime Minister, take their threat that he ordered the construction of a ring of forts around Portsmouth in case France (the new ally but erstwhile enemy) chose to land a force and fire on the dockyard, so preventing the repair of the fleet and undermining the nation's main defence force against invasion. Few projects elsewhere rivalled 'Palmerston's Follies' in their costly nature, but every major nation was brought to realize that vulnerable political and economic, besides coastal, defences were in need of modernization by means of improved armament and strengthened protection.

The need to revise fortifications was as nothing, however, by comparison with the vital importance of radically changing warship construction. Capitalizing upon the experience of the successful floating batteries at the Crimea, the Frenchman Dupuy de Lôme built the first iron-clad warship. Completed in 1859, the *Gloire* was a wooden-hulled, sail- and steam-powered, screw-propelled frigate which had sides faced with iron plates between 4.3 and 4.8 ins thick. With a speed

Above: ***Boydell's steam tractor, with its 'footed' wheel, was a practical attempt to enable vehicles to cross soft, uneven ground and the inferior roads of the 19th century.***

of 13 knots without sail, and an armament of sixteen 6.4-in muzzle-loading guns, it displayed considerable potential, not least in its complete independence of wind and tide when manoeuvring. Its appearance was the signal for an immediate British answer — the laying down of the *Warrior* as the first all-iron warship with a 4.5-in wrought-iron hull, backed by 18 ins of teak to make her proof against 68-pounder shot at 400 yards. Indeed, the advent of these two ships, the latter with a speed of 15 knots and two broadsides of twenty heavy guns each, announced a naval race. Maritime nations drew the correct conclusion that a few of the latest vessels might completely dominate all the old wooden, sail-driven ships in existence 'like a lion among a flock of sheep', as de Lôme foretold. It only needed the ideas propounded after 1855 by the British engineer Cowper Coles, for guns to be mounted on a rotating turntable protected by armour, and all broad-side-arranged ships were fit only for the scrapyard. A complete revolution in warship design to incorporate armoured turrets was inescapable for maritime powers, regardless of the cost.

The key to future development of guns and armour lay, of course, in the science of metallurgy and metal production. The more widespread use of iron for civil and military purposes naturally created a considerable expansion of manufacturing capacity. But iron as armour, and as a means to penetrate armour, had the serious defect of easily shattering on impact; hence the reason for adding timber backing to plates to stop jagged splinters from flying about and causing appalling wounds. The key was given a critical twist by Henry Bessemer, the British designer of a new, spun shell. Discovering that the existing iron-barrelled guns would not stand up to the required pressure, Bessemer began to search for a stronger material and in 1856 announced his way of making mild steel cheaply and in quantity by decarbonizing molten pig-iron with an air blast. Development of Bessemer's Converter method, in parallel with the adaption of the German William Siemen's invention of open-hearth steel-making, contributed by 1864 to a vast expansion in the production of steel, which superseded wrought iron in the construction of much military hardware.

Inventors, their ideas crowding one upon another

as the mood of innovation matched the rewards offered for success, began to overcome the old-fashioned mood which tagged anything undiscovered as 'impossible', and dismissed as 'unsportsmanlike', 'un-British' or heretical any notion or device that looked likely to upset the status quo or threaten vested interests.

It is likely that had machine-guns been offered to Lord Palmerston (as they began to appear in ten or more different forms between 1856 and 1861) he would have rejected them as 'uncivilized' — as he actually described the forerunner of the tank. All the machine-guns of the 1850s were hand-operated and limited in their effectiveness by the existing paper cartridges. But their potential was sufficient to encourage intensive search for improvements, and the day would come when they would introduce their own revolution in war and stimulate yet another — the construction of a working armoured fighting vehicle (AFV) to mitigate the effects of the firepower such guns produced.

James Cowen's suggestion for an AFV in 1855 did indeed have preservation of life in mind as a way to shield men from shell-splinters and bullets. All he proposed was the adaptation of the existing steam tractor of James Boydell (first patented in 1846) to mount cannon behind a dome-like, iron outer casing and travel where other types of wheeled vehicles could not move. The Boydell tractor was a practical example, used in the Crimea, of a portable railway — in this case a vehicle which did not sink into the ground because its wheels were fitted with detached pivoted rails, or feet. These were laid down and then picked up as the wheel turned, to spread the vehicle's load. Had it been built, the Cowen's crews would almost have baked alive behind their armour, yet they might have survived against the enemy in this genuine proposal for a practical, self-propelled vehicle uniquely incorporating mobility, firepower and protection. Certainly the Boydell footed wheel triggered many inventors to begin experiments with 'continuous tracks'.

The Communication Fundamentals

In the vital area of communications, upon which so much depended if forces were to be launched into battle with properly co-ordinated and controlled dexterity, little improvement was made between the Crimean War and the eve of the great American Civil War in 1861. Those communications not delivered by hand of messenger were sent by visual means, with all the accompanying deficiencies in reliability and capacity. Letters spelt out by semaphore, although useful behind the front, were all too easily obscured and interrupted by smoke and confusion on the battlefield. The flashing of morse dashes and dots, as sunbeams reflected from a mirror were interrupted by a shutter, was a system only usable when the sun shone; nevertheless, the heliograph (called the heliotrope when first introduced in 1858) continued in use for many generations. Systems using morse code with flags or shuttered lights were also developed shortly before 1860 by Captain Colom of the Royal Navy and Captain Bolton of the British Army, working together, and by a US Army surgeon, Major Myer.

Telegraph systems, nevertheless, were seen to have most potential after their limited employment during the Crimean War. The rapid proliferation of cable routes soon began to provide world-wide land links, and the introduction of the Duplex method, put forward by Wilhelm Gintl in 1853, enabled simultaneous transmissions to be made in both directions between stations along the same cable. This contributed to a redoubling of capacity and was the first of a number of steps aimed at making better use of existing circuits. Few of these advances in communication were of much use in the tactical handling of ships at sea, of course. Instructions were readily passed to captains while in harbour, but once out of sight of the shore visual signalling alone was available, although greatly enhanced by morse sent at night by light.

Nevertheless, the steadily growing rate of development in communication technology lay at the heart of the revolutionary changes in war — and much more besides. Apart from facilitating the deployment of armed forces, it invigorated the exchange of news and ideas, penetrating every activity of government and society. As the cable network grew into a web, the moment approached when it became far easier than in the past to combine separate states and principalities within the same ethnic groups into compact, though complex, national groups. By the same process, too, smaller industrial units could be merged into greater manufacturing combines. Efforts to hold groups together in the past had broken down because poor communication and misunderstandings laid them open to internal dissent and predatory moves from foreign rivals. The Piedmontese effort to create a fully integrated Italian state and eject Austrian invaders was simply the latest example of the advance of nationalism, made attractive by improved communications and implemented by conflict in argument and battle. The approaching war within the United States of America was very much a struggle between the forces of disruption and unity. And from the heart of these beat the throb of a cause, propagated by the thriv-

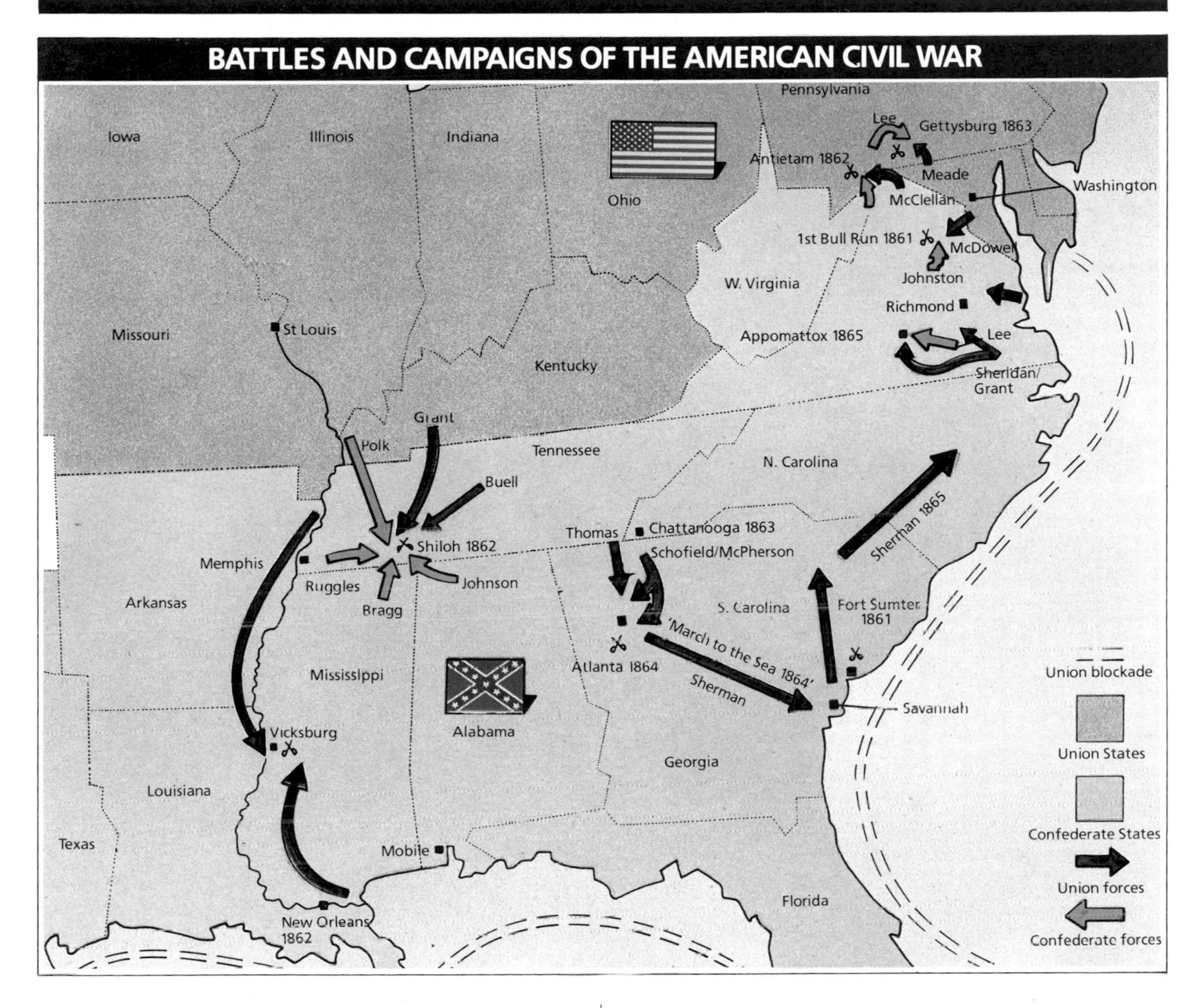

ing communications and publicity systems which persuaded thousands to offer themselves patriotically for slaughter.

The American Civil War

A conflict ostensibly in the cause of individual freedom between those supporting abolition or preservation of slavery in the USA, the Civil War broke out in the summer of 1861. As well as a fight to restore unification, it was also a contest between the diametrically opposed philosophies and economies of the agrarian South and the more highly industrialized and competitive North.

The prowess of the forces initially engaged was geared to that of frontier fighting. Their armament was mostly outmoded. Of the 90 obsolete warships at once available to President Abraham Lincoln in the North, only four were in home waters when hostilities commenced. Not that this mattered since, at first, President Jefferson Davis's Southern Confederacy possessed none at all. Enthusiasm for the competing causes was such that within two months the South had 112 000 men under arms and the North 150 000 — the ever-present discrepancy in numbers temporarily balanced by the far better use to which the South put its officers, who included the best of the pre-war forces. Although the Civil War is best remembered for its land battles in their magnitude, carnage and heroic drama, it is by no means an underestimate to say that the undramatic, stifling blockade at sea, with its monotony and smaller-scale encounters, settled the final outcome. Before anything resembling a clash of armies took place, coastal areas had been the scene of important skirmishes. The taking of Fort Sumter by Confederate forces on 14th April, and their seizing on the 20th

Above: **The steam-driven, armoured** *Monitor* **(centre left) and** *Merrimack* **(centre right), renamed** *Virginia*, **slog it out at close range in Hampton Roads. Out-moded sailing ships burn or stand off and, with the army ashore, await the outcome in trepidation.**

of the Norfolk Navy Yard along with the burnt-out hull of the steam warship *Merrimack*, besides valuable munitions and modern artillery, reflected Davis's concern that his Confederacy should have access to the outside world. Only then could it sell its main product, cotton, and purchase the modern armaments its industry could not manufacture in sufficient quantity. Likewise, the blockade announced by Lincoln on 19th April presupposed that to isolate the South would embarrass, if not eventually cripple, his opponent. Not that the North tried (or was able) in 1861 to enforce the blockade, believing that the capture of Richmond, the South's capital, would settle the issue and end the war by Christmas.

The defeats on land later in 1861, which flung Lincoln's armies back on the outer defences of his capital at Washington, put paid to simplistic notions and gave notice that this would be a long war involving the entire nation. Armies, supplied by expanding manufacturing capacity and imports, would fight for control of resources and communication systems, while navies struggled to enforce or prevent the isolation of the Confederacy. For its part the Confederacy, well knowing that it would never possess the power to strangle the North, pinned its hopes upon inducing a hopeless war-weariness in the North, compelling it to seek a halt.

First Clash Of The Iron-Clads

A vital turning point of the war occurred on 8th March 1862 when the rebuilt *Merrimack*, armed, armoured and renamed *Virginia* by the South, attacked conventional Federal ships which were blockading the James River and the approaches to Richmond. *Virginia* had been protected by two layers of iron, each 2 ins thick, backed by 24 ins of timber; armed with six 9-ins, two pivoting 7-in and two 6-in guns, fitted in two broadsides, five a side; and equipped with an iron underwater ram at the bow. An afternoon's combat ended with *Virginia* in control of the river, her ram having sunk one 24-gun sloop with the loss of 121 men and her guns reducing to a wreck a 50-gun frigate which had run aground, with the loss of 120 men. Her own losses amounted to only 10.

To deal with this lion among sheep the North deployed its latest ship, Ericsson's invention of 1854, the armoured *Monitor* with its two smooth-bore, 11-in guns in a rotating turret. Once called a 'Camembert cheese-box on a raft', it was *Monitor*'s turret, with its eight layers of inch-thick iron, that was revolutionary. Mounted on a flat, unseaworthy hull with only one foot freeboard, it presented a relatively small target and was capable of firing through 360 degrees in azimuth, thus simplifying positioning for shooting by comparison with *Virginia*'s broadside arrangement.

The two iron-clads began their duel on the morning of the 9th March in Hampton Roads, abetted by nearby ships and watched by crowds of civilians and soldiers who knew that the outcome of the campaign on land might be settled there and then. During an engagement lasting over four hours, *Monitor* had twice to withdraw to shallow water to take on more ammunition and make repairs. Thus *Virginia* dictated the action without sinking her opponent. Firing took place at point-blank range as both ships attempted unsuccessfully to ram. *Monitor*, firing 135-lb shot, could shoot only once every seven minutes because her guns had to be retracted into the turret, which was trained away each time for loading. She scored 20 hits from 53 shots. *Virginia*, discharging a far higher volume of shot and shell, scored 23 hits. When *Monitor* made her retirement for repairs, *Virginia*, leaking at the bows and short of ammunition, also withdrew. Neither side had been seriously penetrated; casualties were light; but both crews were exhausted by the clamour of battle and the toil of working the guns and recalcitrant steam engines.

The contest may have been drawn, but it represented a strategic victory for the Confederacy because the presence of *Virginia* blocked the plan of General George McClellan, Lincoln's General-in-Chief, to invade the Yorktown Peninsular via the James River prior to advancing on Richmond. Lincoln feared, moreover, that *Virginia* would next sail up the River Potomac to bombard Washington. So when she emerged in the James River after repairs on 11th April (by which time the Federal Army had landed from the less favourable York River) *Monitor*, now joined by a sister ship *Galena*, was told to stand off on the defensive to guard the landing places and the Potomac.

The fight at Hampton Roads had been a conclusive victory for steam-driven iron-clads, their cannonade sounding the death knell of the old wooden walls. When the Confederacy destroyed the *Virginia* on 11th May, due to the need to evacuate Norfolk in face of the Federal advance by land, the balance of naval power swung irrevocably in the North's favour. Hopeless seaboat as the *Monitor* class was, the armoured ships and river craft which the Federal Navy now built (and which the Confederacy, for lack of facilities, was unable to copy) assumed almost complete command of the sea routes, as well as the rivers which led to the South's heartlands. A few steam and sail Confederate commerce raiders, such as *Alabama* and *Shenandoah*, might brilliantly raid merchant shipping, but they could not break the ring constricting the South or seriously curtail the North's own trade with the rest of the world.

The Peninsula Campaign And Trench Warfare

When the 34-year-old General George McClellan persuaded Lincoln (against the latter's judgement) to leave only 75 000 men guarding Washington from behind fortresses and land more than 100 000 men on the Yorktown Peninsula on 22nd March to strike at Richmond by sea, he sowed the seeds of failure by keeping secret from the President the fact that he was leaving only 50 000 to guard the capital. For when Lincoln discovered the deceit, he withheld 25 000 men from McClellan. By then the General was enmeshed in what amounted to almost constant and costly siege warfare against a series of well-entrenched lines of resistance, dug across his predicable line of advance through the ten-mile neck of the peninsula, but guarded by only 60 000 enemy troops under General Joseph Johnston.

The campaign developed a pattern hitherto unknown in warfare. Excepting sieges of fortified cities, combat in the past had been of short duration, major battles rarely lasting for more than a day. The Peninsula campaign, commencing with the Battle of Kernstown on 23rd March as part of General Jackson's diversionary activities in the Shenandoah Valley, and ending in the withdrawal of Federal forces from the Peninsula in August, consisted of almost ceaseless fighting. Including the siege of Yorktown from 4th April to 4th May, there were no less than six major battles in the Valley and nine in the Peninsula, connected by continual skirmishing and one major raid by a cavalry division. Moreover, the Peninsula fighting coincided with a major campaign in the west, on either side of the Mississippi, where the struggle to control that jugular vein of the Confederacy culminated in the bloody Battle of Shiloh on the 6th and 7th April; the capture of New Orleans by a Federal fleet of 17 warships under Admiral David Farragut on 25th April; and the fall of Memphis to Federal river gunboats on 6th June. Losses were colossal — 14 000 Federal and 11 000 Confederate troops at Shiloh alone. Exhaustion became endemic, halting operations.

Although these heavy losses could partly be ascribed to errors of raw troops, as well as to poor staff work, the underlying reasons were improved technology which had redoubled firepower, and crippling deficiencies in communication which technology had not yet solved. When McClellan advanced from Yorktown in the direction of Richmond, his progress was slowed by an out-numbered Confederate rearguard which gave ground only grudgingly on a wide front. This was possible because no longer did men need to be packed into tight ranks in order to generate sufficient volume of fire to maintain their position against assault. Reciprocally, the thinning out of ranks made them less vulnerable to incoming fire. Such gains were ameliorated further when men took to lying down to shoot or, better still, made a point of firing from trenches or behind cover instead of standing up in the open, as so recently in the past decade. Not that either army was yet able to apply the full devastating potential of modern weapons. Many old, muzzle-loading rifles were still in service, but a sound of the future ripped forth at the Battle of Fair Oaks when on 31st May, within sight of Richmond, a battery of hand-operated Williams machine-guns chimed in to support the first Confederate counter-stroke — a battle which was to save their capital city though it failed, with losses of over 6000 men, to drive McClellan back.

As had been shown in the Crimea and at Solferino, head-on assaults against a well emplaced enemy of equal calibre were no longer profitable operations of war. Even less viable was cavalry against modern artillery and rifle fire. The only chance of making a worthwhile mounted contribution was to ride through gaps in the enemy lines, both for reconnaissance and for raids, into the wide open spaces of the enemy rear. In a country the size of America, and with relatively small forces engaged, there would always be gaps and nobody was better at exploiting these than General J E B Stuart, as he demonstrated between 12th and 15th June when he rode right round McClellan's army, creating havoc in the rear and returning to Lee (given command in the field after Fair Oaks) with invaluable information about Federal dispositions.

The Seven Days' Battle

Ten days later Lee made use of the railways to shift Jackson's army suddenly from the Valley to Richmond. By so doing he unexpectedly produced a superior concentration of force to attack his enemy at

Below: ***Thaddeus Lowe's hydrogen-filled balloon ready to go aloft to 'spot' for the Federal forces.***

Mechanicsville on 26th June — a striking exhibition of the strategic use of railways which might also have been a tactical victory if only the trains had arrived in time for Jackson to take part. As it was, the Federal troops were able to withdraw from the trap and execute a number of steps backwards. Throughout the next week, Lee attempted to outflank or roll over successive Federal positions without ever quite succeeding, but at the end his exhausted forces had suffered over 20 000 casualties, against about 16 000 of McClellan's.

McClellan meanwhile had shifted his base to the James River, safe now after the destruction of the *Virginia* on 5th May when its naval yard was threatened. If provided with reinforcements, he was ready to strike at Richmond again with every chance of success, because by exercising seapower in the James River he could this time advance on either side of the water. That he was denied the reinforcements and later withdrawn to Washington is just one reason why the Civil War dragged on as long as it did. But politics always have a say, and Lincoln's fears played a part when McClellan was relieved of command, presenting Lee with the chance to establish a stronger situation.

The War Of Attrition

The following three years' struggle to exhaustion saw battles the pattern of which had been set by the events of spring and early summer 1862 and which were governed, in the long run, by industrial strength. Resources in manpower counted, but were of lesser importance than in the past because machinery was playing a dominant role both on and behind the battle-fronts. When the South attempted its only major invasion of the North, smashed after three days in a head-on struggle between masses at Gettysburg on 3rd July 1863, it was a sign that Lee's generalship had been matched by the generalship of Lincoln and his leaders. But when Vicksburg surrendered after a long siege to General Ulysses Grant on the 4th July, denoting the North's seizure of the entire Mississippi and the closing of the blockade ring, it guaranteed the defeat of the South. The Confederate economy, let alone its capacity to make war material, was already running down due to the extreme difficulty of purchasing or manufacturing materials of any kind. In contrast, not only was the North turning into a great arsenal, it had taken control of the main systems of communication just when the South's river lines had been cut and its railway system was approaching collapse from lack of maintenance caused by the materials famine. As the war progressed, the inability of the South's railways to move troops would be less of a disaster than their inability to distribute food. Dire shortages resulted in crucial areas of an agrarian state which grew ample food for its needs.

The extension of communication systems, particularly in the North, reached all levels of organization. Along expanded railway routes, troops could be switched from place to place with far greater dexterity in the North than in the South. When Lee moved towards Gettysburg, the Federal Army was able to concentrate against him with relative ease. But when, in the aftermath of Vicksburg's fall, the Federal Army of Tennessee marched on Chattanooga, it not only outmanoeuvred its opponent but in moving by rail, across 1200 miles in seven days, 23 000 men, their artillery and logistic support, it also outnumbered an enemy faltering at the end of a decrepit railway line.

Firepower dominated everywhere. Even with superior numbers and equipment, Grant, by 1864 in control of the Army in the East, was unable to overcome resolute, emplaced Confederate defenders in the approaches to Richmond. He suffered enormous losses in the Wilderness Campaign against infantry and artillery which dug-in as a matter of routine. For an assailant to prevail in these conditions it was no longer permissible for a thousand or more men to push forward shoulder to shoulder on a narrow front without concentrating intensive artillery support on the centres of enemy defence before and during the attack. This sort of fire integrated with movement, which had been only occasionally practised in the past and then rarely deliberately, was yet another form of revolution calling for sophisticated, pre-organized staff work founded upon better communications. Messages from the political centres, written by leaders with an improving knowledge of military realities, could be despatched along multifarious telegraph lines (like the railways, a frequent target for cutting by cavalry patrols and guerrilla bands) to every headquarters in the battle zones and often very near the front. From there to the heart of combat, visual signalling — flag and heliograph using Morse code — would take over; providing, that is, darkness, fog or battle smoke did not intervene. In a rudimentary manner, information and commands could be sent to troops who were engaged in close contact with the enemy. Equally, if not more, important, artillery observers could instantly communicate with guns and mortars lying back at extreme range in order to correct their fire on targets out of sight of the gun positions. This technique entered a new dimension when the Federal balloons of Thaddeus Lowe, tethered (and sometimes towed by a coal barge) lifted artillery observers aloft for them to direct the guns on the other side of the hill.

In effect, a measure of remoteness replaced the unavoidably direct confrontations of old, bringing with it new complications, not least a feeling of hopelessness in the minds of men who saw little chance of survival in the face of machines. One sight of a battlefield, carpeted with dead and wounded by massed fire within a few minutes, was often enough to make even the greatest enthusiast for a cause doubt the wisdom of competing — and such sights were by no means uncommon. The culminating moment, for almost the first time in history, came when the sheer quantity of losses and suffering began to convince one side — the North, which was on the verge of victory — that the struggle should be abandoned. It arrived in the summer of 1864 before Petersburg, where technology absolutely stifled human endeavour and courage.

Trench Warfare — The Lines Of Petersburg

When Grant deluded Lee as to his true intentions after the Wilderness Campaign, managing suddenly to appear in mid-June with massed forces at Petersburg instead of further north as expected, the thinly defended city lay at his mercy. But war weariness and a conditioned caution held back the Federal troops who now approached all entrenchments warily as a matter of course. One quick, determined rush by a Corps on 15th June might well have broken through. Three days later an army of 65 000 was insufficient to overcome the 40 000 men Lee had rushed to the spot by rail. Faced at first by an improvised line, the initial Federal assault failed from lack of co-ordination. Detachments advanced independently, inadequately supported by artillery, and were pinned to the ground by fire of only moderate intensity. By the time a set-piece attack could be launched on the 18th, the volume of defensive fire was annihilating, compelling Grant to call a halt and commence probing the city's southern flank with a view to isolating it. Keeping pace with each Federal sidestep to their left, the Confederates extended their entrenchment to their right, always in time to meet each assault while fiercely contesting Grant's further attempts to cut the line to Richmond or the one running westward from Petersburg. Assault was usually of the battering-ram sort — a blasting of the selected point of attack by artillery and mortars (the latter, with their plunging fire, being particularly suitable for striking at the deeper enemy emplacements) followed by a massed infantry charge.

Historians accuse those in charge of a succession of failed set-piece attacks with bungling. To some extent they are right, although they tend to overlook that the dimensions and ferocity of modern war had produced a complex problem beyond the knowledge and technology of the day to solve. 'In war', said a Prussian officer called Hindenburg, many years later, 'only simple plans work'. In 1864 simplicity could not be adopted. Even if every plan had been perfectly devised, staff work impeccable, communication arrangements faultless and every order executed implicitly, the weather, or the enemy could be expected to disrupt them. But nothing could be perfect in this form of warfare, with masses of men and numerous weapon systems somehow to be co-ordinated. Humanity failed in all its unpredictability. That way chaos and slaughter were assured.

The attack on the Redoubt at Petersburg on 30th July demonstrated in utter confusion the inability of commanders to make human courage prevail over material factors and human frailty. As a powerful augmentation of the, by now, conventional artillery concentration of fire, a mine containing four tons of black powder was to be exploded beneath the redoubt and its defenders. Placed in a cross shaft at the end of a 511-foot tunnel which a regiment of coal miners secretly dug, it was blown at dawn without warning to the enemy. General Ambrose Burnside, whose four divisions of infantry were to exploit the explosion, seems to have relied too much upon the shock effect of the mine; beyond doubt the measures he took to ensure that the troops not only occupied the crater but pressed on rapidly beyond, were ambiguous and unambitious. As for the troops, so staggered were they by the enormity of the explosion, the air pressure of its blast and the scene of carnage which met their eyes when they poured into the crater, that they lost all sense of purpose and stayed there all morning, poking about among the grisly ruins of dismembered men and equipment. On the Federal side leadership came to naught while among the Confederates initial shock was gradually overcome and a counter-stroke launched in the afternoon. Artillery sealed off the flanks of the 500-yard breach in the defences, as infantry rushed to the lip of the crater where they fired volleys into the disorganized mob below. The Federals were flung back with the loss of 3793 men. That day the Confederates lost 1182, including those blown up.

For the rest of the summer and into the fall, Grant strove fitfully to break the deadlock in front of Richmond and Petersburg, creating for the logistic support of his troops a comprehensive conglomeration of base depots, camps and railway spurs leading to within artillery range of the enemy. Facing the Confederate capital the entrenched front was some 37 miles long, manned by 90 000 well-provisioned Federal troops on one side, and 60 000 deprived but fanatically determined Confederates on the other. Try as he would to smash through, Grant was defeated. Likewise, Lee was rebuffed when in March 1865, a last sortie took Fort Stedman but got no further than its ramparts before it was stopped by a Federal counter-attack. In a four-hour battle, the attackers lost twice as many men as the defenders — 4000 to 2000.

Had Grant's exploits comprised the sole Federal effort in 1864 they could well have led to his and President Lincoln's downfall in an election year. The disgruntled General McClellan was campaigning for the Democratic candidacy with a call for an end to the war. He might have won if General William Sherman, taking advantage of the drain of Confederate strength to the east, had not struck the decisive blow in the west.

Chattanooga And The Drive To The Sea

In September 1863, before leaving to take command in the east, Grant had commenced the assault on Chattanooga and the adjacent mountains and ridges. It reached its climax on 24th and 25th November when the strongly-held key heights of Lookout Mountain and Missionary Ridge were taken by surprise — the former with only slight resistance in a battle amid low clouds; the latter an improvised rush to the skyline by Federal infantry who exceeded their initial orders to seize trenches at the bottom of the ridge only. Possession of the route centre of Chattanooga represented the key to the South. Henceforward in the west, no matter how many diversionary operations the two sides chose to mount, they all faded into insignificance when compared to the central thrust of Sherman's army as it set out across Georgia from the vicinity of Dalton early in May.

Sherman's strategic blow, aimed at the economic heart of the Confederacy and carefully synchronized with Grant's advance into the Wilderness, was intended by outright destruction and terror to wreck property, lay waste the land and make 'a hostile people feel the hard hand of war'. More than that, his drive was an act of vengeance upon the people of the South for their past misdeeds, as Sherman and his army saw them. With astute tactical perception, Sherman tried to avoid the weaker armies of his opponent, General J Johnson, in order to make haste at low cost for the key railway town of Atlanta. This he reached on 22nd July after a shrewd delaying action by Johnson and a fierce battle to take Kennesaw Mountain. Here Sherman at first succumbed to the temptations of a direct assault (to be repulsed with the loss of 3000 men against 800) but found final victory through a more laborious flank attack that prised his opponent from the mountain at meagre loss of lives. Atlanta, too, was a tough nut to crack, its entrenchments holding for over a month under heavy bombardment and the effects of flank action. When this cut the railway supplying the city from the east, the Confederates finally withdrew.

There is no doubt that Sherman's celebrated progress through Georgia presented Lincoln with the ammunition he needed to defeat McClellan's bid for the Presidency and attempt to stop the war. It is equally certain that the depredations of Sherman's five columns, marching in parallel on a frontage of more than 50 miles, convinced many civilians in the Confederacy that the game was up, resolute as their tattered soldiers remained in combat. The ruthless destruction of Atlanta and the railway lines radiating from it; and the pillaging of a wide belt of country bordering the sea at Savannah (which Sherman reached after a six-week march of 500 miles) highlighted the impotence of his enemy to prevent further desolation of the countryside. From a logistic point of view it was masterly, in that Sherman harnessed his policy of destruction to the support of his army of 70 000. Cutting the railway behind him, he lived off the countryside until making contact with the fleet at the coast. By carrying his ammunition and medical services in over 3000 vehicles with him, he created a communications void in his rear which was an embarrassment only to an opponent unable to keep up with his movements and incapable of preventing the deep South from being cut off from the main seat of battle at Petersburg and Richmond. A cloud of skirmishers fanned out ahead and to the flanks of Sherman's marching columns, taking what food they wanted, putting the torch to the rest, wrecking the cotton industry and perpetrating damage on a scale which retarded the economy of the South for decades to come.

Yet it was another indication of the changing face of war that the South, with defeat staring it in the face, continued to resist — in part because the persuasive power of propaganda supporting its cause prevented it from publicly admitting to bankruptcy. It was a phenomenon less common than in the past but, in similar circumstances, would be repeated in wars of the future when the apostles of strife were unable in ultimate defeat to abandon the habit of combat. Not until the first week of April did President Davis feel compelled to retire from Richmond and Petersburg because Lee was then being driven back for logistic reasons, apart

A WEEKLY JOURNAL OF PRACTICAL INFORMATION, ART, SCIENCE, MECHANICS, CHEMISTRY, AND MANUFACTURES.

Vol. XL.—No. 24. [NEW SERIES.] NEW YORK, JUNE 14, 1879. [$3.20 per Annum. [POSTAGE PREPAID.]

from pressure by Grant to his front and Sherman moving northwards upon his flank and rear. Even then, struggling in the open with a mere 30 000 scarecrows of soldiers, resistance was prolonged until 9th April when Lee was on the point of encirclement and annihilation at Appomattox.

New Technology — The Lessons Ignored

To the incredulous military experts of Europe, where the art of war was jealously nurtured by professionals who had sedulously studied the subject, the long-drawn-out American Civil War with its death roll of 500 000 was an aberration by amateurs to be dismissed out of hand. They overlooked the basic lesson that new technology had undermined old precepts. In doing so they also lost sight of the fact that the war had ended before even more devastating weapons, already in existence, could show their potential.

Several types of machine-gun had been used, in addition to the Williams which had made its battlefield debut in 1862. The most powerful was Richard Gatling's hand-cranked, seven-barrelled, rotating gun, introduced in 1864, which could ram bullets into the chamber mechanically, pour them out at 3000 a minute and then extract them, also mechanically. The key to Gatling's invention lay in vastly improved ammunition — the light, all-metal cartridge (first designed for sporting purposes by a Frenchman, B Houiller, in 1846) which not only facilitated mechanical ramming and extraction, but also provided tighter breech sealing and, therefore, higher power and greater accuracy. The firepower of the individual infantryman was further enhanced early in the Civil War by the new cartridges in repeater rifles, first produced by Horace Smith and Daniel Wesson in 1852 but brought to rugged practical use by Christopher Spencer in 1860 with his rifle and carbine containing a seven-round magazine in the butt. Paradoxically, the Confederates tended to neglect this rifle because they feared excessive expenditure of ammunition, of which they were in short supply.

Set against the greater propensity to kill was a drive to reduce loss of life. Unprecedented efforts were made by both sides in the USA to remove the wounded quickly from the battlefield and treat them in hospitals organized along the lines laid down by Florence Nightingale. Many lives were saved which in the past

Left: Contemporary illustration of the hand-cranked Gatling gun, showing it on tripods and field mounting.

Above: General von Moltke watches the Prussian Army move into action at Gravelotte in 1870. Aided by primitive telephone control and rail movement, von Moltke's ability to co-ordinate a concentration of forces at the critical point was often faulty, but effective.

would have been wasted, but surgeons operating with unsterilized instruments continued to lose patients. Too late came the discovery by Joseph Lister in 1865 of the power of antiseptics to sterilize and prevent putrefaction in wounds. From his first experiments with carbolic acid (Phenol) a new branch of science would emerge to save countless millions in war and peace and advance the surgeon's skill to unprecedented effect — a process Lister advanced further by producing in 1880 an antiseptic catgut ligature which made safe the operation of arresting haemorrhage.

The Prussian Wars

Between 1864 and 1871, three wars linked to the unification of Germany under Prussian leadership, and provoked by Otto von Bismarck, made use of some new techniques and weapons, but mainly served to convince Europeans that the American experience of war was an anomaly; that trained forces, well armed and directed by experts in strategy and tactics, could win a campaign rapidly and at a politically acceptable cost. The Austro-Prussian invasion of Denmark in 1864 to settle claims over Schleswig-Holstein was determined by a single month-long siege of Dybböl and a quick invasion — the entire, desultory campaign lasting less than six months. Of even shorter duration was Prussia's attack on Austria in 1866, heading an alliance of several German states and Italy, welded together by Bismarck. Under General Helmuth von Moltke, who initially controlled operations by telegraph from Berlin, the Prussians invaded Bohemia and after several preliminary encounters routed the Austrians at Sadowa on 3rd July. The war ended two days later. As it was, the Prussians should have noticed trends akin to those in America. Rifled Austrian artillery dominated the obsolete smooth-bore cannon still in Prussian use and prevented cavalry following up after Sadowa. Moltke's remotely planned, synchronized arrival of his forces on the battlefield broke down after characteristic communication failures and what von Clausewitz termed 'friction'. The effective Prussian rifle, when at last brought to bear against Austrian masses, had a devastating effect and won the battle.

A digressive influence in the Austro-Prussian war was the naval Battle of Lissa, fought in the Adriatic. In this incompetent melée between iron-clad Austrian and Italian steam frigates, the sinking by ramming of the Italian *Re d'Italia* was taken by many naval men as indicating that in future an important weapon in close combat (still assumed to be likely) would be underwater attack by rams. Those who believed this conveniently overlooked that a far more deadly form of underwater weapon, the locomotive torpedo, had already been demonstrated and that the latest guns, mounted in improved turrets, made it extremely unlikely that ships would survive attempts to close.

Similar frictions and miscarriages of intent and execution to those which distinguished Sadowa also featured at the start of Prussia's attack upon France, on 15th July 1870, after Napoleon III had been tricked by Bismarck into declaring war. In the battles which developed along the frontier, both sides often failed to achieve their planned concentrations of force. Each was susceptible to charging emplaced troops, with resultant horrific losses that might have been even worse had the French used their 25-barrelled, hand-operated machine-gun, the Reffye mitrailleuse (literally 'grapeshot shooter') as a short-range infantry weapon instead of medium-range artillery from 1250 yards. As it was, a better organized and trained Prussian Army soon pushed back a French Army of inferior quality whose leaders were split by factions and whose will to fight was lukewarm. The capture of Napoleon along with his main force at Sedan on 1st September was the crowning triumph of von Moltke and set the seal upon Bismarck's policy of formally unifying the German states into one great nation. But Sedan did not end the war as Sadowa had done.

French pride rejected peace overtures. Her armies retreated or were locked up in Strasbourg, Belfort, Metz, Paris and several smaller fortified cities, with every intention of withstanding prolonged sieges. And where the Prussians chose to try smashing forts with their, by now, thoroughly modern artillery, they came up against the perplexing problem which previously had hampered the Americans. At Paris a ring of forts linked by field works was left unassailed; the Prussians were content to let the city surrender from starvation of its large population (as did the garrison at Metz). Belfort, however, with its garrison of 17 600 men, was recognized as a threat that might be removed with relative ease. The assessment was proved sadly incorrect by a French garrison which delayed the Prussian advance to the fortifications and then stubbornly declined to succumb to intensive bombardment and a 105-day siege. For despite the expenditure of 1000 shells to open a breach a mere 30 yards across, a final assault was deemed inexpedient and the siege ended only when the garrison was instructed by its government to surrender. Like the Battle of Lissa, the siege of Belfort would also teach lessons which could be misunderstood.

The 'Arms Race', entering a new leg of the course, now broke from a trot into a canter.

Chapter 3

Firepower and the Masses

1871-1916

Shifts in the balance of international power throughout the latter half of the 19th century were closely related to political instability and an arms race of unprecedented dimensions. As the Ottoman Empire, Austro-Hungary and France slid into decay and as Britain and Russia sought to expand, the newly-unified nations of the USA, Italy and Germany also exhibited positive signs of an urge to expand. Soon they would be joined by a re-united Japan, which emerged from feudalism in 1869 and at once took steps to acquire Western technology, including modern weapons. Science, technology, national commercial and personal ambitions combined, in a predatory atmosphere, to stimulate the development of the arms industry. Simultaneously, an accelerating rise in population, mainly caused by improvements to primitive public health measures and in medical science, created a seemingly inexhaustible pool of manpower from which governments and armed forces could draw, ever more frequently through some form of conscription.

The revolutionary consequences of new technology, so dramatically displayed in the wars of the 1860s, prompted an enthusiastic burst of research and invention, much of it directly related to armaments. Of crucial importance were the latest sources of power and the many new materials revealed in the 1870s and 1880s — above all the means to produce vast quantities of electric power after 1878, when Thomas Edison's first commercial generator was unveiled, and the invention of the internal combustion engine seven years later. None of these would have been feasible had not steel alloys been found for building structures and machines that were previously unattainable. Abruptly, the Industrial Revolution's first phase of coal and iron entered a second and far more exhilarating period, becoming increasingly dependent upon electricity, liquid fuels and steel.

Warships And Fleets Transformed

The principal navies could not avoid taking a lead in modernization after Hampton Roads and Lissa. The majority of warships in commission were out of date and virtually defenceless against any minor naval power which equipped itself with a mere handful of modern warships. Now that steam propulsion of screw-driven ships was established, sails could be abolished and full attention paid to improving power plant. HMS *Devastation*, the first battleship without sails, was launched in 1871 but like all those to follow, she also incorporated the latest gunnery system. Notice had been given that henceforward priority would be paid to ways and means of destruction versus means of protection. In fact, there were two separate but coincidental lines of approach in naval thinking: attack by guns upon hull and superstructure, rivalled by underwater attack by mines and the locomotive torpedo.

As cheaper steel became available in vast quantities and demonstrated superior qualities over iron, nearly all guns were made from it. Simultaneously, the earlier system of 'hooped' barrels was superseded by a method of steel wire wound over the inner tube. And any controversy over the attributes of muzzle or breech loaders was finally resolved in the 1870s by the need to solve problems inherent in loading and firing pieces which grew steadily in weight and length. A British-built 17$\frac{3}{4}$-in calibre gun weighing 100 tons and firing a 2000 lb shell was in service by 1876, while pieces of 40 feet long were already in existence. Lengths were bound to increase, as this proved the best way to

Warship Design And Gunpower In The 1870s

The greatest single factor to influence warship construction in the wake of the iron-clads was the introduction of steel plate, in place of wrought iron, as the main armour protection. Improvements to armour had been necessitated by the phenomenal increase in calibre, and hence weight, of shipborne guns to 38 tons, and then rapidly to 76 tons. Further advance eventually led to Armstrong's 100-ton, muzzle-loading rifled guns supplied to the Italian Navy in 1873. The trend in motive power was from sail/steam to steam only.

Above: The first non-sail capital ship, HMS **Devastation**, was completed in 1871 and this contemporary illustration shows the important elements of her construction.
Top left: A turret containing 4 × 305 mm muzzle-loading guns gives an indication of the armour thickness, which was between 305 to 216 mm at the belt with about 457 to 406 mm teak backing. Subsidiary armament (b) consisted of 6-pounder and 3- pounder guns protected by armour.
Right: The relative penetrative effect of guns against combinations of iron backed by wood, about 1870, shortly before the introduction of armoured steel plate. Comparative diagrams show the difference in penetrative power of British muzzle-loading guns against iron, backed by wood, and German guns against similar materials.

augment muzzle velocity in order to pierce thick armour. The business of manoeuvring a long gun inboard, off-target, in order to push charge and shell into its muzzle was both laborious, complicated and time-consuming. Rates of fire for breech loaders could be twice as great. Their most notorious defect, when first produced in quantity by Krupps and Armstrongs, was a susceptibility to bursting cases or emission of gas and flame due to inadequate sealing between breech and chamber. But by the time of the Franco-German war these faults had largely been eliminated.

Controversy attended the development of the turret, also due to the problems of loading longer guns, but this was overcome once muzzle-loaders were finally done away with. From the 1880s onwards, the turret became the almost universal, self-contained housing for a ship's main armament, its ammunition supply brought up by hoist from magazines located below. Loading and traversing were powered by hydraulics (or later by electricity); and recoil was no longer taken up by friction devices or rope restrainers, but absorbed by steel spring and hydraulic systems (first suggested by the Siemens brothers in England in 1862) which returned the barrel to rest in its firing

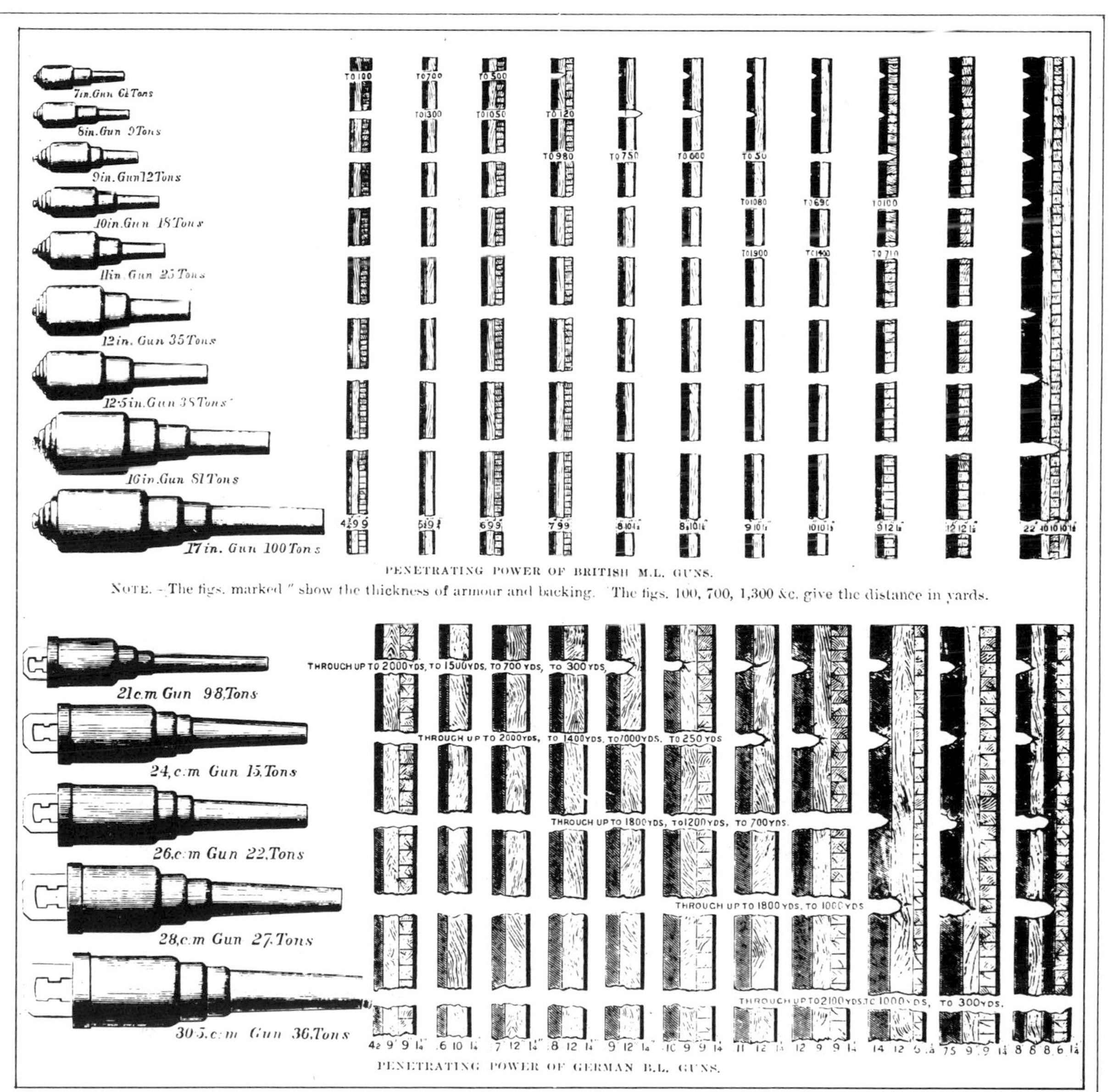

PENETRATING POWER OF BRITISH M.L. GUNS.

NOTE.—The figs. marked " show the thickness of armour and backing. The figs. 100, 700, 1,300 &c. give the distance in yards.

PENETRATING POWER OF GERMAN B.L. GUNS.

The Development Of Underwater Attack In The Aftermath Of Bushnell

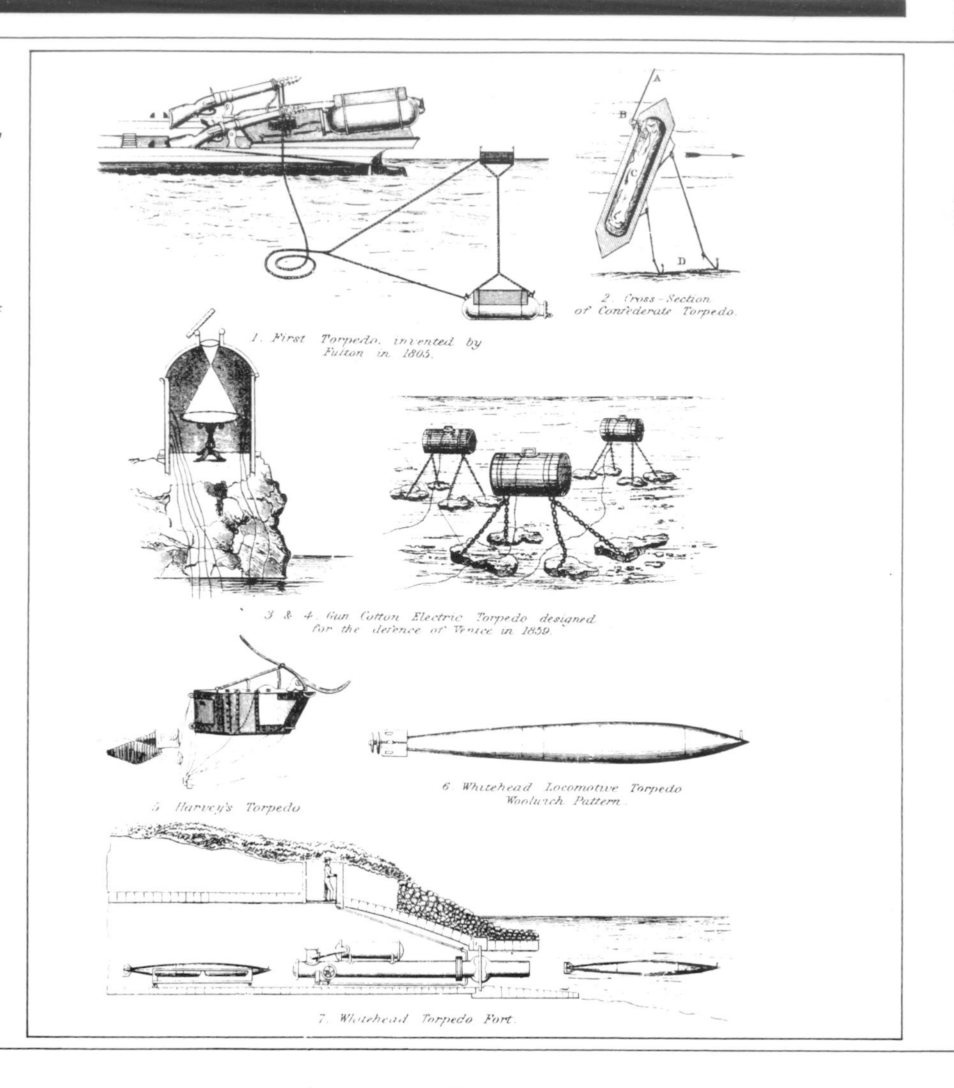

Right: *Initially called a 'torpedo', the uncontrolled moored mines of the early 19th century were fired by means of a gunlock or by clockwork. The idea of Samuel Colt's electrically-fired controlled mine of 1843 became very popular but was overtaken to some extent by the glass-tube chemical contact mines of Nobel in the 1850s. It was the Luppis Whitehead torpedoes of the 1860s, driven by rocket or compressed air engines, which at last gave offensive power to what had previously been a strictly defensive weapon and so revolutionized sea warfare.*
Far right: *Meanwhile, the development of submarines continued, the Rev. Garrett's steam-driven* ***Resurgam,*** *which sank on trials in 1879, being an example.*

position and dispensed with the need to reposition guns after every discharge.

In parallel with the intensive work devoted to guns and turrets went development of new shot and shells to defeat the thicker armour protecting their targets. Solid iron shot failed as an armour penetrator because it broke up too easily on impact. It was superseded by steel shot with a hardened tip, later improved by the addition of a soft steel cap over the nose to take the first shock of impact. Quite as much research and trial went, of course, into improvements to the armour these missiles were intended to smash or penetrate. Naturally, under trial, steel proved superior to iron. When one of the 2000-lb shells from a 17$\frac{3}{4}$-in gun struck iron armour backed by 29 ins of wood at 1470 ft per second, it penetrated; against steel, however, it pierced only 21 ins deep.

More efficient guns depended to a large extent upon improved propellants. The start made by Rodman in the 1850s was steadily built upon by others from many nations. In 1846 C F Schönbein, a German, had created guncotton (nitrocellulose) by adding nitric acid to cotton; subsequently improved by the Austrian Baron von Lenk in the 1860s, it was stabilized in colloid form by Paul Vieille and adopted by the French Army as Poudre B in 1885. This was four times more powerful than existing explosives. Then, in 1875, the Swede Alfred Nobel successfully mixed guncotton with nitroglycerine and in 1885 went a stage further by developing this mixture into ballistite, which was smokeless — a vital advance. Also about this time, Frederick Abel and James Dewar (both British) discovered cordite by mixing nitro-cellulose and nitroglycerine with mineral jelly. These

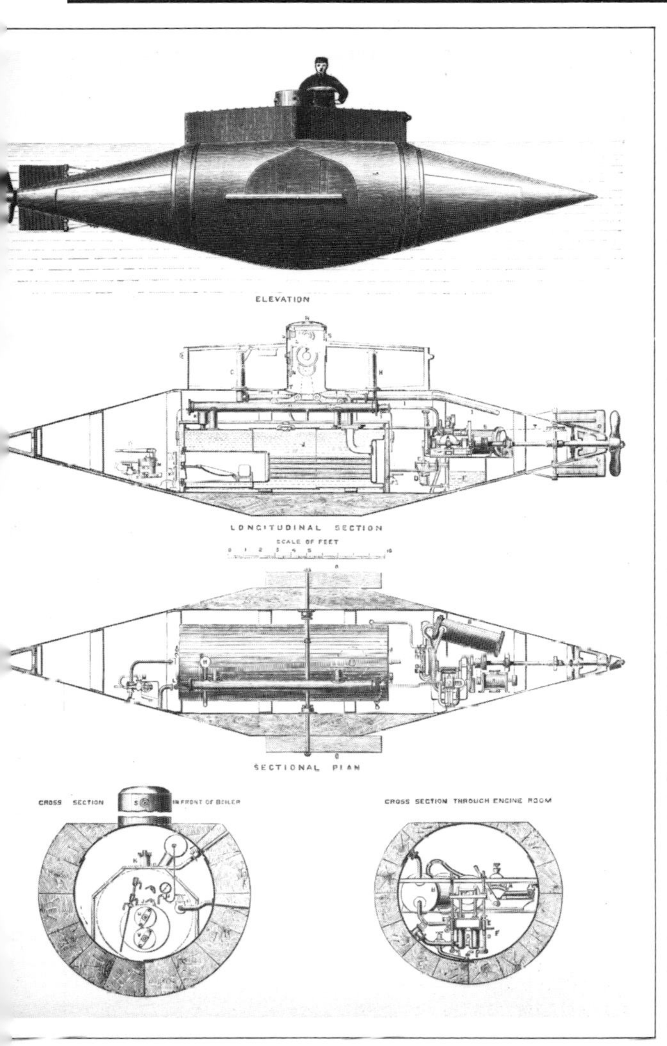

smokeless propellants not only increased the effective range of artillery and led to a new generation of guns and small arms, they opened a new era in shooting techniques and tactics since the gun position would henceforward not be clouded in smoke (as an advertisement of its position) and the gunners would be free of smoke obscuration, enabling them to observe the fall of shot and make positive corrections to aim.

In parallel, significant hardening and strengthening of steel came from the addition of special elements, such as tungsten, nickel, chromium and manganese. Greatest among the metallurgists researching alloy steels was the Briton Robert Hadfield, who invented manganese steel in 1882 and silicon steel in 1885 — the former of immense importance for armour and shot; the latter of great use in the burgeoning electrical industry. Simultaneously, M Marbeau of France invented nickel steel, soon used in America for ships' armour plate. Alloy steels were more expensive than plain carbon steel, but their application by armed forces on sea and land, quite apart from civilian usage, was fundamental and implemented with astonishing rapidity. It was another step forward when in 1876 France launched the *Redoubtable*, the first ship built of carbon steel frames and armour and incorporating a watertight double-bottom and internal subdivision for damage control. It was even more important when, in 1891, the US Navy opted to armour its ships with nickel steel, which did away with the need to back wrought iron or carbon steel with timber.

Attack below the waterline by static mines as a defensive measure had sunk 26 ships during the American Civil War. Offensive underwater operations might have occurred had the US Navy adopted the rocket-propelled torpedo of Pascal Plant, but when his model was tried out before President Lincoln in December 1862, it ran wild and sank a luckless schooner which happened to get in the way. It was collaboration between Giovanni Luppis of Austria and Robert Whitehead of Britain which, in 1864, produced the first practical locomotive torpedo. Luppis demonstrated (as Bushnell and Fulton had shown 90 years previously) that when a gunpowder charge was exploded underwater, the water had a tamping effect. Whitehead put on sale in 1866 a 14-ft, cigar-shaped torpedo, 14 ins in diameter with an 18-lb charge in the nose. It was powered by a compressed air engine driving a propeller, had a speed of 6 knots and a range up to 700 yards. When launched, initially from underwater tubes, but later from tubes carried above water, it ran a set course, controlled in depth by a hydrostatic valve and a pendulum weight working together to activate a pair of horizontal rudders called hydroplanes.

Rudimentary as the Whitehead torpedo was in the 1860s, its impact upon naval thinking and ship construction was immediate and positive. While investigations to improve the existing model went on, foresighted officers began to study changes in tactics which this weapon would impose upon fleet commanders and ships' captains. Among many innovations came the idea of a new class of warship to specialize in attacking with torpedoes. By 1876 the British had built a 33-knot, 19-ton coastal torpedo boat called HMS *Lightning*. Eight years later the Russians had in service 115 sea-going, 40-ton torpedo boats with a speed of 22 knots. Although many old-guard admirals might disapprove of torpedoes, the advent of the torpedo boat could not be ignored. Henceforward battleships would require better protection below the waterline and more watertight compartments, plus extra defensive armament in the shape of quick-firing, breech-loading guns. Escorts, or so-called torpedo boat destroyers, were also called for to supplement fire-power. In 1881, the Royal Navy finally abandoned slower firing muzzle-loading guns and three years later was compelled to take note of a British-built, 386-ton destroyer called *Destructor*, fitted with a triple-expansion engine (the first ship to be powered with this invention by a Frenchman, Benjamin Normand) and with a speed of $22\frac{1}{2}$ knots.

The inescapable outcome of threatening firepower and underwater attack was a race between nations to maintain parity. No nation which looked to the sea for its trade and security could afford to permit rivals a decisive technical superiority. Inevitably increases in weight, power and complexity of arms were called for, together with a considerable rise in costs and a demand for specialists to build and run the machinery and then control the weapons. A wooden ship of the line in 1850 weighed around 3000 tons, mounted some 100 guns up to 10 ins in calibre (with a range of 400 yards) and had its speed controlled by the wind. By the 1890s a steel battleship could weigh 13 000 tons, carry a mere four or six powerful guns of about 14 ins in calibre, with a 10-mile range at a speed of 18 knots.

The overhead costs and logistics of replacing every old ship with types of a totally different kind were enormous. Building and manufacturing called for new industries and the recruitment and training of technicians and labour for unfamiliar tasks, while acquiring new traditions of quality production. So rapid was progress that, in unprecedentedly short periods, equipment was made obsolete by more inventions coming to hand. Warships had recently remained combat-worthy for half a century and more. Now they could qualify for replacement within a decade. Doing away with sail not only placed demands upon engine designers and manufacturers, but also had repercussions in other fields. The rope and sail-making industries were faced with severe cut-backs in production which in turn caused a reduction in the growing and milling of flax and hemp, thus generating problems for farmers. Calls for higher speeds and greater radius of action drove marine engine designers to improve the efficiency of steam engines and forced ship designers to provide far more space for the bunkering of coal. Nations like Britain and France, with world-wide commitments of empire and trade, were compelled to establish coaling bases at strategic points, such as the Falkland Islands, in order to maintain the mobility of their ships. No longer could a fleet be victualled for a year's voyage: movements beyond the range of home ports had to be carefully arranged within a new logistic system, requiring time to assemble, good communications and practice.

Strengthening Fortifications: Artillery Versus Defence

Some similarities of technical impact were experienced by both navies and armies. Bigger guns of harder-hitting power and longer range enabled armies to stand off and destroy targets which previously had to be closely approached by infantry prior to a final, costly assault. Fortresses that had once been proof against bombardment were now dinosaurs needing to be rebuilt or abandoned. Throughout the 1870s and well into the 1880s, the views of those who believed in strengthened fortifications were aired versus those who thought strengthening a costly and wasteful process. Against those who claimed that improvised field entrenchments, such as had been extremely effective in the American Civil War, were better and cheaper than elaborate permanent structures of earth, masonry and steel, the advocates of an expensive modernization scheme for forts held sway — and created as many new problems as they solved.

To begin with, the increased range of artillery no longer permitted a city to rest safely behind a single, close ring of fortifications. It was now necessary to build a ring of several self-sufficient and, if possible, self-supporting forts anything up to six miles from the place they protected. The escalation of cost was so enormous that only truly vital places could be given this protection, leading to the abandonment of many old fortifications at less important localities. Next, the matter of strengthening demanded reconstruction consistent with achieving immunity from whatever large guns might be brought into service. The Germans carried out a trial in 1885, firing 164 rounds from a 210-mm gun at a 430-ft square target, and learned only that accuracy left much to be desired. No hits were scored. Therefore more accurate, heavier guns were ordered from Krupps and from the Skoda factory of Austro-Hungary. A year later the French subjected the old fortress of Malmaison to shells from a 200-mm gun (with thoroughly destructive effects) and established the standards of optimum protection as six to ten feet of earth, concrete and steel. Based on this they began the design and construction of improved forts along their eastern frontier with, eventually, the strongest of all at Belfort and Verdun.

Construction of these forts was largely influenced by a Belgian, General Henri Brialmont, who designed the new defences of Liège and Antwerp deemed vital

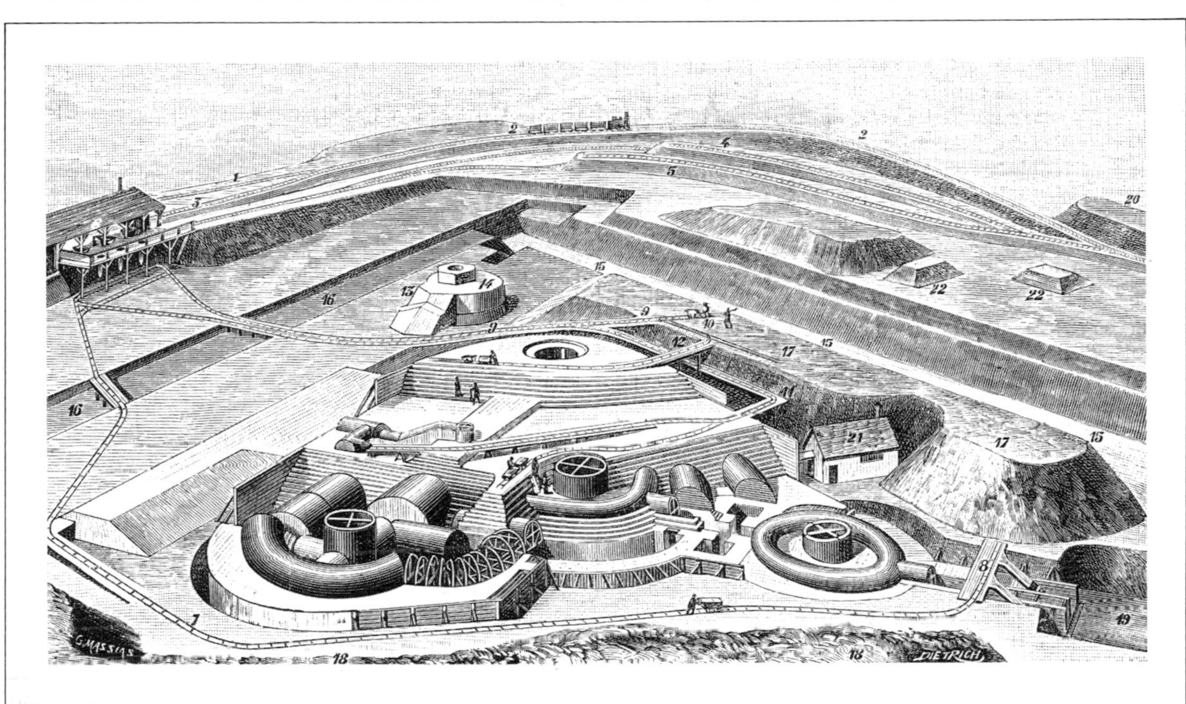

Fortresss Design At The Beginning Of The 20th Century

Right: *The general layout of a Krupp armour-plated, rotating gun turret, similar to many being applied to fortress design in Europe in the 1880 s. In due course turrets would be made to retract for extra protection.*
Above: *The foundations of a typical Brialmont fort.*

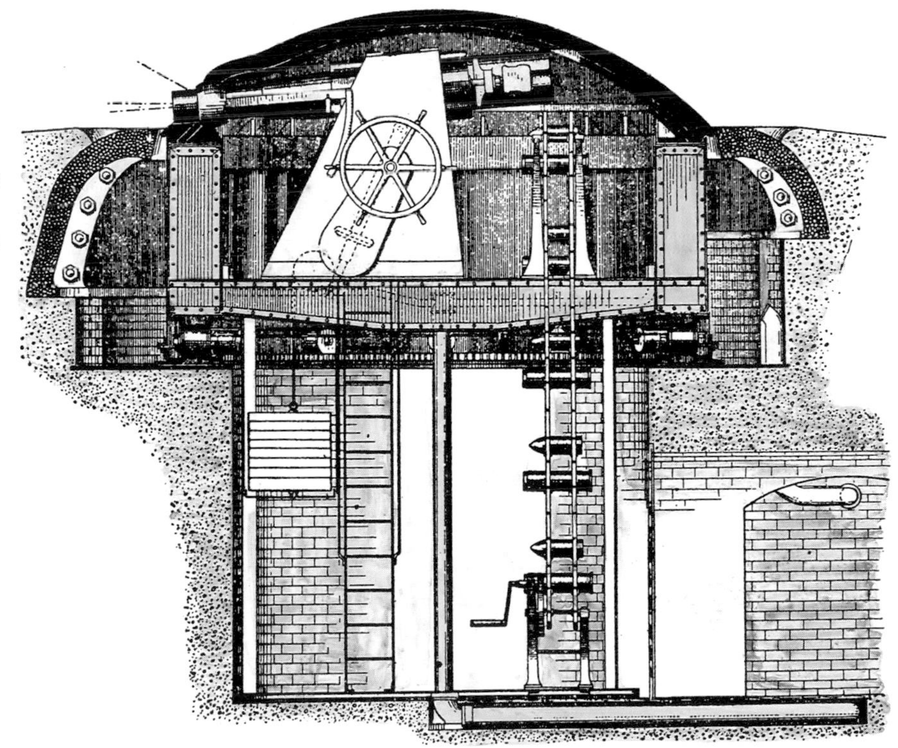

for the defence of his country. Often burrowed into hillsides, they made use, where possible, of natural ground features to strengthen and simplify their layout, and to save cost. Deep ditches, covered by fire from concealed weapons and barricaded by steel, hindered an enemy's approach to the centre where heavy guns and searchlights in retractable turrets and cupolas were emplaced in concrete. Below ground lay accommodation for the garrison, electric power plant and enough food, fuel and ammunition to withstand a prolonged siege. Brialmont was the creator of battleships on land, extremely expensive weapons systems which suffered from the fatally wasteful disadvantage that they were static and so quite unable to influence the battle outside the limited reach of guns which only occasionally had a range greater than those being procured for their destruction. In addition they were often isolated, since at the time of their construction, Brialmont's advice that they must be linked by fieldworks was usually ignored.

Because of its unrivalled destructive powers, artillery continued to be regarded as the dominant weapon, and for practical and political reasons, it was Germany which led the way in the building of heavy pieces, simply because she found herself confronted on all sides by fortifications being modernized as a result of her neighbours' fears — the dread that the bombastic regime which replaced that of Bismarck might repeat the invasions of the 1860s and 1870. And Germany, feeling hemmed in and threatened traditionally by invasion from Russia or France, opted for a pre-emptive offensive strategy should war break out. For this she required weapons capable of cracking the forts that barred her way. Of the super heavy guns, howitzers and mortars she developed, king of them all was the mighty 75-ton, Krupp 42-cm howitzer which could lob a 2052-lb projectile a distance of 15 530 yards. Almost equally effective, and with a higher rate of fire, were the Skoda 305-mm howitzers which would throw ten 846-lb shells every hour to a range of 13 124 yards.

For use in large numbers in mobile warfare, the new generation of cheap light and medium field guns developed in the 1890s were more important than their heavy sisters. Of these the French 75-mm quick-firing gun of 1897 is justifiably famous since it set an entirely new trend in artillery design and practice. The beauty of this gun (made by Schneider) was its simplicity of operation. Like existing naval guns, the piece was carried in slides on a cradle, to which was fitted an hydraulic cylinder. Moving within the cylinder was a buffer piston connected to the piece to control recoil. Weighing barely more than one ton, the 75 was easily handled on a pair of wheels and required only a small 'spade', attached to the trail, to hold it in place against the jolt of recoil, most of which was absorbed by the hydraulic system. The loader opened the breech, pushed home the one-piece round (like an enlarged rifle round), closed the breech, waited for the gun to run back on recoil and return to its position of rest, re-opened the breech (thereby ejecting the spent case) and re-loaded. A well-trained layer and loader, assisted by two ammunition handlers, could fire six aimed 16-lb shells a minute out to 7500 yards.

With either a shrapnel or a high explosive round, this was a deadly rapid-fire gun, the secret of which the French were unable to keep, try as they would. Soon Krupp was producing a 77-mm quick-firing field gun and selling it abroad. Simultaneously Britain, Russia and Austria were also making their own versions.

The immense potential of rapid-fire artillery was recognized by professional soldiers who, like sailors, insisted upon re-equipping their artillery forces in order to maintain parity with possible opponents. The rapidity and accuracy of fire to longer ranges of the new guns were visualized as introducing, above all, far stronger, war-winning offensive power. Optimistically, the leading soldiers persuaded themselves that the negative defensive power of small arms, so devastatingly demonstrated in the 1860s, might be overcome by intensive shell fire. It generated good business for the inventors and manufacturers of explosives and artillery — the Nobels, Hadfields, Armstrongs, Krupps, Schneiders, Erhardts — and it enhanced the burgeoning arms industry to which individual inventors, more than before, had to turn for support with ideas which were growing increasingly complicated and expensive to introduce. By comparison, the makers of the latest magazine-fed, bolt-action rifles were at the lower end of the market, intent upon producing weapons which were more compact and reliable, without endeavouring so much to raise rates of fire, accuracy or range.

The Fully Automatic Machine-Gun, And Barbed Wire

No rush of orders greeted the invention in 1885 of the first fully automatic machine-gun by the American Hiram Maxim, who seized upon the new smokeless powders to actuate a weapon that fully utilised the propellant's energy. With ammunition fed by belt, the cyclic operation of pushing the round into the chamber, locking, firing, extraction and ejection of the spent case, was accomplished (after initial manual cocking of the spring and pressing the trigger button) by gas pressure and recoil. Rates of fire could be between 400 and 500 rounds per minute (rpm) and sustained over a prolonged period with water-cooling of the barrel. But the Maxim gun scored most over its predecessors, such as the Gatling and the Mitrailleuse, by its compact size, light weight and ease of handling. Mounted on a small carriage or a man-portable tripod or bipod, weight was rarely in excess of 50 lbs and preparation for action simple and rapid. Because it was so compact, concealment was easy and made it an excellent ambush weapon. This defensive advantage was understood by few people, so intent were they in seeking weapons mainly for offensive use.

Also unnoticed for its defensive potential was another very simple product which had been on the market since 1874 — single or twisted double-strand wire with barbs attached. The product of cheap steel and new machines with which to work it, barbed wire was introduced as a much cheaper and quicker way for landowners to keep stock in and intruders out. But the barbed wire fence in the hands of farmers was soon the cause of strife, notoriously but not exclusively across the open ranges of the USA. Only slowly was its military usage realized. Eventually its impact on land tactics would be almost as significant as that of the machine-gun.

Above: ***The dynamo, Edison's invention of a simple device to generate cheap electricity through the interaction of electro-magnetic forces.***

The Communication Business

It was not coincidence which caused so many inventions to appear at the right moment to complement each other. The presence or sensing of demand by intelligent entrepreneurs and inventors often triggered trains of thought which led to inspiration. Flourishing international trade called for the fastest possible transmission of information and ideas. Far more complex and sophisticated organizations to handle the financial transactions were generated. To keep pace with the thrust of private enterprise, government had to evolve and, while tending towards greater centralization, had also to seek special office machinery to serve officials. The process infiltrated military organizations as technology burgeoned, and bureaucracy invaded headquarters as never before.

The introduction of duplex telegraphy and the subsequent widespread laying of cable routes lay at the heart of communication expansion, for the networks not only created new industries to run and supply the telegraphic systems, but prompted inventors to seek better ways of recording and transmitting information. A British patent had been taken out as long ago as 1714 for a rudimentary typewriter but it was not until 1867 that the American Christopher Sholes invented the machine which he and others progressively modified until the first practical model was marketed in 1874 by the gunsmith firm of E Remington and Sons. Thereafter modifications to the typewriter followed thick and fast, above all its capability to duplicate through carbon paper.

Newspapers had started to flourish before typewriters appeared to help journalists prepare their copy. Cheaper paper; the introduction of Ottmar Mergenthaler's hot-metal Linotype in 1884 and Tolbert Lanston's Monotype a year later; the installation of John Walter's true rotary mass production press for *The Times* in 1866, all contributed to bulk output. This induced lower selling prices to a public which was becoming increasingly literate and better informed. In the hands of entrepreneurial proprietors, newspapers and magazines, most of the latter well illustrated, asserted a profounder influence than ever

The Maxim Machine- Gun

Hiram Maxim's machine-gun, fully patented in 1885, depended upon the pressure of recoil to thrust the breech block to the rear as the second stage of a firing cycle. The necessary thrust was made possible by the new, more powerful explosives and by Maxim's specially designed, metal cartridge case which enabled complete sealing of the breech at the moment of percussion.

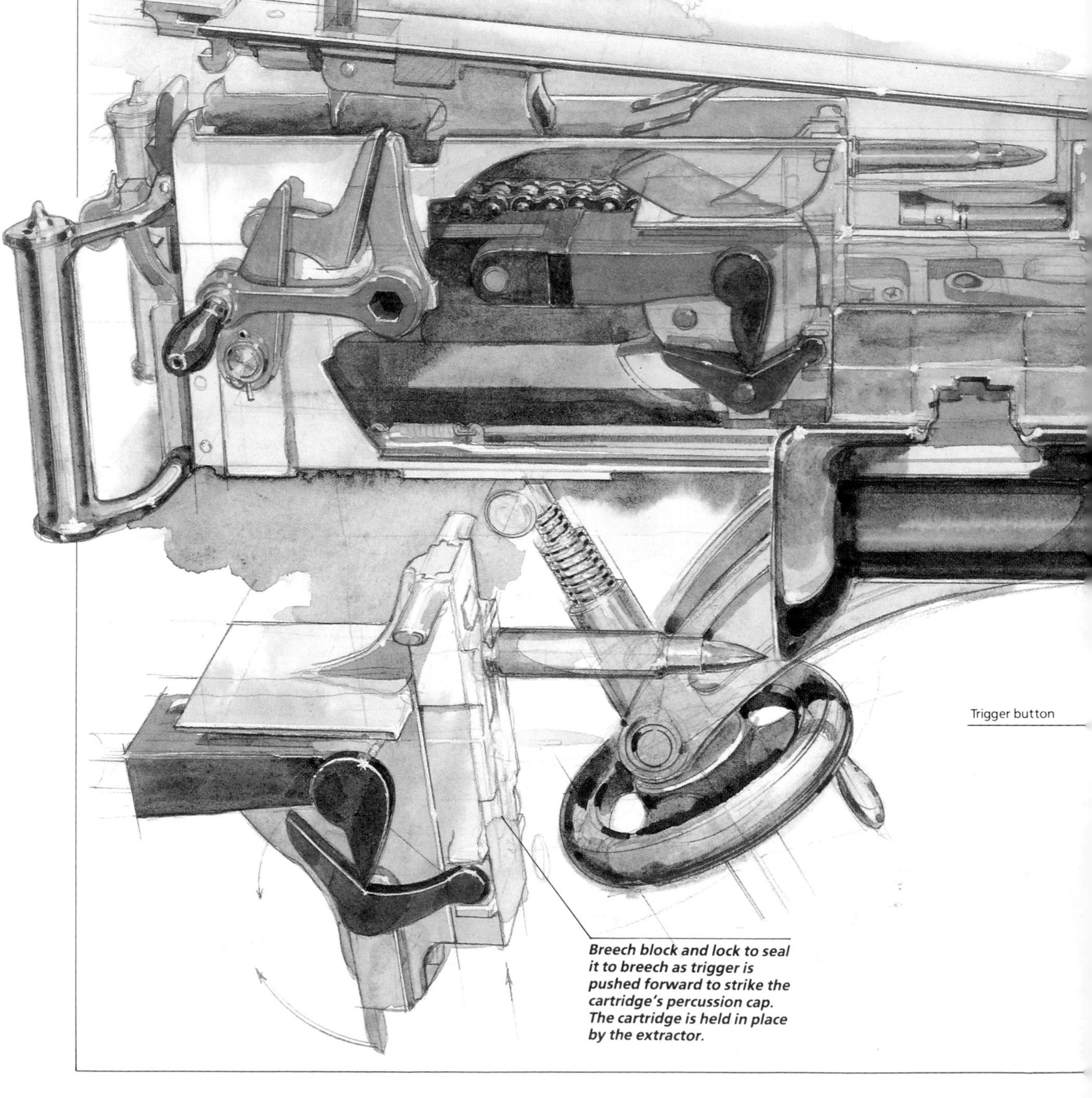

Trigg

Trigger button

Breech block and lock to seal it to breech as trigger is pushed forward to strike the cartridge's percussion cap. The cartridge is held in place by the extractor.

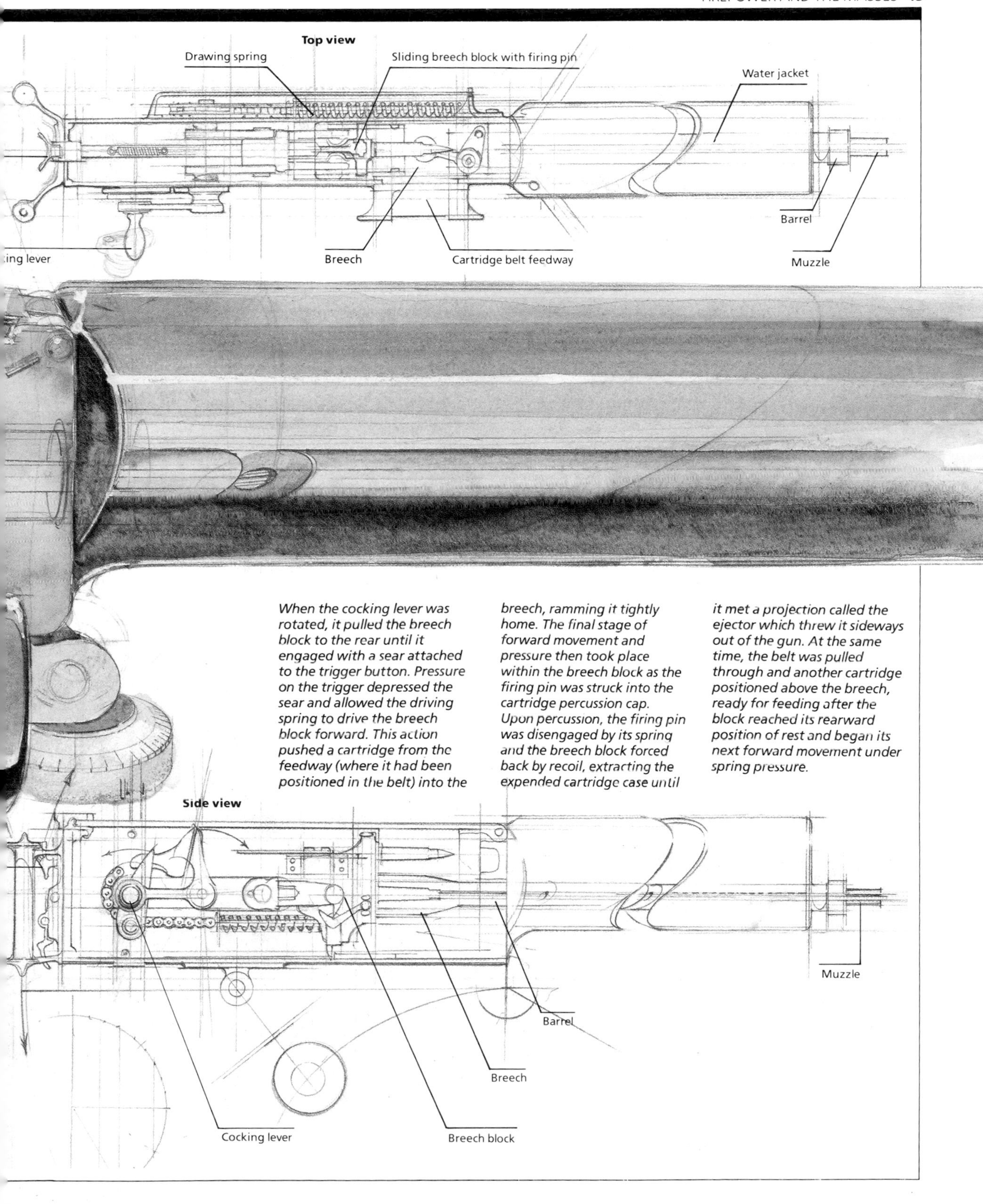

When the cocking lever was rotated, it pulled the breech block to the rear until it engaged with a sear attached to the trigger button. Pressure on the trigger depressed the sear and allowed the driving spring to drive the breech block forward. This action pushed a cartridge from the feedway (where it had been positioned in the belt) into the breech, ramming it tightly home. The final stage of forward movement and pressure then took place within the breech block as the firing pin was struck into the cartridge percussion cap. Upon percussion, the firing pin was disengaged by its spring and the breech block forced back by recoil, extracting the expended cartridge case until it met a projection called the ejector which threw it sideways out of the gun. At the same time, the belt was pulled through and another cartridge positioned above the breech, ready for feeding after the block reached its rearward position of rest and began its next forward movement under spring pressure.

before. Alongside the older style, straight reporting, crusading and advertising newspapers, there appeared popular publications designed to satisfy a liking for sensationalism among the masses, who were prone to have their minds made up for them — to accept without much contradiction calls to patriotism, to arms and to sacrifices for a cause, no matter how worthy or unworthy. Persuasion through rumour and half-truth was boosted as the art of propaganda took a big step forward — and began to break down long-standing social barriers.

From investigations by several inventors into the telegraph emerged the telephone, which enabled speech to be transmitted by electrical impulses through wire. In America in 1876, Alexander Bell (British) was ahead of Elisha Gray (American) by just a few hours in filing his application for patent. Within two years Emilie Berliner and Thomas Edison had invented improved transmitters to turn Bell's instrument into a satisfactory commercial article. Ten years later telephone networks spread a tangle of cables all round the world. In 1885 the Germans, while conducting their trials with heavy guns against fortifications, had forward observers using the telephone for the first time to send fall-of-shot correction to the gun position. Observers in tethered balloons were also soon able to communicate with the ground by telephone.

The telephone introduced a new style of personal military communications for commanders and staffs, but had the severe disadvantage that a network's growth depended upon the slow and expensive laying of cables. Such work could easily be interrupted by enemy fire and saboteurs, and could be utilized only by land forces. The mid-19th century theories and investigations of the electromagnetic spectrum by James Maxwell and Michael Faraday (both British) indicated that electromagnetic energy could be transmitted through space at the speed of light without recourse to cable. Over the closing decades, numerous scientists from industrialized nations progressed towards the moment in 1892 when the British electrical engineer Sir William Preece transmitted a radio signal 400 yards. But it was Guglielmo Marconi, an Italian scientist and entrepreneur who, working in Britain in 1895, demonstrated the immense technical and commercial potential of his equipment: a transmitter and receiver founded upon an induction coil, untuned spark gap and simple aerial system, the signal activated by telegraph key. That same year Marconi transmitted a distance of one mile. Two years later, from a station on the Isle of Wight, he was transmitting morse code to a tug at sea; sending telegrams by radio (or wireless as it was then called); despatching news to liners 56 miles at sea, across the English Channel in

The Communications Revolution

Two unrelated inventions were to revolutionize the transmission of information at the end of the 19th century.

Right: The telephone in use during military operations in Egypt. Communications from ship to shore were connected at the quayside by landline. Radio dispensed with the need for line connection, and was instantly seized on for use by the military.
*Above: Marconi in 1919 with the radio on his yacht **Electra**, giving an impression of the bulk of equipment still in use 20 years after its invention.*

1899 and across the Atlantic, a distance of 3000 miles, in 1901. Improvements upon the original Marconi equipment followed in rapid succession, the year 1904 being distinguished by the first employment of the heterodyne principle to refine enormously the utility of sets; the use of a vacuum diode valve, invented in 1902 by the Briton John Fleming, who coined the word 'electronics'; vast improvements to and miniaturization of receivers and transmitters; and trials with radio direction-finding by those who appreciated that, in a military situation, the location of a sender could provide valuable intelligence of enemy deployment.

Concentrating his initial attention on maritime developments for radio, Marconi nevertheless indicated that the new method had an infinite number of applications in every walk of life. Sailors and soldiers were quick to seize upon these in a world where the pace of their primary means of movement, the railway train, was already challenged and the horse, as the universal carrier and principal means of battlefield mobility, completely ousted. But before that happened there were a few small wars for people to reflect upon; conflicts which gave further notice, to those who cared to think objectively, of fundamental changes ahead.

Signpost Wars

Among the several conflicts which showed the way ahead was that between Japan and China. China's struggle against a hated foreign presence, and above all against Japan, brought on the Sino-Japanese War of 1894, which indicated that Japan was to be the dominant emergent power in the Far East. The war was decided at sea at the Battle of the Yalu River on 17th September — and by superior Japanese gunfire. Slower in the water, better armoured and with bigger guns, the Chinese could not match either the Samurai discipline or the verve of a Japanese fleet of lighter weight cruisers which were well armed with many 4.7-in and 6-in quick-firing guns. These nevertheless failed to sink the two heavily-armoured Chinese battleships even at short ranges between 2500 and 3000 yards. Ammunition and its supply decided the issue. The Japanese had ample and fired an enormous quantity without penetrating thick armour; the Chinese fired sparingly with fewer than 6% hits, but had only 14 rounds per gun. Command of the sea gave Japan the chance to gain a firm hold on the Chinese mainland. Conflict in Korea and Manchuria between the armies rapidly went the Japanese way because her supply lines were shorter and secure; and because her soldiers were better led by dedicated officers — even if one did choose to stride the battlefield in ancient armour, wielding a battle-axe! By March 1895, the Japanese had seized the key ports of Port Arthur and Wei-hai-wei, thus completing the first phase of their conquering ambitions.

America Versus Spain

There was similarity between the Sino-Japanese War and the Spanish-American conflict which broke out in 1898. The Americans, having consolidated the situation within their existing boundaries, were, like the Japanese, bent on expansion overseas. Both had built modern ships and picked on a victim of inferior strength. Attempts by Cubans to rid themselves of Spanish rule induced American sympathy, along with a predatory urge to spread both southwards and westwards and to protect property and stimulate trade. The arrival of the US battleship *Maine* in December 1897 at Havana, to protect US citizens during rioting, and her sinking with the loss of 260 lives in February after a heavy internal explosion, provided the spark for war. Despite Spanish diplomatic attempts to defuse the situation, the American sensationalist press accused the Cubans of mining the ship — although it seems likely that the explosion was of the sympathetic type within the ship's magazine, possibly caused by unstable charges carelessly handled. Be that as it may, the USA declared war on Spain in April and sent Commodore George Dewey with six heavily-armed cruisers to attack the much weaker squadron of Spanish cruisers anchored in Manila Bay, Philippines, on 1st May 1898. In flat calm against immobile targets, several made of wood, others with wooden decks over steel frames, the slaughter was one-sided. The Spanish squadron was burnt and destroyed with 167 dead against a single American who died from a heart attack. Shortly afterwards 11 000 American troops were landed, seized Manila and put an end to Spanish domination of the Philippines — and placed the USA on a collision course with the Japanese in an area into which each would seek to expand.

Near Santiago on 22nd June, the US Army landed on the heels of a blockade set up by the US Fleet, which had four modern battleships standing offshore. As the American troops advanced, driving an enemy superior in numbers before them, a moment arrived when four Spanish cruisers and two destroyers had to make a dash for the open sea. Steaming into a torrent of shot and shell, they were battered to pieces with the loss of 474 killed. The Americans lost one dead and one wounded.

Manila Bay and Santiago were extolled as stunning victories for modern gunnery — but a glance at the

Above: ***British cavalry in the latest khaki uniform resting in South Africa.***

figures indicated to a professional eye that the standard of shooting was deplorable. At Manila Bay only 2.5% of American rounds hit their targets, even at 2000 yards. At Santiago, in similar conditions, not a single 13-in shell found its mark while only two strikes came from the 12-inchers, total hits from all guns amounting to 3% out of 8000 rounds fired. The lesson was not lost on other navies; gunnery methods had not kept pace with weapon improvements.

Britain Versus The Boers

Much was also to be learnt from the South African War, which broke out in 1899 between Britain and the two Boer republics. It started as the result of disagreements over rival expansion plans between the Dutch Boers and the British settlers. Linked to a series of armed confrontations, it led to a Boer ultimatum insisting upon British withdrawal and a tough response from the British. Greatly outnumbered and sparsely equipped with artillery, the Boer militia, organized into Commandos, made up for its loose and informal discipline by a magnificent sense of righteousness in its cause, excellent field-craft and superb marksmanship with the latest repeating rifles. Dressed inconspicuously to merge with the landscape, the Boers had a considerable advantage over the British, many of whom were clad at first in the brighter uniforms of old and whose tactics, better suited to massed battles of days gone by, were still employed elsewhere in skirmishes against primitively-armed tribesmen.

After several costly defeats, British professionalism and numbers inevitably told against the Boers, whose amateur senior commanders failed under the strain of mounting pressure by organized thrusts directed against strategic positions. It was Boer stubbornness and courage that enabled them for two years after the defeat of their main forces to pursue a guerrilla war in the almost limitless space of the veldt, pinning down immense British resources before they were overcome.

The trouble that well-led guerrilla forces could cause was indeed one lesson re-learnt from the campaign. After all, warfare from far back in time had been founded on guerrilla tactics by small units. But the implications for major land warfare of the future were considerably more profound for those who cared to study them thoroughly. Relegating gaudy uniforms of scarlet, blue, green and gold to ceremonial duties and replacing them with drab khakis, greys and blues was the easiest of measures to improve a soldier's survival chances. Radically altering deployment drills, tactical doctrine and logistic systems to match the demands for greater dispersion in a more mobile situation was quite another matter. It meant dispensing with traditional rigidity of command and control in favour of far more flexible methods, which were by no means easy for members of the old school to accept and apply. Ordering thin lines of skirmishers to advance under the impulse of individual initiatives by soldiers taught to think for themselves depended as ever upon intelligent reaction at the lowest level, where junior leaders had as much authority as only their seniors enjoyed in the past. Training officers and men to take cover, move silently and swiftly, fight in the dark, shoot accurately and act on broad instructions, rather than on meticulous orders, were not changes that could take place overnight. But in South Africa the British Army had to do just that in order to prevail. That it achieved a few of these modifications was a sign of desperate expediency; reaction would set in once the emergency was past. But at least camouflage, standards of marksmanship and the application of firepower were significantly improved during this first campaign in which the Maxim machine-gun was used in significant numbers, principally as an offensive weapon.

Another lesson re-learnt and underlined in South Africa was the limited endurance of the horse, particularly when taxed with long-range operations and limited supplies of feed. For every eight hours on the march a horse needs twice as much rest, as well as good fodder. Against the superior mobility of railways in faster-moving operations controlled by cable communications, the horse could no longer compete. A new kind of mechanical motive power was needed and, in primitive form, already existed.

The Petrol-Driven Internal Combustion Engine

When the first successful oil well was drilled in Pennsylvania by Edwin Drake in 1859 it tapped a source of fuel that had long been known but was never before available in what, at that moment, appeared to be almost unlimited quantities. Liquid fuel was more economic than solid fuels to store and handle, and a lot safer than, for example, combustible gases such as hydrogen. Since the time of the first piston, somewhere about 150 BC, inventors had tried to build an internal combustion engine but it was not until 1862 that a Frenchman, Alphonse Beau de Rochas, described a practical solution. He proposed that a piston should suck gaseous fuel into a cylinder, compress it at the top, ignite it at maximum compression to impel the piston downwards and expel the burnt gases with the next upwards stroke, prior to repeating this four-stroke cycle. Employing de Rochas' ideas in 1876, the German firm of Nikolaus Otto and E Langen marketed a heavy four-stroke engine ignited by electric spark — a design dramatically improved by Gottlieb Daimler. In 1883 came Daimler's small engine-powered bicycle; in 1885, a four-wheeled vehicle designed by Karl Benz; and in 1887, a boat.

Above: ***Gottlieb Daimler sits in the car designed on horse-carriage lines by Karl Benz at the start of the mechanized road transport revolution.***

Daimler's engine was among the first to use a refined by-product of oil — petrol (which had been known in the 16th century). From the outset it was seen as a substitute for the horse, its energy rated as 'horse power' and its vehicles — the 'horseless carriages' — often designed by horse carriage-makers whose products were on the eve of extinction. Prone to breakdown if severely vibrated, and sure to give passengers a bumpy ride over existing rough tracks and roads, the new vehicles needed far more responsive and robust spring and shock absorbing systems. At the same time, the solid tyred wheel had to be replaced by the incorporation of rubber pneumatic tubes — an invention patented in Britain by Robert Thompson in 1845 and improved and manufactured by John Dunlop in Ireland in 1888. Daimler's first 3.5-hp car ran at 5 mph. By 1900, cars were travelling at 20 mph and their wealthy owners beginning to call for a considerable improvement to roads — initiating yet another new industry to make smooth and durable surfaces — and the creation of a network of modern roads to serve the fleets of cars and trucks being made. A complete land-transport revolution was under way.

In addition, there was a call for more cross-country vehicles — in some cases, like the Boydell tractor, to negotiate existing roads, but in many others to work on farms and for opening-up undeveloped territory. By 1906 there were several vehicles, mostly steam-driven, with tracks or footed wheels designed to spread the vehicle's weight over the ground. Military applications for powered vehicles at first, however, concentrated upon wheeled, petrol-driven types. In 1899 the British engineer F R Simms demonstrated a four-wheeled motor-cycle, armed with a Maxim machine-gun mounted behind a shield on the handlebars. A year later, for the firms of Vickers and Maxims, he was designing a fully-armoured 'War Car' with a 1½-pounder gun and two machine-guns, 6-mm armour, a speed of 9 mph — and demonstrated it in 1902 without making any sales. Others would improve upon Simms. Armoured cars, indeed, were better business than tracked military vehicles despite an impressive trial carried out by the British Army in 1907 with the first petrol-driven tractor designed by David Roberts and built by R Hornsby and Sons. The most interesting feature of this vehicle was its rather complex and somewhat frail track and suspension. This gave it a truly remarkable cross-country performance and stood up well to trials, without persuading the army of its need for such a vehicle to drag artillery or anything else. The idea of a Major Donoghue, to wrap it in an armoured skin and fit it with a gun as a combat vehicle, was instantly dismissed. So a disappointed Hornsby and Sons sold the rights to the Holt Company of the USA, which adopted the steering system for commercial use.

Flight

If the compact, efficient petrol engine with its high power-to-weight ratio was engineering a revolution for land movement, this was nothing to what it was about to achieve for aircraft. Progress with steerable balloons had been spasmodic and generally ineffectual, whether powered by steam (1852) or electric (1884) engines. The installation of a Daimler 2-hp petrol engine in Karl Woelfert's airship in 1888 proved more successful, if highly dangerous, inflated as the ship was with hydrogen fed by petrol kept at a high temperature with an open flame burner. It was a miracle that Woelfert lasted until 1897 before crashing in flames.

Military aviation had progressed only slightly since the American Civil War. The use of balloons to lift key people and micro-filmed messages out of and into a besieged Paris in 1870 had been sufficient in scale to prompt the German Army's request to Krupps for an anti-balloon gun. It resulted in the mounting of a 37-mm, high-angle cannon on a pedestal in a horse-drawn carriage and, later, in a boat. But the most important developments in lighter-than-air craft were with non-rigid airships. Once more metallurgists provided the key to progress. Isolation of the light-weight metal aluminium in 1825 by the German Hans Oersted led to its public display in 1855 and the partnership in 1889 of Carl Berg, who was casting, rolling and stamping aluminium, and David Schwarz, a Hungarian timber merchant without engineering experience who dreamed of building an aluminium rigid airship. Inflated with hydrogen and powered by a 16-hp Daimler engine driving four propellers through an unnecessarily complicated arrangement, the Schwarz airship was barely aerodynamic and far from gas-tight. Quantities of the 114 700 cu ft of gas stored in the all-aluminium cylindrical envelope leaked from the seams. Although a series of malfunctions caused it to crash when it first left the ground in 1897, Schwarz's craft acted as an inspiration, notably to Brigadier Count Ferdinand von Zeppelin.

Von Zeppelin had served with balloons during the American Civil War and in 1887 became alarmed by the French military, non-rigid airship *La France*. Powered by an electric engine, it had flown at 11 mph and demonstrated a capacity to reconnoitre at long range and to carry a load of bombs. The Prussian War Ministry, failing to comprehend the potential of the new rigid airship von Zeppelin now proposed, declined sponsorship until he had built one. Royal patronage encouraged the Count to proceed. At last, in 1900, came the flight of LZ 1, a 420-ft long, cigar-shaped, twin-engined Daimler ship which trials showed was capable of flying at 17 mph. It would be 1908 before the German Army commissioned its first Zeppelin (the Z 1 with a speed of 27 mph) and by then positive defensive measures were under development, notably in Germany. In 1906 the firm of Rheinmetall produced a high-angle, 50-mm anti-aircraft gun mounted in a turret on a lorry chassis — the first of several such mobile forms of equipment to appear over the next decade. Krupps, for example, produced a

Air Warfare In The Early 20th Century

Above: *The 1911 version of Count von Zeppelin's successful hydrogen-filled, rigid aircraft. With a speed of 30 mph, maximum altitude below 12 000 ft and use of inflammable hydrogen gas, its operational capability was severely restricted by its vulnerability to ground fire and heavier-than-air craft.*
Right: *Weapons were rapidly developed to counter attack from the air. Derivations of the Ehrhardt anti-balloon gun were developed by the Germans after the siege of Paris, 1870–71. Despite the problems of aiming, they were made in large numbers to counter all forms of aircraft.*

65-mm mobile anti-aircraft gun in 1909 and a year later a much more sophisticated 75-mm model. By then, indeed, the need for ground forces to defend themselves against air attack was pronounced, for in 1903 an event at Kitty Hawk in the USA had introduced a flying machine of demonstrably greater military promise than the airship.

The brothers Orville and Wilbur Wright were but the latest of hundreds who had attempted to build a successful powered heavier-than-air flying machine. Their *Flyer*, which made its first 120-ft hop on 17th December 1903, owed its freedom to fly at a speed of 30 mph to a 12-hp engine driving two propellers by chains. The machine created interest without immediate commercial or military excitement. A pause between invention and production of so revolutionary a machine was inevitable. Time was needed to acquire experience, to set standards of safety and operation, to obtain appropriate materials and to train designers and craftsmen in a brand-new process — let alone crews and mechanics to operate the machines. Peculiar stresses affected all aircraft, allied to an insistent need to build airframes and power plants which were both strong and lightweight — a conflict rarely encountered before in such degree. To begin with, all types of aircraft were made of wood, fabric and wire, frequently assembled by recruits from the furniture industry. Schwarz's use of aluminium introduced the vital material, however; indeed, without it the rigid airship would not have been feasible. The discovery in 1909 by the German Alfred Wilm of duralumin — aluminium alloyed to 4% copper and smaller parts of manganese or magnesium to harden and strengthen it — made the essential step forward with a metal superior to steel in terms of strength to weight. Incorporated in von Zeppelin's LZ 26 in 1914, this alloy would gradually supplant most other materials in aircraft construction and call, also, for special skills and machinery in its manufacture and utilization.

Given this background, it seems not unreasonable that it was 1907 before the US Army issued the world's first specification for a military aeroplane, and 1909 before the Wright brothers delivered for trials a test machine. By then, world-wide interest had been generated and in all the industrial countries machines were being built and aircraft performance raised. Speeds of about 50 mph were being recorded, and distances of over 100 miles covered non-stop as lift capacity steadily increased to carry greater payloads.

Europe led the way with military air development of a passive kind when, in 1905, a British balloon received radio signals in flight; in 1909 the Germans fitted a radio set into a zeppelin. Trials of aggressive measures were concentrated in the USA. In 1910 Americans dropped a dummy bomb upon the outline of a battleship, fired a rifle from an aircraft and launched a machine from the deck of a warship. They followed this in 1911 by landing on a ship and taking off and landing on water, besides dropping high-explosive bombs. That year, too, the Italians dropped a torpedo. In 1912 attention focused rather more on armament. The Americans fired the new lightweight, drum-fed, gas-operated Lewis gun from the air and the British went one better by loosing off a 37-mm gun without damaging the aircraft's frail structure. That year, too, the British carried out experiments from aircraft and explored their use for the detection and attack of submarines, while the Americans began launching aircraft from ships by catapult. Three years of development showed beyond doubt that a new weapon system of another dimension had arrived.

All this progress stemmed from the invention of the lightweight internal combustion engine which burned lightweight liquid fuel. Similarly, by 1909 the French decided that oil fuel was vastly superior to coal for warships. Not only would the exhausting and time-consuming business of coaling ships be dispensed with, but heavy oil was also thermally more efficient and much more easily bunkered in previously inaccessible or unused parts of the ship. Thus the demand for fuel oil increased by leaps and bounds, and the parts of the world from which it was drawn — initially the USA, the Middle East, Romania, Burma and Sumatra — began to assume overriding strategic importance, making them targets for potential predators and objects for defensive measures. Expansionist nations began jostling for position and influence, not only on their own borders but also in remote, undeveloped, defenceless lands which, as the result of improved transport systems, now lay ripe for colonization and exploitation.

The Boost To Expansion

The outward thrust of the more advanced industrial nations was, of course, a by-product of their superior modernized manufacturing capacity linked to the search for new markets. In 1900 Britain, France and the USA took the lead in the race to expand, followed closely by Italy and Germany, with Russia, owing to inherent social inertia, falling behind. At the root of modernization lay the entrepreneurial urge linked to technical expertise and education. Britain was in the forefront here because of the start she had gained at the beginning of the first Industrial Revolution and the

Above:** The Wright brothers'* Flyer ***makes one of its first flights. Designed around a 12 hp IC engine, it was the result of meticulous design and engineering, backed by tests in a home-made wind tunnel and intensive testing of gliders, while the brothers taught themselves how to control the machine safely.

wealth of experience and resources since accumulated. That her industrialists managed this without much support from government and the universities — most of whose members were classicists at heart and despised and feared commerce and technology — is a commentary on the virtues of their enterprise. But it followed that to lose their inventive and entrepreneurial drive would soon destroy competitiveness. And at the start of the 20th century all entrepreneurs worth their salt could recognize from whence future competition could be expected, and felt bound to seek protection. Inevitably, the arms manufacturers stood to benefit from this attitude. Remorselessly they sought to adapt new inventions to a military purpose, and did so at a considerably greater rate than in the past — particularly in Britain, where the resources of research, development and mass-production were better founded than in any other country.

Britain, for example, built nearly 80% of the world's ships and led the way in marine technology. As the supreme example in this activity, she was in 1900 inaugurating yet another revolution in propulsive power. In 1889 Sir Charles Parsons had taken the design of existing water and steam turbine engines a step further by fitting several stages of blades to the shaft in order to make further use of steam, improve energy efficiency and increase the power delivered. His original aim was to provide cheaper power for electric dynamos, but the fitting in 1897 of one of his engines to drive a 44-ton boat called *Turbinia* was of farther reaching importance. For *Turbinia* could move at 34 knots and did so in public before the assembled British Fleet when it lay at anchor for Queen Victoria's Diamond Jubilee Review.

With steam turbine engines, a 550-ton torpedo boat destroyer, HMS *Viper,* was built in 1899 to a specification demanding 30 knots — and the day was not far removed when turbine engines, greatly enlarged and infinitely more reliable than the reciprocating type, would drive nearly every ship in service, including the latest battleships. Prior to that, however, the ships then in service with reaction engines were to be put to the test of battle on a scale large enough to expose their strengths and frailties and illuminate the future shape of naval technology and tactics.

The Russo-Japanese War

The emergence of Japan as a modern power in the 1890s posed Russia with a serious challenger to her own long-term expansionist aims in east Asia. Instantly Russia had expressed her hostility by compelling Japan to give up Port Arthur and Wei-hai-wei — the latter's gains from her recent war with China. Quite apart from the loss of a foothold on the mainland, an unforgivable humiliation was thus inflicted on a proud ruling hierarchy. The construction of the trans-Siberian railway also provoked Japan into counter-measures since it would enable the Russians rapidly to strengthen their local forces from the west. Predatory ambitions focused on Manchuria (occupied by Russia) and Korea (under Japanese guarantee). Determined to take possession of both countries and without unduly concealing her intentions, Japan built up her naval and land forces to the state that, at the end of 1903, they were locally superior in strength to those of Russia in the Far East. Paying scant regard to protocol, the Japanese began hostilities on 8th February 1904 with a surprise attack on the Russian Fleet at Port Arthur, followed a week later by the unopposed landing of an army in Korea.

The war which ensued was dependent on sea power and strongly influenced by the latest weapons and communications technology. At once both sides resorted to underwater attacks, with Japanese torpedo boats entering the harbour at Port Arthur to torpedo a cruiser and two battleships, and both sides laying defensive and offensive minefields in attempts to bottle each other up in port or sink ships in shallow waters. On 13th April the Russian flagship *Petropavlosk* struck a mine which detonated the ship's magazines and sent her to the bottom in two minutes, carrying the C-in-C and 600 crew with her. A month later, two Japanese battleships hit mines which had been laid in waters they were known to frequent. Again, there was a catastrophic sympathetic explosion on one while the other sank later.

Guns achieved nothing like the success of torpedoes and mines. When Admiral Heihachiro Togo led six battleships and nine cruisers into Port Arthur at the outbreak of war and exchanged a welter of fire at 7000 yards with a Russian force of almost equal strength, the results were anything but impressive except for the noise and spectacle of shooting from 240 Japanese guns and about 140 Russian guns of between 12- and 4.7-in calibre. The percentage of hits scored was extremely low owing to the inability of gunnery officers to observe and correct fall-of-shot amidst a tumult of splashes and explosions. And when hits were scored they did little damage, failing in every instance effectively to penetrate armour and resulting in only 124 Russian casualties with still fewer among the Japanese. There was a somewhat similar story to tell on 10th August when the fleets met at the Battle of the Yellow Sea. On this occasion the issue was settled during a long engagement by a mere four shots out of the thousands of all calibres fired and the dozens of hits scored. Just two 12-in hits on the Japanese flagship *Mikasa* seriously impaired the fleet's communications and gunnery, while two 12-inchers landing on the Russian flagship *Czarevich* killed the admiral, produced disorder and led to a precipitate Russian retreat, several ships choosing internment in neutral ports rather than honour in Port Arthur.

At root of the inability of shells under 12-in calibre to inflict fatal damage or sufficient hits lay a fundamental misappreciation by admirals of basic gunnery techniques. A profusion of fire which did not inflict serious damage did not necessarily undermine enemy

morale, as had been hoped. And existing gunnery methods, principally devised by the British and superior to those of the Americans for single ship-versus-ship actions, were still hopelessly inadequate in a fleet action. Direct hits from the heaviest guns were essential; near misses counted for nothing. The existing optical range-finders, devised by the British firm of Barr and Stroud and first fitted to British ships in 1892, were not the complete answer. Direct hits were likely only at very short ranges where range-finding, due to the flatness of projectile trajectory, was of less importance.

Above: ***US Navy torpedo boat in 1906, launching its compressed air-driven, gyro-guided weapon from tubes carried amidships.***
Below: ***The turbine-powered*** *Turbinia* ***shows her paces at 34 knots before the assembled Royal Navy in 1897. Surrounding her are ships now out-moded by her speed and reliability.***

Nevertheless, the Japanese admirals, through superior handling of their ships and excellent use of radio to send orders and report enemy movements to their ships at sea, were able to obtain a tactical advantage, to win command of the sea and prevent serious interference with their invasion of the mainland. They built up land forces which outnumbered the Russians by 280 000 men and 870 guns to 83 000 men and about 200 guns. Likewise, superior generalship also enabled the Japanese to win a series of bloody victories and drive the Russians back into Port Arthur. At the beginning of August the city was isolated and the port under artillery fire, no longer safe for the fleet, which was compelled to break out and fight the battle of the Yellow Sea before falling back to rejoin the hard-pressed garrison.

The fighting on land went the way of all other encounters since 1853. When armies of sturdy courage equipped with weapons of massed firepower came into conflict, casualties were enormous. Strive as each side might to manoeuvre for advantage against the enemy's flank and rear, the moment almost invariably arrived when one had to make a headlong and costly assault upon the other. During six major assaults on Port Arthur subsequent to 7th August, slaughter was fearful. Delayed in opening a major bombardment through the loss of heavy guns at sea, the Japanese could make little headway at first against modern steel and concrete fortifications, lit by searchlights, linked

by entrenchments and defended by artillery and well-sighted machine-guns. Day and night between 19th and 24th August, General Nogi lost 15 000 men for only minor gains and 3000 Russian losses, his dead and wounded piled up by machine-gun fire against barbed wire fences the artillery had failed to cut. Trying again in the final fortnight of September, the result was much the same and the key position of 203 Meter Hill remained in the defenders' hands. The arrival of heavy artillery, including 18 11-in howitzers, added significantly to the bombardment; but mining and sapping operations, like those at Petersburg, linked to successive assaults, simply piled up the dead until 5th December when, at last, 203 Meter Hill fell and the Japanese artillery observers could see the Russian ships at anchor and begin their systematic destruction by gunfire. The total cost to the Japanese when the Russians surrendered on 2nd January amounted to 58 000 killed. To the Russians it was 31 000, plus their entire equipment and stores and the destruction of most of the ships still afloat.

Much of the haste which attended Nogi's costly assault was provoked by public knowledge of the transfer of the Russian Baltic Fleet to the Far East. He might have worried less had he known that the commander, Admiral Zinovi Rojestvensky, had neither confidence in his ships nor in the prospects of recovering command of the sea. Moreover, although the seven-month voyage to the Far East gave ample opportunity to train the Russian crews, it would also provide Admiral Togo with time to refurbish his ships and prepare a welcome. When the two fleets met in the Straits of Tsushima on 27th May 1905, the four top-rate Japanese battleships with their total of 16 12-in guns were, on paper, inferior to a larger force of Russian battleships with 41 10-in or 12-in guns. But while Rojestvensky at once threw his fleet into confusion with imprecise orders, Togo manoeuvred with precision and skill (having been warned by radio of the enemy approach) and succeeded more than once through superior speed in crossing the enemy's T. That is, he managed to steam across the head of the enemy column, subjecting each ship in turn to devastating broadsides from all his guns, allowing the Russians only to bring their forward-facing armament to bear. Togo, moreover, was able to pick the ideal range of between 5000 and 5500 yards for his gunners with the result that his better-trained crews were able to overwhelm an opponent who never settled to the task of shooting back methodically.

Every Russian attempt to escape was foreseen and blocked by the Japanese so that, one by one, their ships were wrecked and capsized by heavy 12-in shells. As one observer remarked, 'shots from 10-in pass unnoticed, while for all the respect they instil, 8-in or 6-in guns might just as well be pea-shooters'. Then as nightfall approached, Togo unleashed his torpedo boats among the shattered enemy while himself moving ahead again to block the way to Vladivostok. It seems that some 100 torpedoes were launched, of which only seven struck home: but they sank two battleships and two cruisers with the loss of only three torpedo boats. Next morning, all that remained of the Russian fleet, except three ships which slipped through, either retreated into internment in neutral ports or hauled down their flag to Togo. In addition to losing cruisers and smaller ships, along with 4830 killed and nearly 6000 prisoners, eight of their battleships were sunk and four surrendered. In this, one of the most complete naval victories in history, the Japanese had but 117 killed and 583 wounded.

It was the culminating moment of a war which had now to be brought to an end through exhaustion on both sides. At Mukden, between 21st February and 10th March, Japan had won the final land battle, although at an appalling cost in casualties — some 100 000 Russians and much of their equipment and 70 000 Japanese. Beyond doubt, Japan emerged as a modern power to be reckoned with, one which unflinchingly assimilated all that Britain, the USA and other advanced nations could teach but also stood back and drew its own original conclusions about what was needed next in the armament competition and on the battlefield. The Japanese had learnt what was free for others to copy; that on sea and land firepower was vital, but that an apparently pulverizing bombardment did not necessarily subjugate the enemy. That, time and again, well-trained and motivated troops would survive, stand up and fight back despite heavy losses. They had seen that subtle tactics and accurate fire by a relatively few heavy weapons could win at sea; and that against wire and machine-guns on land there was no hope for the massed attacks of old. Instead it was necessary to play hide-and-seek with the enemy, to destroy his guns by better fieldcraft and shooting, employing superior strategy, tactics and surprise.

Few among the other nations registered the lessons learnt in the Far East, perhaps dismissing eye-witness accounts *because* it was the Far East and therefore unlikely to be relevant elsewhere.

***Right:* Japanese torpedo boats in a night action against Russian battleships. The naval action at Tsushima was possibly the earliest instance of searchlights being used in an encounter at sea.**

Dreadnoughts And Submarines Take Over The Naval Race

German ambition fired the starting pistol for the next heat of the naval race, beginning with her acquisition of naval bases in the Far East and Pacific in 1897 and announcement of a large warship-building programme that was doubled in 1900 and supplemented in 1904.

Diplomatic attempts to curtail the arms race had made no headway at the First Hague Peace Conference of 1899, and such was to be the fate of all subsequent efforts made in the coming decade. Ambition and jealousies overrode prudence. New weapons proliferated, led by battleships as major symbols of security, prestige and strength. Thinking along parallel lines, British and American naval constructors separately devised the inevitable next generation of dominant warships. The keel of the first so-called dreadnought battleship, named after the 18 000-ton British version, was laid in 1905, completed a year later in a truly remarkable building record and called *Dreadnought*. The brainchild of William Gard, she was innovative in almost every respect: some 3000 tons heavier than any predecessor; powered by 13 000-hp Parsons steam turbines driving four propellers to give a speed of 21 knots, with unheard-of reliability; adequately armoured to defeat 12-in projectiles; and armed with ten 12-in guns in twin turrets, plus 27 3-in anti-torpedo-boat quick-firers. But her most potent claim to omnipotence was the technology and technique used for centralized direction of salvoes from the main armament. Developed chiefly by the British Captain Percy Scott and American Captain William Sims, this fire control system combined range-finding devices, plotting machinery and electric communications, allied to precise calibration of guns and ammunition as related to spotting and interpreting fall-of-shot. Controversial as the method was, proof of its superiority over existing systems and control by independent gun layers was provided by the British trials. Those which took place in 1904 registered only 42.86% hits, while two years later, using the new method, the total had risen to 71.12%.

Dreadnought made all other battleships obsolescent. She and her successors, it was reasoned, could defeat all other threats without themselves being seriously harmed. The battery of 3-in guns (soon improved to 4-in) would keep torpedo boats beyond the 3000-yard range of existing torpedoes. The latest tethered mines in shallow waters would be cleared by minesweeping boats whenever she put to sea. Objections to the construction of such a radically-advanced warship came mainly from those who deprecated the financial cost and would have preferred to build more ships which were individually cheaper. Few, however, paid much heed to the latest and most potent threat to the dreadnought, the torpedo-armed submarine, and

***Below:* The streamlined *Holland* submarine of around 1900 embodied all the design characteristics for future non-nuclear powered submarines: electric engines powered by storage batteries which were charged when on the surface by a petrol-fuelled engine; periscope for vision while submerged; and torpedo armament on the eve of the invention of the so-called fast 'heater' torpedo.**

in this critical omission the pundits were both myopic and wrong.

A practical submarine was not feasible until a reliable electric motor, powered by rechargeable storage batteries, was available. In 1886 Lieutenant Isaac Peral of Spain built such a boat, powered by 480 batteries driving two motors and propellers. But it was the French, Russians and Americans who took the lead in developing cigar-shaped, streamlined boats, powered by electricity and armed with torpedoes. Step by step, battle-worthy submarines evolved. Stephan Drzewiecki's 18-ft, four-man boat of 1879 carried four externally-mounted torpedoes. The appearance of the periscope and its mounting in M Romazzotti's *Morse* in 1899 provided commanders with a vital vision device, enabling them to conduct operations while submerged. And in 1900 came the use by the Irish-American John Holland of a petrol engine which both drove the boat on the surface and powered a dynamo for recharging the batteries. So was engineered a practical self-contained submarine, an advance on Maxim Laubeuf's *Narval* (the first vastly to increase its buoyancy with a double hull) which had used, as several previous designs had done, an operationally impractical steam engine.

The *Holland* boat, accepted by the US Navy, was also adopted by the British. From this moment submarine development accelerated, even if its tactical use remained clouded by scepticism and lack of experience. Speeds were steadily raised to over 10 knots on the surface and 8 beneath; endurance lengthened to several hours; and diving depths increased through improved structural strength. With these advances, submariners and those of real insight visualized the feasibility of closing unobserved upon battleships and sinking them with torpedoes. The only existing anti-submarine measure was a ship's ability to ram — if it was lucky enough to find a submarine close to or on the surface. Moreover, the range and reliability of submarines was considerably increased when, in 1904, Laubeuf's *Aigrette* substituted a heavy oil-fuelled engine for the dangerous petrol engine employed by Holland. This, indeed, was the first major military use for the compression ignition engine (which replaced spark ignition) invented by the German Rudolf Diesel in 1892. Soon it would be introduced to both sea, land and air forces as a source of power which, for many purposes, was safer, more reliable and provided greater fuel economy than its petrol-fuelled rival. Detractors pontificated that the world's oil supply would not long bear the drain upon it — but that problem the new oil industry quickly sought to solve by stepping up exploration and production, thereby converting itself into a massive economic force as provider of a crucial commodity.

Essential to the future of the submarine and naval war in general was the torpedo, which was about to be given a capability far in excess of the uncertain 3000-yard range upon which the battleship's supporters based their calculation of survival. By 1910 the Royal Navy, shortly and inevitably to be copied by others, was in possession of an improved Whitehead torpedo. The so-called 'heater', steam-driven torpedo, introduced about the time *Dreadnought* entered service, had a range of 15 000 yards. It enabled attacks to be made well outside the effective range of big guns, let alone those specifically mounted to deal with torpedo boats. Simultaneously, the incorporation of a gyroscope, invented by the Austrian Ludwig Obry in 1895, provided more flexible guidance. This important gadget could, by activating the hydroplanes, counter unwanted deflection of the torpedo by external forces and so keep it on course. But the gyroscope could also be programmed to turn the torpedo onto a desired heading (up to 90°) after launching, thus helping an attacker to a variety of tactical options. No longer could it be expected that the submarine would confine itself to defensive operations, as some assumed and with whom Admiral Sir John Fisher (the driving force behind modernization, improved gunnery and the *Dreadnought*) violently disagreed in a paper of 1904. He visualized submarines travelling at 18 knots and penetrating enemy ports to sink ships without warning, even before outbreak of war. 'I don't think it is even *faintly* realized — *the immense impending revolution which the submarines will effect as offensive weapons of war*', he wrote.

Ten years later, as the cost of building battleships escalated rapidly with every attempt to retain their dominance, the latest, much cheaper, undersea weapon system was about to demonstrate its ability to do all and more than Fisher had foretold.

The Balance Of Forces In 1914

The causes of the First World War in all their complexity and controversy have long been debated, without incontrovertible conclusions being reached. Obscure treaties; national and personal suspicions and rivalries; vengeance for past humiliations; the struggle for markets; the vast expansion of armed forces and the political influence of general staffs all contributed to the explosive situation of Europe in June 1914, when the assassination of the Archduke Franz Ferdinand of Austro-Hungary provided the spark of ignition.

Above: ***The Royal Navy at Spithead in 1914. On view (from the left) are a dreadnought battle-cruiser; a cruiser; a line of dreadnought battleships and, approaching from the right, a torpedo boat.***

Undeniably, a contributary factor was the existence of strong armed forces committed to ambitious plans of conquest at the outset of a war everybody expected. Undeniably, too, the state of people's minds in approaching war in a mood of euphoria can be laid at the door of politicians and the propaganda fantasies of the sensationalist press, who raised war fever to pitches of optimism and hatred in excess of reason. Enthusiasm for battle might have cooled significantly had the same politicians and press propagandists been aware of the reality — that the immense destructive powers of the navies and armies were self-cancelling and that belief in firepower and massed manpower as arbiters of decision through offensive action was a delusion. In fact, all they achieved was a crushing defensive effect, smothering efforts to attack and so inhibiting decision. The clash of surface fleets did not occur, as expected, within days of the outbreak of hostilities; and only one man, a Polish banker called Ivan Bloch, envisaged in print the stalemate which locked fast all the battlefronts on land. In essence, the latest weapons of mass destruction were portrayed not as deterrents but eulogized as the means to quick victories — and thus misled, instead of discouraging, those responsible for taking the fatal steps. One fundamental reason for this almost total incomprehension of reality was that virtually none of those in government had the slightest technical education, and thus were in ignorance of the technological revolution which had taken place since 1850. In their ignorance, those who sought war on the false assumption that a quick decision could be attained without prolonged fighting or immense damage, were in fact risking the destruction of civilization and society as they knew it.

Power At Sea: Fleets of The Opposing Nations

A comparison of the combined fleets of Britain, France and Russia with those of their opponents, the fleets of Germany, Austro-Hungary and Turkey, provides not only a measure of the overwhelming advantage of the former, but an indication of the enormous industrial effort put into ship construction over the preceding two decades. In addition to proportionate numbers of cruisers, these main contenders could send to sea:

	Britain and Allies	Germany and Allies
Dreadnoughts	32	17
Battle-cruisers	11	6
Pre-dreadnoughts	54	45
Destroyers and torpedo boats	420	178
Submarines	179	44

The battle-cruiser, another of Fisher's innovations, was simply a ship of the same weight and armament as the original dreadnought, but more lightly armoured and capable, at 25$\frac{1}{2}$ knots, of a higher speed. The price difference between the two dreadnought types was small; the improvement in speed, according to Fisher, though important from strategic and tactical aspects, would do little to increase protection since modern gunnery could easily hit fast-moving targets. But the price of thinner armour would, in due course, be paid for in another way. And so, too, would lax precautions and arrangements regarding ammunition. Carelessness in handling explosives (instances were recorded in 1912 of cordite bags being dried out before a battleship's galley fires and inspections of magazines being made by exposed candle-flame!) was matched by insufficient awareness of the dangers of flash being transmitted from the turrets to the magazines below. No procedure was laid down for dealing with such an occurrence, and this, too, would prove expensive.

Wide though the disparity in numbers was, and important as it was to the British to retain a margin over the Germans as well as a hold on their world maritime trade, there was some amelioration of the German disadvantage. When it came to confrontation on the continental shelf separating Britain from Europe and the German North Sea ports, the dreadnoughts, upon which sea power was deemed to be founded, forfeited some of their importance. For in waters that were mostly shallow, the mine posed a lethal threat to all types of ships; and at the distances prevailing, the short-range submarines possessed a potency lacking in more open waters. From the start, therefore, both sides tended to keep their dreadnoughts safe in port, taking them to sea only under the most stringent protective arrangements which hampered their tactical freedom. In other words, a crude, cheaper form of new technology had already imposed severe restrictions on an only slightly older, sophisticated, expensive weapons system.

The Strength Of Land Forces

It was different on land. The numbers of men committed to battles on four fronts — that is, in the Balkans, on the Russian front, in Western Europe, and in the Middle East (when Turkey entered the war in October 1914) were roughly equal. Some 2 857 000 belonged to the Franco-British-Russian-Serbian Alliance (the Allies); about 3 135 000 to the German-Austro-Hungarian-Turkish Central Powers. As with the war at sea, strategies and tactics employed by those in command and the advantages and disadvantages incurred were to a considerable extent decided by technology and its exploitation. Since soldiers were almost unanimous in believing that modern weapons favoured the offensive, the fullest scope and precedence were given to artillery of all kinds to destroy enemy defences and help the cavalry and infantry to advance. Similarly, transport systems, and above all the railways, were adapted to move masses of men and materials from pre-ordained mobilization centres to the front to execute meticulously pre-planned advances aimed at the enemy heartlands.

There was no holding back of forces on land. Every man and every piece of equipment was committed to a collision course. But since the vast majority in all armies were short-service conscripts or recently-recalled reservists, standards of marksmanship failed to match those of their weapons' accuracy. Only trained shots, as in the small, long-service regular British Army, were able to hit men beyond 300 yards. Shooting in the other armies was usually profligate, nervous and wasteful beyond 200 yards, a typical engagement consisting of rapid fire which had deterrent if not fatal effects until the enemy reached point-blank range. Mass manpower, while facilitating the replacement of losses, did not necessarily compensate for deficiencies of training — and often its existence merely encouraged careless despatch of men to futile destruction.

With hindsight it is not difficult to see which side had the initial advantage in technical terms. The vastly superior heavy artillery of the Central Powers had the surprise capability to destroy nearly all the Brialmont-type forts guarding the frontiers and key nodal points within the hinterland of the Allies. In field artillery there was very little to choose between the numbers and efficiency of the Anglo-French armies and the Germans and their Austro-Hungarian allies, but the Russians, suffering from the paucity of their industrial base, were woefully deficient of modern pieces and, like all the other contestants, short of sufficient stocks of ammunition or the means to make it. Indeed, matters of supply to the armies, and the logistic system in the field, represented the almost universal weaknesses in the forces involved. Despite the vast increase in men and material and the long-studied formulation of expansive plans for conquest or defence, because neither side had visualized the nature of combat, neither had established a true basis from which to calculate the supply demand or foresee the unprecedented load placed on transport.

Logistics

While the Germans made admirable preparations to mobilize their large forces and transport them to the frontiers by a railway system specially laid out in peacetime to satisfy these conditions, they underestimated the problems of supplying so many men, horses and guns when they reached the extremities of a long march into Belgium and France. Founding their plans upon bulk supply by railway, the old-fashioned but flexible system of horsed transport by road was, while carrying considerable quantities, relegated to a secondary role in which it was supported by a small number of lorries. The potential of lorries was recognized, but their production had yet to be put on a massed basis; they were relatively expensive and still somewhat unreliable. Horses, too, were unreliable but took precedence over lorries, not only because they were available and familiar, but also because scrapping all the facilities related to horsed transport could no more be granted by treasuries than could the complete

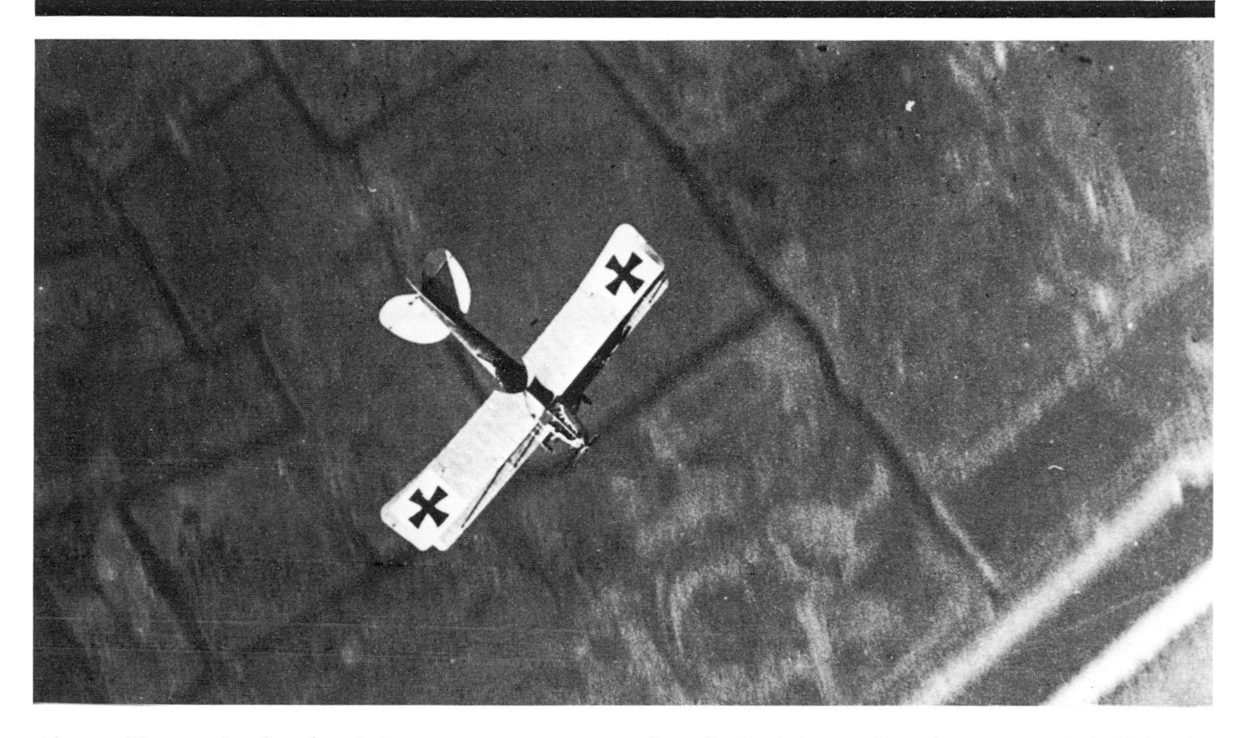

Above: ***The eye in the sky. A German reconnaissance biplane photographs enemy land forces below. This development made imperative the fighter aircraft as a counter measure.***

re-training of men to drive and service mechanized vehicles be accomplished. Mechanization on such a vast and fundamental scale had, by the nature of economic conditions, to be gradual. The degree of change would be dictated by the restrictions of peace-time accounting and the urgencies of crisis and war.

In The Air

In the realms of the imponderable stood the, as yet, virtually untried weapons of war: aircraft and radio communications. True, the latter had proved helpful — perhaps decisive — in Japanese hands in the recent war with Russia. Similarly, the use of aircraft in 1911 by the Italians against the Turks in Tripolitania had been promising. Nine primitive aircraft and two non-rigid airships had provided useful information from reconnaissance flights and had avoided harm so long as they remained above 3000 ft and therefore out of range of small-arms fire. But bomb dropping had failed and in fact proved politically counter-productive in the train of a vitriolic propaganda campaign unleashed against airmen who, allegedly, had dropped bombs on a hospital. This, the first of many such mistakes and misadventures in the generations to come, only highlighted what was to be a long-standing difficulty — how hard it was to navigate and identify the target, let alone hit it if found.

Air forces in 1914 were regarded mainly as instruments of reconnaissance, despite several pre-war demonstrations of their ability to drop bombs and carry machine-guns. Air-to-air combat, in fact, was hardly contemplated; partly because of the technical difficulties involved; partly because there were several among an international brotherhood of airmen who thought it would be unsporting to kill those with whom they shared similar perils; but also because only a minority foresaw any need to do so. And yet a very realistic assessment by a naval airman, Captain Murray Sueter, to a sub-committee of the British Committee of Imperial Defence in 1912 made it perfectly clear that '. . . war in the air, for the supremacy of the air, by armed aeroplanes against each other', was likely. He went on to declare: 'This fight for the supremacy of the air in future wars will be of the first and greatest importance, and when it has been won the land and sea forces of the loser will be at such a disadvantage that the war will certainly have to terminate at a much smaller loss in men and money to both sides'.

There spoke the self-interested enthusiast in search of support for his own particular branch of the

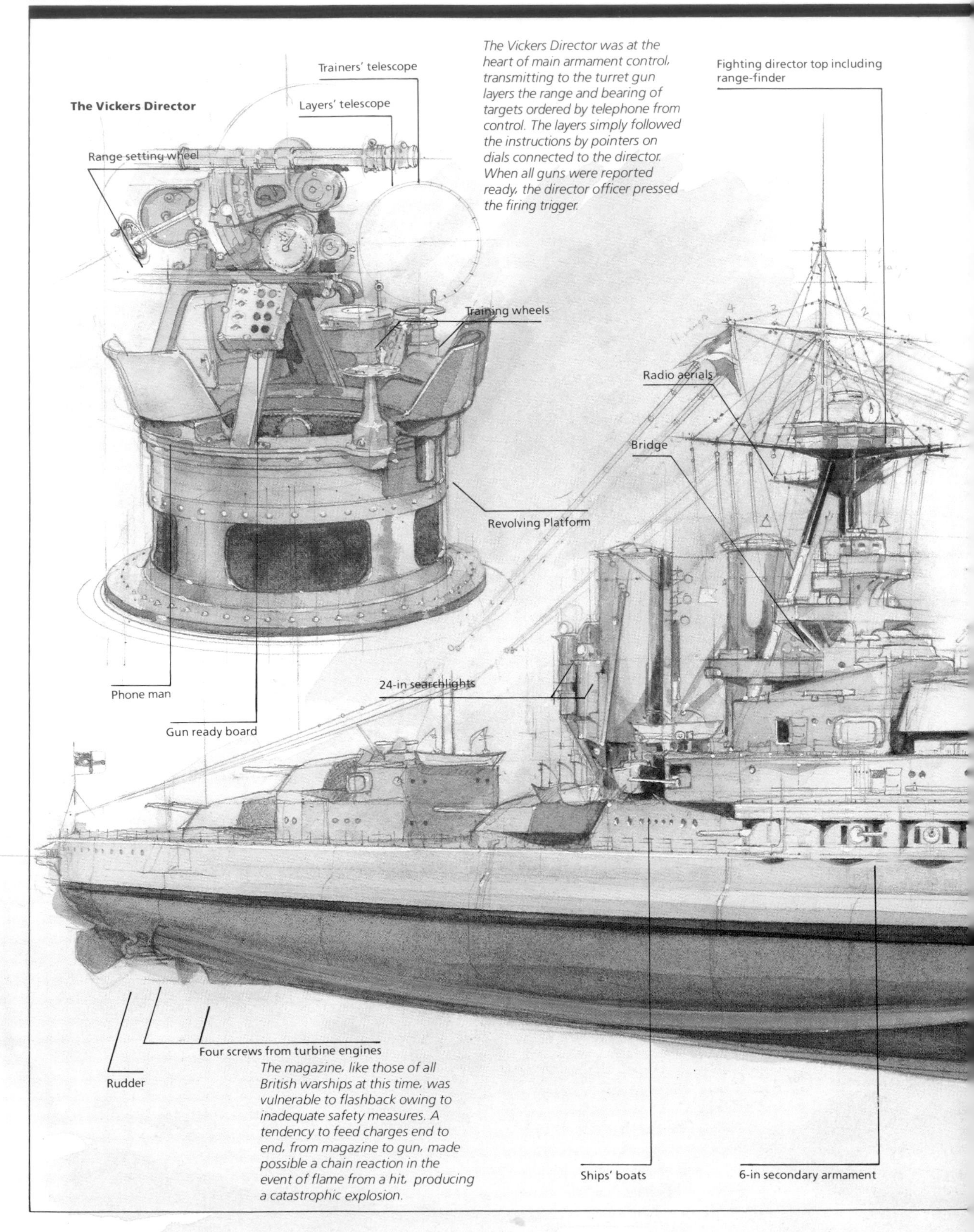

The Vickers Director was at the heart of main armament control, transmitting to the turret gun layers the range and bearing of targets ordered by telephone from control. The layers simply followed the instructions by pointers on dials connected to the director. When all guns were reported ready, the director officer pressed the firing trigger.

The magazine, like those of all British warships at this time, was vulnerable to flashback owing to inadequate safety measures. A tendency to feed charges end to end, from magazine to gun, made possible a chain reaction in the event of flame from a hit, producing a catastrophic explosion.

The Dreadnought Battleship HMS *Iron Duke*

HMS *Iron Duke*, launched in 1912, was a 21 250-ton battleship armed with 10 × 13.5-in guns. Their fire was controlled by instructions from the range finder and director mounted in the fighting top. With four, shaft turbine engines producing 31 000 hp, she had a maximum speed of 21 knots and was coal-fuelled with all the filthy, hard labour involved. Her 12 × 6 in guns, for use against torpedo boats, were mounted in sponsons; neither they nor the 12 × 24 in searchlights were director or centrally controlled.

The proliferation of radio aerials indicates the growth of this form of communication from ship-to-ship and ship-to-shore. But its restriction to morse code made it of only limited use for quick tactical reactions. For speed and security, much ship-to-ship contact continued to be by light signals and semaphore from around the bridge.

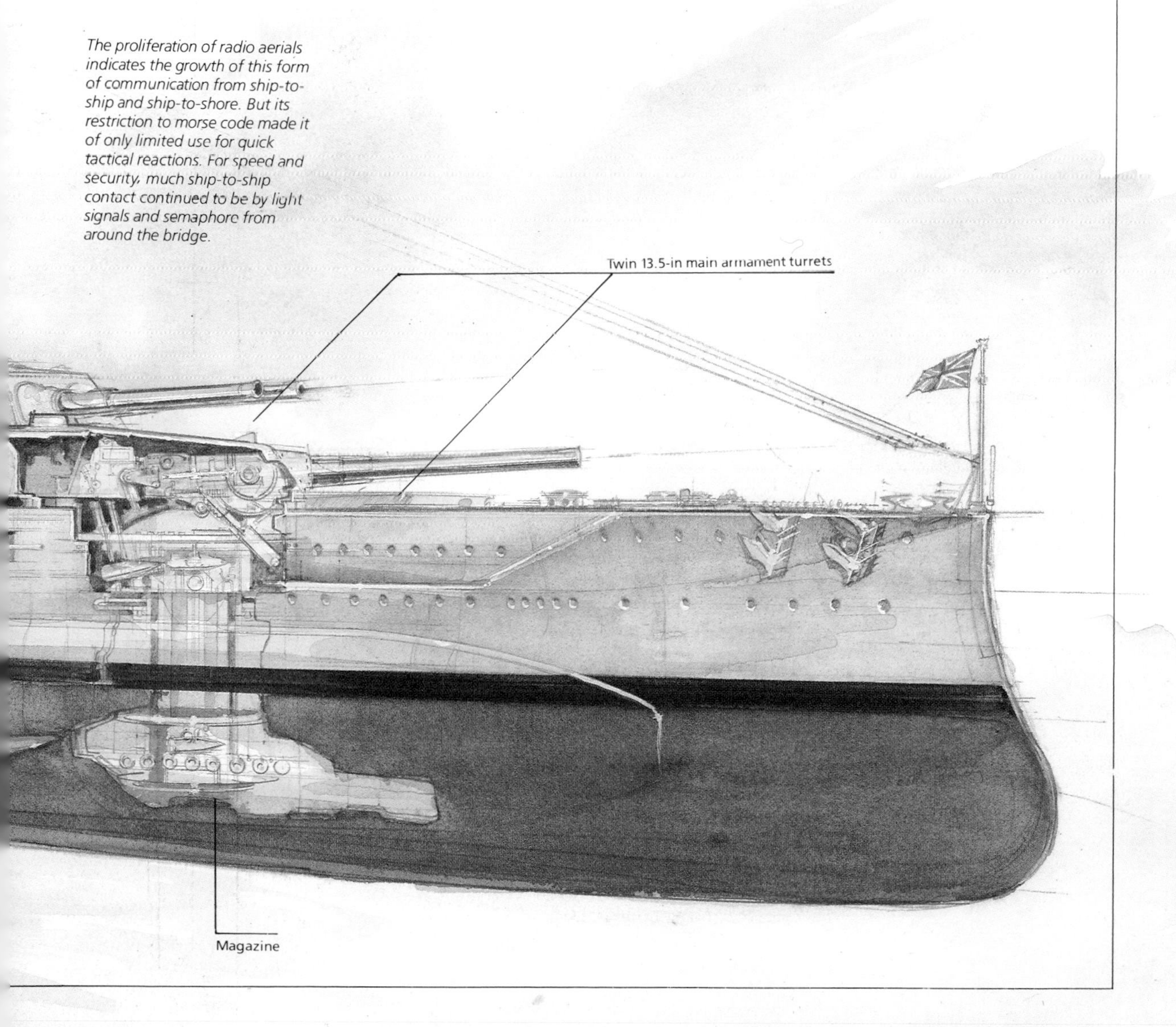

services. Arguments in similar vein had frequently been proffered by advocates of new, all-powerful weapons in the past and increasingly would be deployed in the future — and not exclusively by airmen. The fact that Murray Sueter's ultimate prediction has yet to be achieved is far from beside the point. In 1914 it was ridiculous with the equipment available. For one thing, Germany had only 246 aircraft and Austro-Hungary a mere 35 against 160 in France, 110 in Britain and 25 in Belgium. For another, their top speeds were rarely as high as 70 mph; their ceiling, at best, 13 000 ft; and their endurance four hours at the most. Only airships could carry out prolonged flights at long range, but they, with their inflammable gas, were vulnerable at speeds of only 50 mph and a maximum ceiling rarely in excess of 12 000 ft. In any case, Germany possessed only 11 of the latest rigid types which were superior to an equal number of lighter-than-air craft in French and British hands. Though airships would drop bombs at the very start of the war, they were chiefly committed to long-range strategic reconnaissance, in which their effectiveness was mainly hampered by the need to descend low in order to penetrate cloud cover while searching for information.

Only a few in authority fully appreciated the threat posed by disclosure of plans through deep penetration reconnaissance. No longer was it merely the front line under observation from a tethered balloon which had to be concealed from view. Henceforward, troops in back areas had to be hidden, and all the more cunningly when the results of the latest aerial photographs came to be studied. Those taken with difficulty by Gaston Tournachon on plates over Paris in 1858 had represented a notable 'first', but now vital was the ability to take shots with the latest cameras of all kinds of military installations and terrain. Leo Warnerke's first use of spool film in 1875 (vastly improved by George Eastman) and the first portable roll film camera, marketed in 1888 by Kodak, were firm steps towards the introduction of simplified cameras, developing and printing processes. They enabled British Army airmen in 1914 to photograph clearly the entire Isle of Wight and the fortifications of the Solent from a height of 5600 ft in one day, developing their negatives in the air ready for printing on landing. Neither navies nor armies could ignore such threatening spies in the sky.

Out of sheer necessity, measures to shoot down aircraft from the ground, if not from the air, had to be taken with some urgency. A so-called mobile balloon gun (soon renamed an anti-aircraft gun), mounted on a lorry, was offered in Germany in 1908 and was but the first of many such weapons introduced into all navies and armies in the years to come. From machine-guns to 3-in and 4-in pieces (many just adaptations of field guns on pedestals and high-angle mountings) a whole range of anti-aircraft guns were put into service to tax artillerymen with the entirely new problem of how to hit a fast-moving target in space. For the problems of securing a stable reference point and of ranging were insoluble before the invention of special devices for the purpose. Gun-layers were left with only one option: that of hose-piping with machine-guns, or firing off great volumes of shells in the direction of the enemy in the extremely optimistic hope that perhaps one in several thousand might hit. Such tactics placed an immense drain on scarce ammunition supplies.

Radio War

Strangely enough, in military communities where resistance to new technology has been recorded as proverbial, there was pronounced willingness to make maximum use of radio, and the Germans placed considerable reliance upon it for passing messages. A typical mobile radio detachment, horse-drawn, would be bulky and weighed down by batteries and charging equipment. And while the large sets mounted in General Headquarters and fortress towns might have a reliable range up to 675 miles, those in the field were doing well to transmit to 125 miles. Volume of traffic with morse code was restricted to the speed of the man on the key — rarely faster than 20 words per minute and often broken by discontinuity of communications. The Germans, for example, used a bizarre procedure which laid down that out-stations were 'responsible for establishing touch with central control,' a most difficult task for the operators concerned if the network was busy, as it always was. Furthermore, the Chief of Signals was not informed of the war plan and therefore was quite unable to make prior arrangements for radio, telephonic and telegraph communications with the armies as they plunged deep into enemy territory. This gulf in understanding showed that the signal arm lacked the confidence of the General Staff and was to lead to serious consequences.

Most vital of all, however, was the question of security in connection with 'open' radio communications that could be overheard by both friend and foe. Before the war all nations created codes and cyphers to guard security; but these, particularly among the Russians, were unwieldly, not properly understood by the operators and frequently ignored — to the detriment of war plans. From the outset the British, French, Austro-Hungarians and Germans were at

Communications in the land war

Top: The bulky German field radio station with its teams of horses, heavy storage batteries and large aerial arrays. Such inefficient working arrangements contributed to the breakdown of command control during the land battle in 1914.

Left: The more vulnerable and sometimes less reliable pigeon messengers, of slow speed and operational inflexibility, were still used extensively. Motor cycle dispatch riders replaced the horse for delivering lengthier commands.

Above: Dogs were often used both for carrying messages and for paying out telegraph cable under fire.

pains to monitor enemy transmissions and make attempts, with success, to break their codes. Thus radio technology became a two-edged weapon and one which spawned yet another science, that of the intellectual codebreakers who would employ mathematical deductions and, from the start, mechanical calculators to serve their purpose.

Secret plans were in jeopardy, particularly when ships and armies were divorced by distance from using secure land-line links. Moreover, the introduction of radio direction-finding made it possible to discover the location of transmitting sets and thus plot the movements and intentions of major naval units and land formations.

In effect, the nations went to war in 1914 at unprecedented strength with forces of enormous power which, nevertheless, were hampered or endangered by vital technical defects and corrupted technologies. The people manipulating them based their judgements and plans upon knowledge of the past without properly attempting to project into the future the latest advances in weapons technology.

The Campaigns Of 1914 — A Short-Lived Mobility

The course of the fighting from the start of the war, at the end of July, was largely dictated by German strategy. When General Graf von Schlieffen became Chief of the German General Staff in 1891, he began formulating plans which stipulated that, in the likely event of simultaneous war against Russia and France, everything would be staked on knocking out France while assuming the defensive against Russia. That these plans, in 1914, did not wholly accord with those of Austro-Hungary, which decided to concentrate everything upon conquering Serbia and making a preemptive attack upon the Russians before their mobilization was complete, was, in the event, conjunctive to the German intention. For while the Germans kept only one army in the east to guard Prussia and threw the remaining seven armies against the Belgians, French and British, the activities of the Austro-Hungarians did help to relieve the strain on the Germans when two Russian armies flooded into Prussia on 13th August. For by then the bulk of the German Army was committed to its advance into Belgium and France.

The moves which culminated in the complete defeat of the Russian Second Army at the Battle of Tannenberg at the end of August will not be described in detail here. Suffice it to say that after initial Russian successes threw back the German Eighth Army and led to the replacement of its commander by General Paul von Hindenburg, a technical failure on the Russian part enabled the Germans to concentrate all their forces against the Second Army and bring about its destruction. Meanwhile the other, more slowly advancing, half of the Russian forces was temporarily left to its own devices.

The confidence which allowed Hindenburg and his Chief of Staff, General Eric Ludendorff, to take this apparent risk was based on positive intelligence of Russian intentions gleaned from monitoring insecure radio links, and confirmed by reconnaissance from airships and aeroplanes. Confusion among the Russian radio operators often led them to ask each other in 'clear' which code they were using. Also, staffs' insistence on speed in preparing transmissions frequently led to the sending of the officers' names and place names in clear and, in the case of two vital messages on 24th/25th August, the almost complete (if somewhat garbled) text of orders to the Russian First and Second Armies. With exact knowledge of enemy movements, it was no stroke of genius on the German part to concentrate nearly everything against the Russian Second Army as it advanced, incompetently led and increasingly out of control, into the complex terrain near Tannenberg. Without the gift of priceless information from misuse of technology on the Russian part and its exploitation by the Germans, Hindenburg's and Ludendorff's reputations might have been tarnished and the war diverted to a different course. For already the panic ensuing from the Russian advance into Prussia had caused General Helmuth von Moltke (the German Chief of Staff and nephew of the victor of Sedan in 1870) to send troops destined for the advance on Paris to help defend Prussia — where they arrived too late for the Tannenberg battle.

Von Schlieffen had always set his face against subtracting resources from the attempted knockout blow against France. The essence of his plan was an overwhelming concentration of forces on the right wing to enable an advance across Belgium to swing wide to the south, preponderantly overlapping the French left wing to the west of Paris and thus enveloping the enemy from flank and rear. When von Moltke took over as Chief of General Staff in 1906, he progressively weakened this concept by gradually withdrawing troops from the right wing, allocating some to the centre and some to the left wing as a precaution against correctly-anticipated French attacks into Alsace and Lorraine. Nevertheless, the right-wing forces were immensely strong — some 320 000 in the First Army (which had furthest to march round the decisive right flank) and another 260 000 with its neighbour, the Second Army, whose role was to conform as the wheel into France began in the vicinity of Brussels. But first the Belgian forts at Liège had to be taken, and quickly, to prevent any delay to the smooth progress of the main advance, the sheer size of which was a close secret faithfully kept from Allied intelligence.

The Great Bombardment Begins

The spearheads to attack the forts were the secret, mighty pieces of heavy artillery which had been built specially for this kind of task — the rapid destruction of supposedly impregnable bastions. Kings among the big pieces, of course, were the 420-mm Krupp 'Fat Bertha' 2052-lb shells. But of equal importance for tackling the 12 separate Belgian forts were a number of Skoda 305-mm howitzers on loan from the Austrians, firing 846-lb shells. It was hoped, however, that the forts might first be isolated by a cavalry sweep and then seized by infantry *coup de main* or simple diplomatic persuasion. That way lives and, above all, time would be saved. But on 4th August diplomacy was rejected; then the cavalry sweep fizzled out (as most cavalry drives would when faced by small arms fire); and the attempt to penetrate the forts in darkness largely failed when illuminated by searchlights and raked by fire from Belgian garrisons, whose confidence was far from shaken except in a few isolated instances. Progress by the Germans was, for the next week, discouragingly slow. Minor gains only were made with the help of light artillery, while the big guns were brought forward by rail and road and laboriously emplaced.

The effect of the great shells was swift and catastrophic, with one in nine accurately striking and cracking apart steel cupolas and concrete roofs. The out-dated Brialmont forts, overtaken by the latest technology, failed their garrisons. Morale broke as explosions like earthquakes ripped their shelters apart and, in places, detonated the magazines. In less than three days the last bastions collapsed. The way was open for the avalanche of men, horses and guns to pour through Belgium.

Until the principal French and German armies collided on a line running north-westwards from Luxembourg to Mons, both sides' plans kept moderately close to schedule. Only as their communications were extended and the anger of battle began to pressurize commanders at all levels were inherent defects revealed — and most prominently in Belgium where the Anglo-French and German armies met at a long distance from their railheads. Elsewhere, in Lorraine, where the French attacked the Germans (who withdrew slightly), the main test was battle in which the effect of firepower against men in the open was every bit as disastrous as it had recently been in the Far East. Shattering their ranks against prepared German defences, the French exposed themselves to a German riposte: von Schlieffen had always set his face against such a move, but von Moltke permitted it in response to the pleading of ambitious subordinates who scented glory in attack. Inevitably, the Germans seriously damaged their formations by reckless assaults upon fortifications. At Nancy a single French corps, under command of General Ferdinand Foch, administered to two German armies a stiff dose of artillery and small arms fire such as they had previously endured themselves.

In a crisis, von Moltke was unable to impose his will upon subordinates. Once a mobile battle became joined in the Ardennes and along the River Sambre, the strain thrown on the links between von Moltke and his army commanders increased — a strain which was exacerbated by von Moltke's insistence on remaining in a headquarters which stayed farther and farther to the rear. Depending for the passage of orders wholly upon telephone, telegraph and radio, amplified by visits from liaison officers who sometimes enjoyed plenipotentiary powers, the German leader became divorced from reality through failing to see for himself what was going on at the front and impressing his personality, man to man, upon his generals.

On the other hand, General Joseph Joffre, the French C-in-C, was repeatedly visiting the front and demonstrating a remarkable ability to arrive at the right place in moments of crisis, to stiffen his subordinates' morale, and give orders as appropriate to the situation as possible. Joffre was not enslaved by technology, but even had he conducted operations only from his headquarters he would still have been more effective than von Moltke. For not only did the French benefit from being forced back upon their bases and reserves, but also from operating in their own country where the civil communication system was dense and still functioning. Men and material could thus readily be moved about, almost at will. Messages could be passed in quantity with only the slightest delay, permitting quick reactions to changes of situation. The Germans, on the other side of the lines, were entering territory which was not only damaged by the fighting but hostile to their very presence and thus subject to disruptive attacks upon communication systems. Because of routine demolition of bridges and railways by the retiring Franco-Belgian armies, the large-scale movement of stores and

THE SCHLIEFFEN PLAN AND THE RACE TO THE SEA

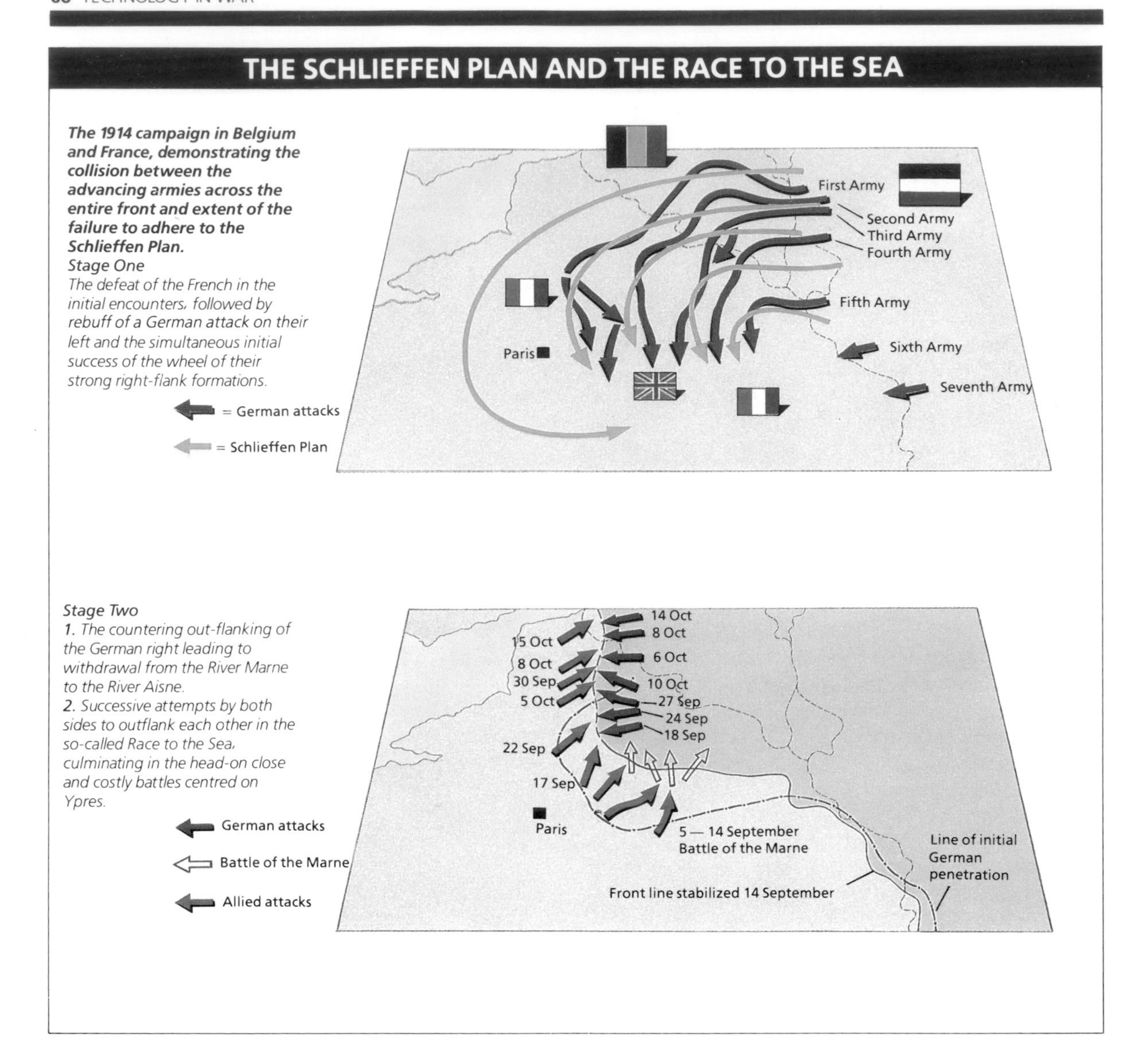

The 1914 campaign in Belgium and France, demonstrating the collision between the advancing armies across the entire front and extent of the failure to adhere to the Schlieffen Plan.

Stage One

The defeat of the French in the initial encounters, followed by rebuff of a German attack on their left and the simultaneous initial success of the wheel of their strong right-flank formations.

Stage Two

1. The countering out-flanking of the German right leading to withdrawal from the River Marne to the River Aisne.

2. Successive attempts by both sides to outflank each other in the so-called Race to the Sea, culminating in the head-on close and costly battles centred on Ypres.

reinforcements forward from the frontier was throttled, compelling men and horses to trudge almost the entire distance in dust and heat. The cutting of signal cables; breakdown of the civil communication network in the wake of invasion; and the unreadiness and inability to lay land lines at the same pace as the marching troops placed an unbearable load on the already over-worked and inefficient radio links. It has been recorded that the noise level within the French GHQ was quiet, but very loud inside the German counterpart, where officers were attempting furiously to remedy faint and tenuous telephone conversations by the raising of voices and loss of tempers.

Lacking positive and sure central control, those commanding the First and Second Armies of the extreme German right began to veer from their pre-ordained paths. When Second Army, for example, became embroiled in battle on 21st August with the French Fifth Army along the Sambre, its call to First Army to come to its aid was unchecked by von Moltke. That key force was thus deflected inwards, diverting it from the full radius of its wheel towards the west of Paris and effectively diminishing the physical and psychological impact of the intended wider sweep. True, the Germans proved superior to the French in the encounter battles, but this was more to the credit of junior commanders taking the initiative in seizing fleeting opportunities than it was for those who failed to combine the action of artillery with the movement of infantry and cavalry. Time and again losses were far heavier than necessary, owing to infantry advancing without artillery support. And nowhere was this more pronounced than at Mons on 23rd August, when the German infantry assailed cleverly-emplaced British infantry without shellfire to soften them up first.

At Mons the Germans suffered terribly from aimed rifle fire, backed up by machine-guns and artillery, fired by professional soldiers whose marksmanship was at a peak never again to be rivalled. Indeed, had not the French Fifth Army on its right been withdrawn, the British could have held their ground much longer. But now, at last, Joffre came to realize the overbearing weight of manpower falling down upon his weakened left flank from the north. Withdrawal in a long retreat towards Paris was imperative. For both sides this was to prove an exhausting experience in which, although the Germans appeared on the verge of triumph for much of the three weeks of marching and sporadic battles, they were often suffering more serious difficulties than their hard-pressed opponents.

The distances covered — several hundred miles on foot by most troops — took a toll that led to the forfeit of opportunities. For example, on 31st August when a gap opened in the French front through which the German cavalry could have ridden, the horsemen (whose activities had been circumscribed since 12th August by respect for enemy fire) reported they were unable to go farther until they had reshod their mounts. Already the German Army had been deprived of sufficient information about enemy dispositions in depth due to lack of cavalry penetration of the enemy lines. In fact, the traditional reconnaissance role of cavalry was better performed by aircraft, slight though their contribution was at this moment. But it was not only the frustration of German cavalry as a reconnaissance and disruptive force which helped the French (whose cavalry was also of only limited value) to contain the German offensive. They, too, made only rather uninspired use of aircraft but benefited greatly from having broken the German codes within 48 hours of the outbreak of war and so could read German radio traffic with the same ease as the Germans were reading that of the Russians.

Use (or misuse) of the latest technologies was decisive in blunting the German offensive. As the drive continued southward, its weakening logistic state and diversion to the east of Paris instead of the west granted the still-intact Allied armies freedom of manoeuvre to prepare a counter-stroke. Denied use of the railway system, the German lines of communication failed to supply the full needs of the troops at the front or move them without long marches. The French meanwhile retained their main route pivot at Paris and were able to move troops easily from the less threatened right flank to the left, where they out-flanked the out-flanking Germans. Improvised employment of Parisian taxis to transport a division to the front demonstrated mechanization as a vital factor in future war — even if, on this occasion, the commercial panache of the drivers, who overtook each other in happy abandon on the way to the front, threw all into confusion!

Because of their weakened right wing, the Germans were unable to bring in fresh reserves to confront the French manoeuvre and had to extract troops from the wilting drive on Paris. In so doing they opened a gap in their front into which French and British poured almost unopposed. This Allied manoeuvre compelled a complete German retirement from the River Marne to the River Aisne. Superior use of command facilities gave victory to the Allies as the German system collapsed. When at least one German cavalry commander resolutely declined to keep in contact with his radio station, it was but symptomatic of a general decay. It is not surprising that the actual order to retreat was given not by von Moltke in person, but by a Colonel with plenipotentiary powers who had been sent to the front to assess the situation — a task which would have been better performed by von Moltke had he not become tied to a central organization hampered by inefficient techniques.

Yet in defeat the Germans at once demonstrated their adaptability in assuming the defensive while mounting a counter-thrust. On the heights above the Aisne they quickly assembled defences similar to those used by the Russians at Port Arthur: a line of entrenchments, defended by barbed-wire fences covered by fire from dug-in machine-guns and artillery, which could not easily be stormed or destroyed. Simultaneously, they sought to out-flank the Allied out-flanking moves by side-steps northwards, a process enabled by the repair of telephone and radio communications and of railway lines. It then became possible to react and move forces with ease from Germany and other parts of the front to new battlegrounds. The so-called 'Race to the Sea', as each side endeavoured to out-flank the other, culminating in a final collision of forces in the vicinity of Ypres during October and November, engineered stalemate on a vast scale. With the resources at their disposal, neither side was able to overcome even the semblance of an artillery, machine-gun, wire-entrenched position. The technology of defence defeated the current technique of attack, and with quite appalling loss of life to reinforce the lesson.

Armoured And Air Warfare, 1914 Style

A veil was lifted from the future during the opening moves of the Race to the Sea — a revelation across a wide tract of open terrain standing thinly defended between the besieged entrenchments of the port of Antwerp and the gradually encroaching main armies extending their flanks from the Marne. Defence of the Channel ports, opposite England, was one excellent reason for the British involvement here but important, too, was the need to prevent the Germans establishing air bases to threaten London with bombing. With the intention of keeping the Germans at arm's length, Winston Churchill, who as First Lord of the Admiralty was also responsible for air defence of Britain, sent a force of sailors and marines in motor-cars, armed with light cannon and machine-guns, to prevent incursions by enemy cavalry into Flanders. And it was the sailors who quickly saw the advantage of imitating some Belgians, who armoured their cars with 6-mm mild steel plates to give the crews a modicum of protection. Until the Germans dug ditches across the roads, the effect of the armoured cars (the protection of which was soon improved by giving them 8-mm armour plate) was out of all proportion to their numbers. Apart from artillery, the Germans had nothing which would penetrate the armour, and the artillery was lucky if it scored more than one hit in 30 above 100 yards. The cars were able to move rapidly from place to place, countering each enemy thrust and mounting attacks of their own. Until the trench lines precluded all mobility, both sides rated each car equivalent to a company of infantry and therefore a very economical force.

Below: **A German shipborne seaplane, lifted on and off by derrick, adds scope to the search for enemy forces in vast ocean spaces. It heralds the day when aircraft, armed with surveillance equipment, bombs and torpedoes would dominate sea warfare.**

At the same period, three British naval aircraft, single engine Avro 504 biplanes with a range of 100 miles, attempted from Antwerp to raid airship hangars at Düsseldorf but were thwarted by thick mist over the target; all, that is, except for one aircraft which dropped three bombs, none of which exploded. (Also, in September 1914, a Japanese seaplane joining action against the German garrison at Tsingtao had dropped a naval shell adapted with added fins onto a minelayer and sank it.) A second British attempt at Düsseldorf on 8th October saw hits scored and an inflated zeppelin destroyed. These bombing raids, several more by aircraft of most nations from land bases, and a British attempt to attack airship hangars using seaplanes launched from carriers escorted into the Heligoland Bight on Christmas Day, removed existing doubts that aircraft were not fighting weapons. Before the end of the year both Germans and French had formed squadrons specifically designated to carry out bombing missions, the German unit being based on a railway train to enhance its mobility. Strategic attacks were launched against enemy cities, ostensibly against military targets but frequently striking civilians. And on 5th October, the gentlemen's agreement against air-to-air combat ended when Sergeant Joseph Frantz, firing a Hotchkiss machine-gun from a French Voisin V89, brought down a German Aviatik.

The tactical value of reconnaissance was itself ample cause to make the destruction of prying aircraft imperative. No commander could afford to have the enemy freely observing his deployment and seeking out his intentions. Nor could he allow enemy aircraft, unopposed, to direct artillery fire against his own guns. The effect on morale alone could be disastrous. By the same token, nobody with Home Front responsibility could permit bombing of cities for fear of the effect on morale, with all its political repercussions. A zeppelin had bombed Liège on 6th August, Paris on the 30th, Antwerp on 2nd September and Warsaw on the 24th — all of them fortress cities, although only Antwerp was under siege at the time. Each attack, along with the propaganda pamphlets also dropped, was intended to destroy morale, for the bombs were more noisy than dangerous.

Far more serious was the deliberate decision by the Germans in January 1915 to attack British and French cities with zeppelins by night. There could be no pretence at deliberate aiming of bombs and little defence either. Some damage was caused and people killed. The zeppelins easily avoided anti-aircraft guns or flew too high for their reach. Not until 6th June was the first German raider brought down in air combat, and then by a small monoplane struggling up to drop six 20-lb bombs on it from an altitude of 11 000 ft.

The characteristic combination of panic demands by politicians, cries for help from fighting men, response from innovators, and profit-seeking by entrepreneurs produced an immediate and remarkable addition of technical improvements to aircraft which were already foreseen but not yet seriously acted upon. In April 1915, at the start of the campaign to seize Constantinople via the Dardanelles, a balloon equipped with radio directed ships' gunfire against an encampment and later against a Turkish battleship; in August a seaplane successfully dropped a torpedo from the air to sink a merchant ship. And in France during the Battle of Neuve Chapelle between the 10th and 12th March, as part of the initial unsuccessful attempts by the Allies to drive the Germans out of northern France in 1915, aircraft for the first time played a deliberately integrated part in the land attack. They proved their worth by directing artillery fire against enemy guns (using the 'Clock System' of fire control in which a celluloid disc, divided into concentric circles and segments, helped the airborne observer to plot fall-of-shot in relation to the target at the middle of the circle); by bombing railway junctions to interrupt the movement of German reinforcements to the front; by flying 'contact patrols' at low level to discover and report the extent of friendly infantry advance (since, in trench warfare, commanders on the ground had no way of telling how far their men had got); and by using French methods of aerial photography and the latest box-type Camera (A) to obtain clearer, annotated pictures of enemy positions for use by artillery and infantry assault parties. Zeppelins were also trying out improved methods of navigation by cross-bearings upon radio transmissions, and optical bomb sights such as the German Görz model and the gyro-stabilized sight produced by the USA firm of Sperry. But these sights, in assisting the aimer to measure his speed over the ground and make allowance for drift, were of limited use only, particularly since navigation to the target remained at the mercy of weather and human error.

Aircraft losses were far more frequently the result of accidents caused by weather, mechanical failure or pilot error. Gunfire from the ground managed only a marginal improvement in predicted shooting, and hit very infrequently — despite the French introduction of salvo shooting controlled by a central post which attempted to track the target and give the gunners the target's height and speed. Obviously, the best method of shooting down one aircraft from another was by a machine-gun firing straight ahead. All the pilot had to do was aim his machine at the target, but the configura-

The Single-Seater Fighter Aircraft

Even before 1914 it was clear that the best way for one aeroplane to aim a gun was by pointing the attacker at its target. With a pusher type, such as the Vickers Gun Bus of 1913, it was easy to arrange, but with much more efficient tractor machines, such as the Niueport Scout, it was necessary to mount a drum-fed gun on the top wing in order to clear the propeller.

Anthony Fokker's EI monoplane (right), 1915. Although it was to incorporate the first practical interrupter gear, allowing a belt-fed gun to fire through the propeller arc and thus simplify accessibility to the pilot and increase fire power, the aircraft was itself unremarkable. Its design was strongly influenced by the French Morane Saulnier monoplane and was none too robust, with a tendency to wing flutter and disaster during violent manoeuvres. But as long as the opposition lacked a fighter with an effective forward-firing machine-gun, the Fokker E monoplanes, with a speed of 80 mph, were able to shoot down their victims with impunity and seize control of the air.

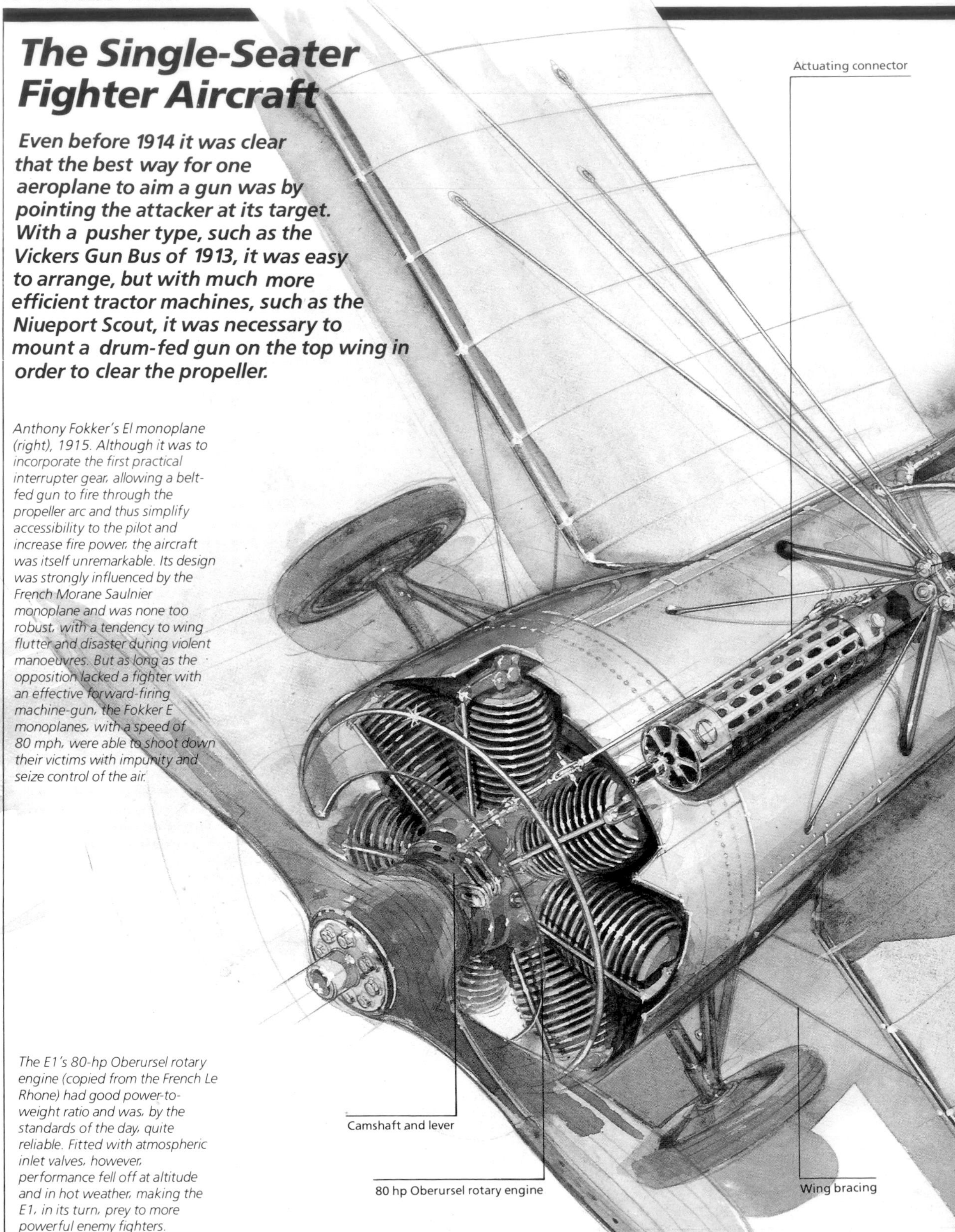

The E1's 80-hp Oberursel rotary engine (copied from the French Le Rhone) had good power-to-weight ratio and was, by the standards of the day, quite reliable. Fitted with atmospheric inlet valves, however, performance fell off at altitude and in hot weather, making the E1, in its turn, prey to more powerful enemy fighters.

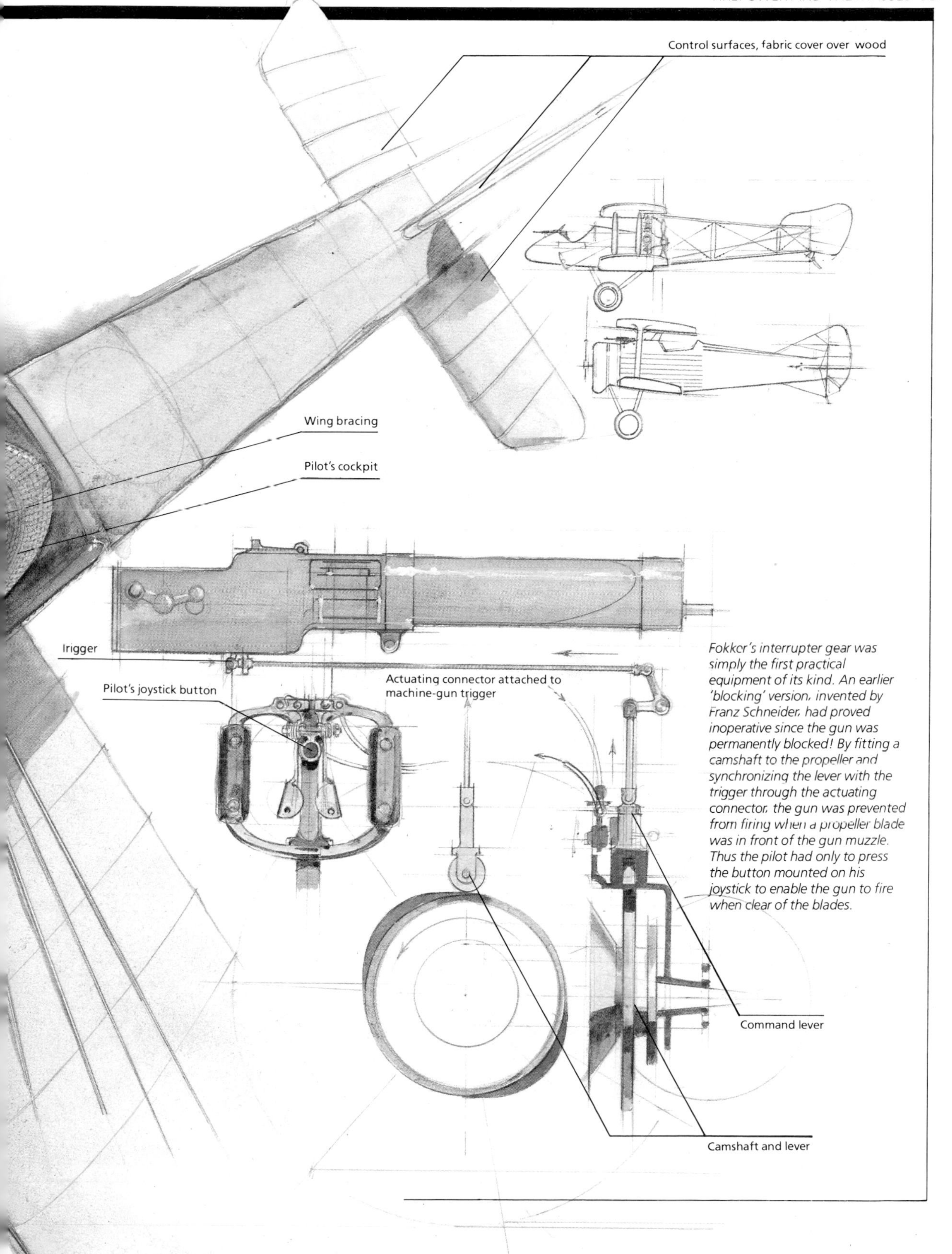

Fokker's interrupter gear was simply the first practical equipment of its kind. An earlier 'blocking' version, invented by Franz Schneider, had proved inoperative since the gun was permanently blocked! By fitting a camshaft to the propeller and synchronizing the lever with the trigger through the actuating connector, the gun was prevented from firing when a propeller blade was in front of the gun muzzle. Thus the pilot had only to press the button mounted on his joystick to enable the gun to fire when clear of the blades.

tion of single engine aircraft set problems. Only 'pushers', with propeller in the rear and the pilot with his gun in the front, were then suitable. True, the Russian Lieutenant Poplavki had attempted to make a Maxim gun fire through the propeller arc of a 'tractor' aeroplane in 1913, and in spring 1915 the Frenchman Raymond Saulnier had fitted deflector plates to the propeller to allow Roland Garros, in a Morane monoplane, to surprise and shoot down three German aircraft. But it was not until the Dutchman Anthony Fokker perfected a mechanical interrupter gear (based on Saulnier's) and fitted it for the Germans in the E1 monoplane fighter of his own design, that the first fighting machine with a fully synchronized system could be operated with safety. This system depended, incidentally, upon the use of consistently manufactured ammunition, without which the chances of striking a propeller blade were high. On 15th July, when the first victory was scored by a Fokker E1, the day of the specialized fighter aeroplane had dawned, making inevitable the ensuing struggle for air supremacy over sea and land.

Gas, Flame, Explosives And The Battle Of Verdun

In the year 1915 the war expanded as more nations joined in, and the battlefronts, with one exception — Russia — remained locked in stalemate. The apostles of artillery power ruled like gods. It was a year which also witnessed the growth of mutual hatred generated by sensational, and almost wholly false, stories of atrocities by the enemy purveyed, most vividly of all, in the French and British press. The news media were now able to generate far greater punch with their stories by the use of photographs, shot in the forefront of battle with the latest cameras of reduced size and easier operation, using paper and celluloid film. To this could now be added moving pictures taken at the front, the cinematograph having become common-place reality at the turn of the century. In the cinema, the mass of civilians could watch battles unfolding before their eyes — and the effect on their emotions was both conflicting and powerful.

Notwithstanding a censorship which prevented accounts of the worst horrors reaching the people, they were able to judge the rigours and ferocity of the fighting from scenes of ships sinking after mine strikes during the attack on Gallipoli. They could see intense artillery fire engulfing trench lines, mixed with glimpses of the strain on men's faces under fire, in successive and appallingly costly and abortive Allied offensives — at Neuve Chapelle, Vimy Ridge, Gallipoli, Aubers Ridge, Champagne, Loos and northern Italy (Italy having joined the Allies), to name but the main venues of desolation. Among the new 'horrors' given most prominence, and quite predictably as part of the hate campaign, were bombing attacks upon civilians, the sinking of merchant shipping by German U-boats and the use by the Germans of flame throwers and poison gas. Popular indignation was easily aroused by these manifestations of the latest technology — as had often been the case, though less well publicized, whenever some new and frightening weapon had been introduced.

German interest in gas stemmed more from unfounded rumours in 1914 that the French were about to use it than from strong conviction that it would break the war's deadlock. But her rapid creation of the necessary chemical substances was the direct result of pre-war encouragement to science in industry by the state through the universities and, above all, the Kaiser Wilhelm Institute for the Advancement of Science, headed in 1911 by one of the most brilliant chemists of the century, Fritz Haber. Although Dr von Tappen instigated the production of liquid xylyl bromide tear gas shells, and enthused the General Staff about their potential, it was Haber (who detested war) who led the drive to increase the supply of explosives and headed the team which produced bottled, asphyxiating chlorine gas. Of the two projects, easily the most important was that connected with explosives, which was based upon his vital synthesis of ammonia from the air (from nitrogen and hydrogen). This made possible a far cheaper manufacturing process; further relieved Germany of dependence upon the import of nitrates from which ammonia, up to then, was made; revolutionized the making of nitric acid, the essential raw material for explosives, and strengthened the outstanding German chemical industry in its ability to produce agricultural fertilizers.

Since lack of explosives and a looming shell shortage most seriously worried the German General Staff in 1914 (and its opponents too), the new Chief of General Staff, General Erich von Falkenhayn, was not greatly enthusiastic in agreeing to the use of gas, which was contrary to the Geneva Convention. Nor were the results encouraging. Xylyl bromide failed in the extreme cold when launched against the Russians at Bolimov on 3rd January 1915; and although chlorine, half-heartedly released at Ypres on 22nd April, caused panic among some Allied troops, it also failed when the wind changed and blew the gas back in the Germans' faces. Thereafter gas, the effects of which would

usually be nullified by equipping each man with a respirator, became merely one of several new weapons.

Bizarre innovations aside, the nature of fighting was transformed in the apparently immutably entrenched conditions of siege warfare. While artillery inflicted wholesale destruction on anything which seemed worth hitting, troops at the front shot at each other from behind armoured loopholes, often using periscopes to scan the enemy lines. Precise long-range shooting by marksmen fell into disuse as the masses fought it out below ground, armed with magazine-fed machine-guns (such as the American Lewis) which sprayed bullets; light trench mortars which could lob bombs the four hundred yards separating one trench line from the other; and hand grenades which scattered a shower of splinters. None of these weapons on their own were likely to recreate the conditions of mobile warfare wherein lay the key to victory. For a start, not even the most intense artillery fire could guarantee to destroy barbed-wire entanglements, which grew thicker and thicker as time went by. Nor could the

Above: ***Canadian versus German infantry at Ypres, 1915, showing Vickers/Maxim machine-guns and Lee-Enfield rifles shooting Germans caught in the wire.***

placing of mines through tunnels provide other than local effect, kill men though they would by the hundreds and thousands, along with all the other associated weapons.

Massed killing, as part of the wearing-down process later known as attrition, became the motif of the generals. It reached a climax of insensitivity when General von Falkenhayn planned an attack in February 1916 at Verdun — a key, fortified sector of the line which, he correctly reasoned, the French would feel compelled to hold unyieldingly for emotional, nationalist reasons quite apart from military necessity. The commitment of nearly 1400 pieces of artillery, supplied by 2.5 million shells on an eight-mile front for which less than a score of infantry divisions could be allocated, indicated the manner in

which material was being regarded as of greater importance than manpower. This line of reasoning was embodied in Falkenhayn's plan when he initially laid down that the infantry were to be spared losses by advancing cautiously in small numbers, as the guns put the French through 'a mincing machine'. Capture of ground was of only minor importance. It was killing that counted, abetted by the French in their own way.

In the belief that even the best-protected of forts were made useless by the German heavy artillery, the French had stripped theirs of almost all armament and instructed their garrisons to occupy field-works rather than risk their lives inside concrete and steel boxes which would be crushed — despite the evidence of an exploratory bombardment by 420-mm guns in 1915 which had failed to crack Douaumont, the keystone fort of a ring guarding Verdun. In addition, they had reduced the defenders of Verdun to a mere six divisions of which the majority were inferior militia. Ironically, when on 21st February 1916 the Germans began systematically to drench the French defences with high explosive shells and gas, and as their long-range guns hit communication centres and bridges in the French rear, they might actually have broken through at once had they tried. Instead, they merely pushed forward strong detachments to probe the extent of the damage and identify those positions which had not been hard enough hit. Then, just to accelerate the task of killing Frenchmen, the French generals indulged, out of doctrine, in counter-attacks and lost heavily in so doing. Only gradually were the German lines pushed forward, one quite extraordinary success coming their way on 25th February when a mere handful of their infantry, probing ahead in a snow storm, found a way into Fort Douaumont and, without loss to themselves, took prisoner, also without bloodshed, the 56 French artillerymen remaining.

The German press represented this lucky *coup de main* as symbolic of the defeat of France and the church bells were rung in honour. The French, on the verge of collapse, sacked a few leaders, put General Henri-Philippe Pétain in command at Verdun and further satisfied the German aim by pouring more and more men into the cauldron to hold rigidly to those forts still intact. The difficulties of supplying reinforced artillery were overcome. The French artillery began almost to match that of their opponents. Attrition became a double-edged weapon as the German infantry, too, began to suffer, particularly when, in order to retain the initiative, it was found essential to abandon the probing method and commence full-scale assaults. By 31st March, after six weeks of pounding, German losses amounted to 81 607 dead against 89 000 French. The time had come to halt — as Falkenhayn seriously considered doing, and dismissed, at an early stage. It was not simply that it would be impossible rapidly to try an assault elsewhere. It was a question of clashes of will-power, public relations, and the prestige of the principals involved. The press had proclaimed a victory and the German Fifth Army now attacking Verdun was commanded by the Crown Prince. To stop at this point and admit failure might do irreparable damage to the monarchy and the army — and to Falkenhayn. In any case, all commanders drove on with optimism and determination. To falter was to admit weakness, even if to continue butting against a solid wall of French defence also amounted to obstinacy. It was a vicious, intractable perpetuity.

The slaughter continued, with each side attacking and defending by turn in a narrow rectangle of ground where the shell holes were lip to lip and the profit and loss evenly shared by the contenders. By 1st June the German artillery had increased to 2200 guns; that of the French to 1777. In June a new type of gas — phosgene — was used by the Germans but had only local effect. Just as the Germans were trapped by a false concept into determination to destroy the French, so the French for nationalistic, political and personality reasons were committed to hold. When Pétain stated that withdrawal was necessary, he was ordered by Joffre to stand firm. Joffre, for his part, applied all the pressure he could upon his British and Russian allies to begin offensive operations quickly on their

***Left:* A gassed man in the front line has a field dressing placed on his leg.**
***Below:* An operating theatre dug into a casualty clearing station in Gallipoli.**

respective fronts in order to draw off German troops.

Battles waged for the sake of killing ran out of control because politicians and military leaders were unable to withstand the adverse publicity of the press should they appear to weaken. It was a dreadful impasse for which none were adequately prepared and in which runaway technology always seemed to be offering a solution it never quite achieved.

Impasse At Sea — The Travails Of Blockade

By the end of 1914 the oceans were completely dominated by the Allies. Those few ships of the Central Powers not sealed in port were either scattered and hunted raiders or blockade runners. This isolation had been foreseen, and provided Germany's sharpest incentive to make herself self-sufficient in essential materials — in the ability to make nitric acid from ammonia, for example.

Combat at sea had run its predictable course. Numbers and sheer weight of metal doomed German ships to elimination, except when they were fortunate or skilful enough to create a temporary local superiority — as happened, for example, when Admiral Graf von Spee's force of two armoured cruisers out-gunned and out-ranged two obsolescent British cruisers at Coronel on 1st November only to be caught, out-gunned and sunk off the Falkland Islands by two British battle-cruisers on 8th December. Closer to home, the Germans were able to launch two hit-and-run bombardments by battle-cruisers on English east-coast towns towards the end of 1914. But when they tried again on 24th January 1915 and were intercepted by five British battle-cruisers off the Dogger Bank, the price to the Germans was the armoured cruiser *Blücher*, sunk by concentrated gunfire.

In this battle the British, through breaking the German codes, enjoyed the initial benefit of knowing the times of German sailing, coupled with insight into the composition and movements of their flotillas. But although the Germans had a long-range air reconnaissance capability with their zeppelins, the only one of these aloft and able to watch the fight failed to contribute information. Both sides groped their way into contact in much the same tentative way as in battles of old, and secured the same low percentage of hits. Despite the use of Director Control and the spotting of salvos, a mere 73 British hits were scored (of which 70 were upon the *Blücher*) out of 958 shots: and 14 German out of 1276, but shared more evenly between two British battle-cruisers. These were nothing like the results obtained in training at 14 knots against targets which did not change course — an index of the difference between war and peace. Men under tension made more mistakes, compounded by the difficulty of identifying salvos amidst several flurries of explosions and through billows of smoke from the funnels and guns of so many ships. Conditions were also exacerbated by spray from ships at full speed which obscured the layers' sights. The effects of successful hits were, however, quite impressive, mainly because battle-cruiser armour was easily penetrated by battle-cruiser guns, at ranges of 16 000 yards after fire had been opened at 22 000 yards. But while hits above the water-line were locally devastating, causing serious fires, a near-catastrophic magazine explosion in one of the German ships and putting the British flagship *Lion* out of action at a crucial moment, not one ship was penetrated below.

Yet attack below the water-line was by this time dreaded, and had a more profound influence on tactical decisions than gunfire. Already far more warships, including battleships, had been sunk or damaged by torpedoes and mines than by guns, and the pursuit of the German battle-cruisers at Dogger Bank might well have ended in annihilation had not a false submarine sighting led to a change of course which diverted the chase. The seriousness of this threat was emphasized two months later in the Dardanelles. The attempt to force the Narrows with a line of battleships systematically bombarding the Turkish shore batteries was brought to an abrupt end by a small, surprise minefield, which sank three battleships and severely damaged one other.

Blockaded by the Allies and fearful, for more than a year after Dogger Bank, of seriously challenging the British in the North Sea, the Germans were compelled to concentrate their efforts upon underwater attack against merchant shipping. But damaging as this was and hard to combat, the effect of this entirely new form of blockade was also counter-productive because it fuelled the campaign of hate and turned neutral nations (most important of all the USA) against the Germans. Loss of neutral sympathy might have been acceptable had the means to create a total blockade of Britain and France been available. But the gross inadequacy of existing craft, weapons and techniques made this impossible. Fast surface minelayers had no hope of making repeated sorties into protected sea lanes so close to enemy ports. A few small submarines of limited range could not sustain an all-out offensive. Paucity of intelligence and lack of reconnaissance facilities made the locating of targets a matter of improbability. In these circumstances the only way to break the British blockade and impose their own was for the Germans, under a new C-in-C, Vice Admiral Reinhard Scheer, to attempt to wear down the British at sea by a kind of attrition.

Commencing in February 1916, Scheer mounted a series of sweeps, first by light forces, later by the High Seas Fleet, whose gunnery and tactical handling had reached a high pitch of efficiency after exercises in the Baltic Sea. Hoping to bring an action against only a portion of the British fleet, temporarily unsupported by the rest, Scheer's concept admirably coincided with the willingness of the British C-in-C, Admiral Sir John Jellicoe, to accept action providing the terms were not dangerously adverse — and in communications and intelligence the British had the edge. For while their communications were mostly secure, the Germans continued to transmit messages in their compromised code and clearly spelled out their intentions. Again, their use of zeppelin reconnaissance in no way compensated for the British acquired insight. Rarely could zeppelins find their quarry through cloud or in darkness, and on the occasion of the major Battle of Jutland they were unable to fly at all due to gusty winds. So once more the collision of main forces, in gloomy weather, provided surprises to both sides.

The Battle of Jutland

Scheer's plan was to lure the British battle-cruisers to sea by feigning a bombardment of Sunderland with his battle-cruisers, unsupported by the battleships of the High Seas Fleet. Hoping to have his U-boats ambush and sink a few battle-cruisers with torpedoes as they left Rosyth, he took the High Seas Fleet to sea but fooled the British into thinking they were still in port by anchoring a ship in the River Jade to transmit signals purporting to come from his flagship. This part of his plan succeeded. Nevertheless, so good was overall British intelligence that their ships left port two hours ahead of the Germans on 30th May and were thus within striking distance long before expected — despite warnings from a U-boat which failed to ambush but did signal vital information that Scheer, out of incredulity, discounted.

The clash of battle-cruisers at 1520 hours on 31st May therefore came as a surprise to Vice-Admiral Franz Hipper, commanding the five German ships, but he reacted correctly by turning away and racing towards the 16 battleships under Scheer, some 60 miles to the southward. Eagerly, the nine faster battle-cruisers under Vice-Admiral Sir David Beatty followed in hot pursuit, gradually overtaking until superior technology, in the shape of excellent long-base Zeiss optical range-finders, enabled the Germans quickly to pinpoint their targets and hit first. This disturbed the British gunners and diverted the pursuit as Beatty felt obliged to change course. Fire opened at 1547 hours at a mere 16 000 yards because of poor visibility and the usual funnel smoke. Hits on Beatty's flagship, *Lion*, narrowly failed to destroy her when a mortally wounded officer ordered the closing of a magazine door in the nick of time. But a few minutes later the weakness in the battle-cruiser's magazine safety was crucially exposed when, within a period of twenty minutes, two of these great ships were hit repeatedly and torn to pieces in vast, internal detonations caused by flash igniting unprotected charges. And at this moment, as the British at last began to hit the Germans, they frequently failed to strike fatally because their shells, equipped with an over-sensitive fuse, exploded on impact (unlike the German shell which detonated *after* penetration of armour).

The gun duel, nevertheless, began to shift the British way and decisively so when four of the latest fast, heavily-armoured *Queen Elizabeth* battleships, with their great 15-in guns, moved within range and at 19 000 yards out-classed their opponents' best 12-in guns. Not even a series of torpedo attacks by the destroyers of each side could drive the battle-cruisers apart as the British, with superior speed, began to dictate the exchanges with their far heavier weight of metal. In the event only one torpedo hit was made on a

Above: ***Admiral Sir John Jellicoe. His belief in the vital deterrent value of large surface units later blinded him to the requirements of anti-submarine warfare and the need for the convoy system.***

The Battle Of Jutland

The only major clash between the British and German fleets seemed to end in a draw but, in effect, lost the war for Germany through blockade.

The converging of the fleets from Scapa Flow, Rosyth and Wilhelmshaven. The British are fully aware that the Germans are at sea; the Germans largely uninformed of the British approach.

14·00-15·00 hrs

First contact between cruisers scouting ahead of the respective battle cruiser squadrons as a preliminary to contact between the fast dreadnoughts.

15·00-16·00 hrs

Fire opened between the battle-cruiser squadrons as the Germans turn southwards to lead the British onto the guns of the advancing High Seas Fleet.

16·00-17·00 hrs

Two British battle cruisers sunk as the chase leads to the meeting of the British battle cruisers with the High Seas Fleet. The British retire northwards and, in their turn, begin to lead the unified German fleet towards the advancing British Grand Fleet.

17·00-18·00 hrs

The British battle cruisers make contact with the Grand Fleet and sweep round to lead it across the T of the oncoming German columns.

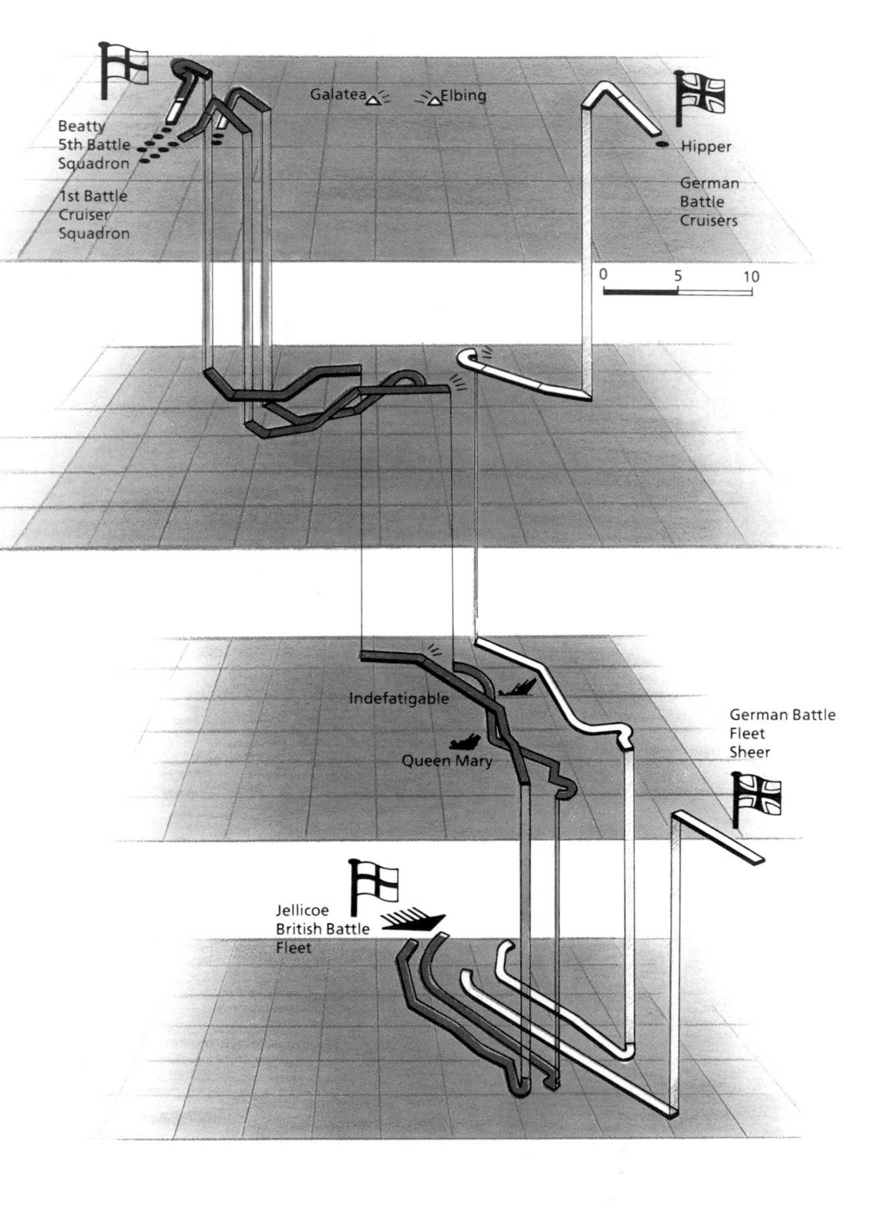

18·00-19·00

The British battleships deploy from columns to form a long line of dreadnoughts, headed by the battle cruisers (who lose one more of their number) and cross the German T. The Germans turn away to the south-eastward, enabling the British to revert to column on a southerly course and interpose between the Germans and the safety of the gaps in the minefields protecting their home ports.

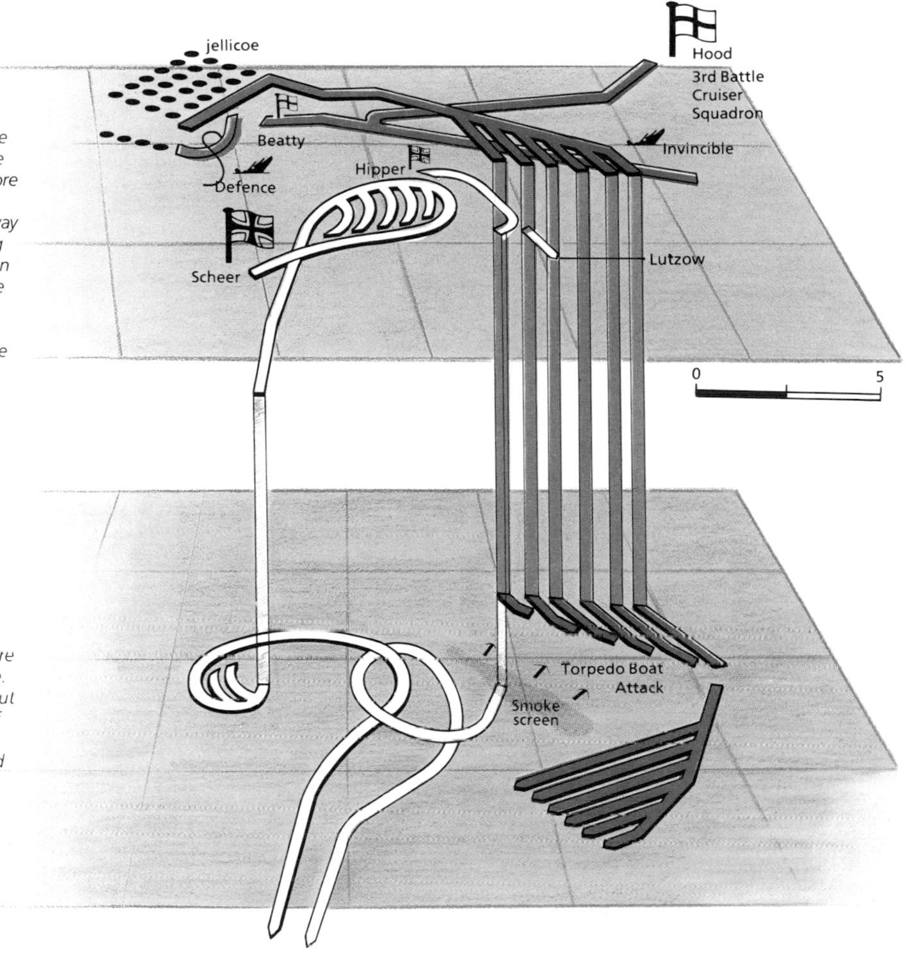

19·00 hrs-Nightfall

Attempts by the Germans to break through the British ring are thwarted by a maelstrom of fire. As night falls the Germans are cut off from home and in danger of annihilation next day by a vastly superior and largely undamaged opponent.

The Night Action

The German High Seas Fleet passes through the British light force, across the rear of the Grand Fleet, with the loss of only one battle cruiser (damaged in the earlier fighting) and a single pre-dreadnought battleship sunk by torpedo boat. Only the British light forces challenge the Germans and they suffer losses through failing to inform their C-in-C of what is happening in their midst. Come the dawn and the Germans are safely home, leaving the British with command of the seas for the rest of the war, depsite having sustained heavier losses.

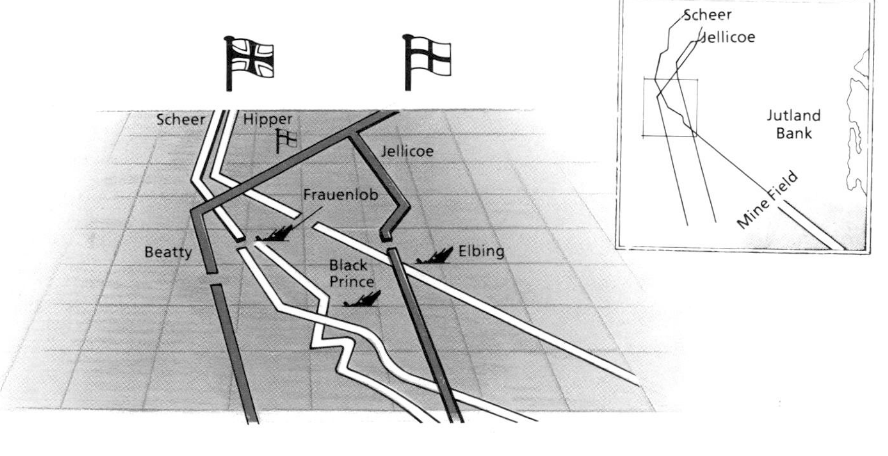

German battle-cruiser at this stage. Things might have been serious for Hipper had the support of Scheer's battleships not been close. It undeniably came as a surprise to Beatty when his scouting light cruisers reported the presence of the line of German battleships — for now it was his turn to run before the storm, seeking to avoid damage and draw the Germans into the lethal embrace of Jellicoe, whose 28 battleships of the Grand Fleet were rapidly closing the scene, quite undetected by the Germans.

At 1735 hours the British trap was sprung. Beatty, knowing he was almost within sight of the Grand Fleet steaming towards him from the north-west, suddenly turned east to head across Hipper's course, thus crossing his opponent's T and bringing to bear upon the Germans a deluge of fire. That from the 15-in guns of the *Queen Elizabeth* ships, their thick armour having withstood severe punishment, was quite devastating. Hipper was forced to turn away as damage to his battle-cruisers mounted. In doing so his attention was attracted to the east, following Beatty's movements. Neither he nor anybody else saw Jellicoe approaching. The Germans were entering an ambush at the very moment that the British were at last making full use of their technology. Hipper later commented on the way his enemy's shooting 'demonstrated how carefully the British had eliminated all the factors that increase the spread of guns firing as a battery, and how thoroughly their fire control installations are perfected'.

The radio silence imposed for good security reason upon the British fleet, and the consequent total reliance upon visual signals, gave Jellicoe several anxious minutes. It was not until 1801 hours that he could see Beatty's *Lion*, and another 13 minutes before, in reply to two interrogations by signal lamp, he at last received positive information that the enemy battle fleet was 'bearing SSW'. Based on this, Jellicoe could confidently deploy his six columns of battleships into line, enabling him to take station behind the battle-cruisers (but ahead of the *Queen Elizabeth*) as they headed east. He therefore presented a moving wall of 35 dreadnoughts, 7 miles long, steaming across the bows of Hipper and Scheer in another classic crossing of the T, at a range of 14 000 yards. The effect upon the Germans of the sudden deluge of fire engulfing them was shattering, not only from the shock of surprise and the damage steadily inflicted, but also because in the smoke and gloom they could hardly catch sight of the fleeting targets from whence the shells came. The British, on the other hand, enjoyed the full benefit of seeing their targets silhouetted against a brighter background. A flurry of hits from good German shooting caused a catastrophic internal explosion to another battle-cruiser; and two older, armoured cruisers by mischance blundered to within 7500 yards of the High Seas Fleet and either blew up or were reduced to scrap metal; but otherwise several of the German ships' guns were silent for lack of targets. Meanwhile, the British battleships shot steadily and virtually unscathed at 14 000 yards, where they were safely out of torpedo range. During this phase of the battle not a single hit of the 36 landed by the Germans struck a battleship; whereas 12 out of the 34 British hits were on enemy battle-cruisers.

Had not Scheer devised and rehearsed a special turn-away manoeuvre for use in the event of having his T crossed, the High Seas Fleet would have been crippled and thrown into confusion there and then. As it was, Hipper's battle-cruisers were now in desperate straits, with all but one severely damaged. His own *Lützow* was taking in water, forcing him to leave the line, and he had to transfer to a destroyer to keep touch with the battle. Torpedo boats were again beginning to make their presence felt. Already the battle-cruiser *Seydlitz* had been hit by one British torpedo and now it was the turn of the German torpedo boats to launch a co-ordinated attack upon the British, without scoring hits but compelling Jellicoe to take standard evasive action by turning towards the threat, thus losing contact with an opponent who vanished into the gloom.

Both Scheer and Jellicoe were at this moment poorly informed. Indeed, Scheer, totally lacking news of the enemy from his scouting forces, based his decision again to reverse course and return to the fray, in the hope of catching Jellicoe by surprise, upon intuition alone. But Jellicoe, bothered by conflicting false reports of enemy submarines, linked to news of a battleship being hit by a torpedo (probably fired from a German cruiser) did have priceless accurate information from his own cruisers of the enemy's exact position and approach. Thus he could confidently prepare a hot reception and remorselessly maintain his course southwards to place the Grand Fleet across the enemy's retreat to home ports through defensive minefields. Again, the High Seas Fleet entered a maelstrom of fire Scheer had not expected. Once more he had to turn about to escape annihilation from an enemy which clearly saw him, but which he could hardly pick out. In desperation, Scheer ordered his limping battle-cruisers, and those torpedo boats in a position to do so, to charge the enemy. The torpedo boats were then to lay smoke to cover a retreat. A crushing fire hit Hipper's ships before Scheer relented and called them back from their sacrificial death-ride. And the only effect on Jellicoe was to make him temporarily turn away to avoid 33 torpedoes before continuing southwards, clear in the knowledge that he had only to ride out the approaching night until, next day, he could destroy an opponent who was cut off.

Yet at nightfall the comprehensive British numerical and tactical supremacy was nullified by a particular technical superiority on the German part, as Scheer, with the determination of desperation, sought to make his way to safety round the rear and flank of the Grand Fleet. For although British main armament gunnery was at least as good as that of the Germans, they had, despite plans for implementation, neglected to develop centralized control of their secondary armament and of remotely controlled searchlights. They had also failed to produce a satisfactory illuminating (so-called 'star') shell. An abortive search for a solution to *all* requirements had been remarkably self-cancelling. Since Jellicoe could not contemplate fighting a night action, all the advantages were with the Germans, who possessed the vital instruments of darkness; and who, as a bonus, received prior knowledge of the British signal challenges of the night.

The tale of the German drive through the rear flotillas of the Grand Fleet, among a screen of British cruisers and destroyers and with the loss of only one pre-dreadnought battleship and two cruisers, was one of utter confusion. Jellicoe was made only vaguely aware of events through intermittent flashes of battle edging down from the north to his eastern flank. Repeatedly, throughout the night, the Admiralty was in possession of enemy signals which indicated the German intention. Still more frequently, the British light forces were in action with an enemy whose course was known. Never was enough of this vital information relayed to Jellicoe to enable him to draw the correct conclusions and make the necessary counter deployment. By 0100 hours on the 1st June the High Seas Fleet was clear of danger, except from a few mines laid in its path which damaged one battleship.

Intense activity ensued after the return of the fleets to repair the damage, rectify technical faults and improve training. For example, the British at once sent an unexploded German shell to Hadfields to discover why it was superior to the British design, and in a searching examination sought to prevent a recurrence of magazine explosions. In the aftermath, controversy raged over who had won the battle. The losses tabled below indicate that the Germans slightly had the edge.

These figures reflect the heavy loss of life caused by no less than three British dreadnoughts blowing up. But they also obscure the widespread and serious damage dealt to all the German battle-cruisers, bar one, and the harm done to several battleships — all calling for extensive repairs before the High Seas Fleet was ready for sea again. The Grand Fleet merely needed to replenish coal and ammunition and patch a few holes before it was ready, within a week, to flaunt its presence in the North Sea.

The true mark of British victory was to be found in the subsequent unwillingness of the Germans ever to risk a major encounter again — though they might tentatively put to sea on occasion. An admission of defeat appeared in Scheer's report to the Kaiser when he wrote that England would not be compelled to make peace through a sea battle and that only U-boat action against English commerce, crushing economic life, offered a not too distant victorious end to the war.

Thus Jutland, with its uses and misuses of technology, was summed up by several critics as a 'battle of lost opportunities'. It was to propel navies into a feverish race to improve their gun power (while increasing resistance to it) and to step up underwater attack, without immediately taking counter-measures to defeat it. Basically, naval warfare made overtaxing demands on research, design and industrial resources without achieving an instantly decisive effect. Trench deadlock on land could not be broken that way. Sea power could apply only indirect pressure, by diverting industrial effort from supplying armies and air forces with weapons which were also incapable of forcing a quick conclusion.

	British	German
Battleships	nil out of 28 (2 hits on 1)	1 out of 22 (25 hits on 4)
Battle Cruisers	3 out of 9	1 out of 5
Armoured cruisers	3 out of 8	nil
Light cruisers	nil out of 26	4 out of 11
Destroyers	8 out of 77	5 out of 61
Casualties	6748 out of 60 000	3058 out of 45 000

The Battle Of The Somme And The Evolution Of Attrition

To relieve the French from the strain of the German pressure at Verdun, four Russian armies under General A A Brusilov attacked the Austro-Hungarian armies on 4th June and made fine progress. On 1st July the British, under General Sir Douglas Haig, attacked close by the River Somme and advanced hardly at all. Neither offensive made use of striking tactical innovation nor, at first, of technology. In Russia it was Austro-Hungarian weakness and the vast open spaces available which enabled the Russians to prevail in the assault and exploit success in the gaps which opened up. On the Somme, the British were confronted by strong German field works manned since 1914 by a resolute and well-trained enemy. Additionally, they tended to ignore the lessons of trench warfare experienced throughout 1915 and at Verdun. They were persuaded by the gunners that a heavy, prolonged artillery bombardment was in itself sufficient to destroy and cow the enemy, leaving it for the infantry to walk forward, heavily laden with packs and equipment, to occupy the devastated area. They shrugged off their shortage of super-heavy and heavy artillery, which alone might wreck deep enemy trenches and neutralize gun positions. They underestimated the effect of medium and field gun fire, overlooking, too, that although a serious shell shortage of 1915 had been largely overcome, many of the shells were of inferior manufacture with a nasty habit of exploding prematurely in, or just beyond, the gun barrel — and not always detonating at the desired terminal point of impact among the enemy. They also failed to recognize that the deep dug-outs so thoroughly constructed by the Germans (who rarely did anything by half-measures) were proof against all but the heaviest shelling.

The result was nearly 60 000 casualties for only minimal gains on the first day of the offensive, followed by prolonged losses each day until mid-November, as the British persisted in their attempts to wear down the Germans, but without striking specifically at any recognizable strategic objective. This was the outcome of allowing technologists to dictate. Faced with a deadlock, generals can only turn to the weapons on hand for an immediate solution; in this case to aircraft, artillery and machine-guns, of which they had some knowledge. Since aircraft and machine-guns offered only supporting assistance, the artillerists, among whose officers (along with the engineers) were to be found the vast majority of the scientifically trained, took charge and promoted their enthusiasm for overwhelming use of high explosives.

One essential aid the gunners required was air power, as part of the perennial need to occupy commanding heights both in observation and combat. Indeed, the most apparent sign of change in the nature of warfare was to be found in the air, where the battle for supremacy centred upon Allied attempts to build fighter aircraft capable of beating the 'Fokker scourge'. It was more than a matter of designing a machine which could fire straight ahead; indeed both French and British had produced examples of interrupter gears without copying the fractious Fokker model. Satisfactory solutions were eventually found in a pusher machine, the British DH 2, with its fixed belt-fed gun firing from the cockpit, and the French Nieuport tractor biplane, with a drum-fed Lewis gun mounted on the top wing firing above the propeller's arc. The crucial problem was how to produce faster machines which could defeat the none-too-aerodynamically-sound Fokkers by out-climbing and out-turning them. Then pilots had to be trained to shoot down the enemy by surprise from the rear, or by means of a circling contest in which the better machine turned inside the circle of its opponent to bring its gun

Below: ***Typical trench fortifications along the French lines at Verdun, 1916. This bunker is constructed of timber, earth and sandbags, in contrast to the steel-reinforced concrete constructions being erected by German engineers.***

to bear at short range. This the Nieuport and DH 2 pilots achieved, nullifying the initial German advantage over Verdun and virtually driving the Germans from the skies over the Somme before 1st July.

Air supremacy was recognized as an almost essential prerequisite for land forces, particularly if they contemplated an offensive. Without it they could hardly conceal their preparations or protect their own reconnaissance and artillery-spotting aircraft from an all-seeing opponent. They would also be increasingly vulnerable to bombing attacks on all manner of targets, spreading from the immediate front line to their main concentrations of communication and population. True, the efficiency of bombing had improved little since the beginning of the war and by mid-1916 the zeppelin raids had been all but defeated by higher-climbing fighter aircraft equipped with machine-guns. Their incendiary ammunition made short work of these highly inflammable craft. But the threat to morale of scattered bombing prevailed, especially since longer range attacks usually took place at night in order to avoid interception. Night bombing had an even greater effect on morale because of its interference with rest. Raiding by night, usually in moonlight, stimulated the development of night-fighting techniques — of early warning systems using forward observers, for example; and a demand for night-fighting aids, such as better searchlights and anti-aircraft guns, take-off and landing aids, and communications systems between ground and air (usually at this time in the form of lamp displays and visual beacons) to guide defenders towards approaching bombers. The comprehensive systems of air defence, which had to be organized wherever air attack threatened, were yet another stimulus to invention and industrialization, a further drain upon resources to the traditional fighting arms and a stage in the deeper involvement of an entire nation in the war.

Chiefly, of course, the Battle of the Somme (like that of Verdun, which the Germans tried to end early in July) focused attention upon artillery tactics. Throughout 1915 the British Army had experimented with a considerable number of techniques that prior to 1914 had simply been interesting debating points. By the end of that year the Royal Artillery had centralized command and control, coupling it to standardized techniques with new and improved equipment. Applying the known performance of each type of gun and ammunition to a carefully calculated bombardment data, the indirect shooting of several gun batteries could be combined to control linear barrages of 'creeping' shells as they shifted ahead, bound by bound, to a timed programme and achieve concentrations against pin-point targets such as enemy gun positions, communications centres and headquarters. Infantry and cavalry were invited to integrate with these plans in order to benefit fully from fire support as they closed with the enemy. Employing sound rangers and flash spotters, the artillery were able accurately to locate enemy guns for counter-battery fire and, with a plethora of intelligence from other observers (mainly in the air), could actually begin hitting a freshly-arriving enemy battalion as it de-trained in rear and harrass it, by day and night, as it marched to the front. As few as 65% of its somewhat demoralized strength would eventually enter the trenches to experience the firepower of the many other artillery weapons.

Scientific Control Of Firepower

Of these other weapons, the sustained fire, Maxim-type machine-gun was also making a wider contribution than that of defence alone. Centralized control of these weapons under a newly-formed Machine-Gun Corps enabled scientifically calculated barrage-firing by machine-gun batteries to project a hail of fire against targets up to 2000 yards behind the front. The same task might otherwise have been dealt with by the artillery, with undesirable cratering of the ground. Machine-guns mounted on tripods, shooting on fixed lines of elevation and azimuth, could, as with the artillery, keep targets under steady attack by night and day without observation. This method, especially for the artillery, was enhanced by the use of meteorological data for wind-speed, temperature and humidity; but, if not fully compensated for, it also led to unacceptable inaccuracies. By the end of 1916 it became possible to adopt what was known as 'unregistered fire'. That is, shooting from guns which were precisely and individually 'surveyed in' to their positions, firing standard ammunition on a calculated bearing and azimuth with a fair certainty of striking the target area — and with the highly desirable elimination of ranging by observed fire. Unregistered fire promised a tactical revolution: the ability to hit the enemy without warning by ranging and to restore to commanders that factor of surprise by which, alone, they could hope to demoralize and paralyse the foe.

Devastating though firepower was to both sides at the Somme (and wherever else it was used with unbridled ferocity and ever-improving technique) the trench lines were never broken and the cavalry went on awaiting a gap that never appeared. The soldiers needed an entirely new weapon which could crush the

Above: ***General Erich Ludendorff controlled the German war effort and underestimated the threat of new technology, notably the tank.***

barbed wire, cross the broken ground and bring direct, accurately-aimed fire from behind some sort of protection against the enemy. Frequently in the past it had been civilian inventors and entrepreneurs who had come forward with a viable solution. This time — and as a prime example of how the travails of war concentrate minds — it was a scientifically-trained Royal Engineer, with a bent for writing history in excellent English, who proposed and pushed through to battle on the Somme on 15th September, a genuine armoured fighting vehicle (AFV) which satisfied, in temperamental form, all the requirements needed to breach the trench lines.

Colonel Ernest Swinton simply proposed the adaptation of the Holt track, which before the war had been so improved by the Roberts track, to carry a steel box armed with cannon and machine guns across no-man's land, through the trenches and among the enemy artillery, shooting machine-gunners and infantry on the way and so dominating the enemy positions that the AFV's escorting infantry could reach their objective almost unmolested. From Swinton's specification came what was known, as a way of disguising its nature, as the tank. It was the product of combined design and development by a team of inventors of whom the naval architect Eustace Tennyson D'Eyncourt, and the motor engineer Walter Wilson (both of whom had been involved in 1914 with naval armoured cars), along with William Tritton, managing director of the firm of Fosters, were pre-eminent. Slow and clumsy, the tank nonetheless had a proven ability to cross broken ground and wide gaps, thanks to the genius of Wilson in designing its rhomboidal shape and the entirely original track link which was far in advance of the Holt type. The new vehicle made an encouraging battlefield debut by doing, with a mere 32 machines, everything that was required of it and a lot more than some people expected. One did break through to the gun lines. Those which reached the enemy trenches had rolled flat the wire and were followed by infantry whose casualties were significantly reduced. The enemy panicked and, had it not been for some resolute shooting from a few 77-mm guns, might have suffered more harm than they did.

The majority of tank losses came from mechanical breakdown, with artillery and the armour-piercing bullet, introduced at an earlier stage to defeat trench armour, sharing the credit. Unreliability was to be expected from an untried improvisation of a machine which, apart from the special track invented by Wilson, incorporated existing materials and weapons. The engine came from Holt artillery tractors and the guns and armour from the navy, whose contribution was made rather reluctantly. But tactics similar to those devised by Swinton for co-operation between tanks and infantry proved sound (and basically remain unchanged to the present day) while the achievements of recognizably crude machines were sufficient to ensure their admission to the armoury of war.

On the face of it, attrition at the Somme produced no more positive results than elsewhere. An advance of six or seven miles in four and a half months for 450 000 British and 195 000 French casualties was nothing to boast about. Yet the 650 000 losses inflicted upon the Germans, plus those they suffered on other fronts, were of such severity as to compel the newly-appointed team of Hindenburg and Ludendorff, who now took charge of the German war machine, to concede that the army was incapable of offensive operations in 1917. It was the Russians who were to crack, their human masses smashed and discouraged by an enemy whose training and weapons overrode blind courage and obedience. In March 1917 revolution broke out, the Tsar was toppled from his throne by a popular uprising of the people and the navy and army's resistance crumbled, allowing the Germans to do much as they pleased. Russian industry had failed to keep the men well-armed, and their leaders made virtually no contribution to advancing the technique or technology of war because they had invested so little intellectual or industrial effort in it.

Chapter 4

A Shift in Dominance

1917-1930

It was the year 1917 in which the new technology boosted by the war began to supersede the old, and in which military and industrial leaders on both sides began to apply fresh techniques in profusion to practical advantage. Over sea and land, aircraft made an increasing impression, but over neither element could even their keenest protagonists claim that they played a fundamental role in initiating what, within two decades, had become established revolutions in the art and conduct of war.

The initial German attempt in 1915 to impose a blockade of the British Isles with submarines had been called off. First, from a shortage of submarines, with only 20 available in February at the start of the campaign; and second, because of worldwide indignation, whipped up among the neutral nations by intensive propaganda when merchant ships were sunk without warning and when civilians were drowned. The full fury of American wrath fell upon Germany when the liner *Lusitania* was torpedoed on 7th May with heavy loss of life, including many Americans. In fact, German U-boat commanders often gave warning of sinking and helped to save crews, not only for humanitarian reasons but to prolong their voyage by conserving what few torpedoes they carried. It was more economical to surface and board the vessel to open its scuttles, or to use the gun with its ample supply of cheap ammunition, and not very risky if, as was usually the case, their prey was alone and unarmed.

In the absence of detection devices, a U-boat was vulnerable only if it surfaced within sight of a hostile ship or struck a mine in shallow waters. Once beneath the surface it could be tackled only by ramming or by explosive sweeps dragged through the water. Suggestions from outside the Admiralty to train seagulls to sit on periscopes or seals to bark at the presence of a U-boat were no more ridiculous than those adopted by the Admiralty itself; such as issuing a picket boat with a canvas bag and hammer with a view to slipping the bag over the periscope prior to hitting it with the hammer. The invention in 1915 of a depth charge, which could be set to explode at a required depth, was of only marginal help since the first hydrophone location devices were found to be useless; establishment of precise bearing was impossible beneath the surface where all kinds of ship's noises and varying densities of water intruded.

When the new German Chief of Naval Staff, Admiral Henning von Holtzendorf, managed to persuade the Kaiser early in 1917 that the only answer to the complete Allied blockade was to starve Britain out of the war, he based his case upon a statistical study which, like all exercises of its kind, was subject to imponderables and the unknown. The basic proposition that sinking 600 000 tons of Allied shipping a month and dissuading 40% of neutral ships from entering British ports would bring Britain to her knees in five months was more contentious than the claim that a U-boat fleet, enlarged to 105 units as the result of monthly increases of eight or nine against losses of two or three, might achieve that result. The Germans did attain this latter goal, and on occasion exceeded it. The underrated risk lay in the political gamble of an unrestricted campaign of sinkings, regardless of warning or nationality — an almost total break from the established blockade Prize procedures of the past. The reason given was fear of enemy defensive measures which, the figures showed, tilted against the survival of surfaced boats, and did so because a combination of aircraft, surface ships (including armed merchantmen) and mines was slowly beginning to make itself felt. Twenty U-boats had been lost in 1915: about 30 in 1916. The decision to begin the offensive not only discounted the apparent risk of America entering the war against Germany within five months, but also the introduction of defensive measures which might prevent the U-boats meeting their target of sinkings.

Submarines Versus Convoys — Statistical Analysis

Politically, the U-boat campaign which opened on 1st February 1917 was a disaster. America joined the Allies at war two months later. Technically it acted as a spur to defensive measures. The latest U-boats, some weighing 900 tons and carrying 42 mines or 18 torpedoes, were able to stay at sea for up to four weeks and had useful ocean-going characteristics. Along with the smaller, coastal types, they could seek their prey on the recognized sea lanes and in the immediate approaches to ports where the pickings were good. So long as Allied ships sailed individually, as in peacetime, there was little need for U-boat commanders to call for special intelligence or to co-ordinate their attacks with each other.

This freedom of action was prolonged by a lack of adequate challenge, as the British and American admirals were so unwilling to change their methods. They refused to allocate enough escort vessels, particularly destroyers, to guard shipping and hunt U-boats, and declined to institute a universal convoy system. At the root of this apparent intransigence lay the instincts of generations of sailors who based their decisions on intuition and resisted scientific methods. Not until 1917 did the Royal Navy form a Staff College to educate their staff officers, although the Americans had started theirs in 1884. Unlike the Germans, the British Admiralty managed without a statistical branch and therefore had no clear measure of problems or their trends. Moreover, they consolidated the system by transferring Admiral Jellicoe from command of the Grand Fleet to First Sea Lord, thus ensuring that many destroyers would be kept idle in a Grand Fleet which was unlikely to be called upon to fight another major action. Jellicoe clung to existing methods, arguing against convoys as an unmanageable organizational problem in relation to the 5000 ships which he believed entered and left port every week, and because of the lack of available escorts.

Close examination of Ministry of Shipping records by Commander Reginald Henderson proved, in an early example of scientific operational research, the falsity of Jellicoe's intuitive guesswork. Weekly ocean-going sailings numbered only 120–140. Those few vessels which were usually escorted, such as warships, colliers and troop transports, suffered hardly any losses because, it was shown, submerged U-boats under threat could not aim their torpedoes accurately. But it took a direct appeal by Henderson to the Prime Minister, in the teeth of Jellicoe's objections, to try a convoy system in May 1917 and prove from demonstration the veracity of his calculations. Losses of 869 000 tons in April, 593 000 in May and 683 000 in June dropped sharply as soon as the convoy system was fully introduced. Within a year they fell below the 300 000-ton mark, even though by May 1918 the number of U-boats at sea had risen to 55.

Statistical analysis provided the key to countermeasures, the collection, evaluation and presentation of material highlighting the results achieved in corroborating or refuting hypotheses. The transfer of Grand Fleet destroyers to convoy duties could now be justified. Study of known U-boat movements to their hunting grounds could be converted into blocking those routes with defended mine barrages, such as made the Straits of Dover a death trap for boats on or beneath the surface, by day or night. Relating losses to

Right: ***Captured German mine-laying U-boat, with mine in front of the conning tower. This type of boat, built in sections, was sent to Flanders where it was assembled for operations in the English Channel.***

ML 121
U
UC 5
34
32
30
28
26

their causes indicated which offensive measures were most worthy of promotion — and of these, the figures for sinkings in 1918 showed that the depth charge (22) and the mine (19) were most successful out of the 59 achieved. Nevertheless, the use of 2000 depth charges, out of 4000 produced, to put down only two of the 14 U-boats sunk in May 1918 was hardly a good return. The location of submerged U-boats continued to elude the scientists, although the invention by the British and French in 1918 of a sonic-beam detector called Asdic (later known as sonar and actuated by piezo-electric effect) pointed the way to a considerable improvement.

Statistical evidence also aided the German U-boat commanders, although those who survived had no need of figures to realize that finding a few convoys was more unlikely than intercepting a plethora of individual ships; that surface engagements had to become the exception rather than the rule; that torpedoes were not wholly reliable; and that the enemy had a depth charge — although not until May 1917

Above: ***A destroyer, about 1917, showing torpedo tubes (amidships), depth charges and launcher rails astern. In such a primitive form, without any device to locate a submerged submarine, the use of the depth charge was haphazard in the extreme.***

were they aware of this. As a counter to centralized enemy control, Commodore Hermann Baur proposed (and had rejected) the idea of a large cargo-carrying submarine, fitted with wireless, controlling and supplying a pack of U-boats. Such craft would not only give and receive orders, but would monitor Allied messages and provide intelligence of shipping movements. Radio was used, however, to guide individual and paired U-boats towards their quarry, but their messages were being read by the Admiralty and a reception prepared when they reached their target area. Indeed, the only major, informal concentration of a pack on 9th May 1918 led to disaster. Intercepts and cross-bearings on U-103's messages to four other boats, placing them in the path of a convoy, were translated into an ambush. U-103 was sunk by ramming; U-72 stalked on the surface by the British submarine D-4 and torpedoed.

The attempt to starve out the Allies failed because the original German calculation was flawed. Primitive defensive means, belatedly developed to combat the primitive offensive weapons, were more than enough and Allied ship-building capacity was sufficient to replace losses. But there was a blurred image of the future for those who could recognize that underwater attack had immense potential. Ninety per cent of Allied shipping had been sunk that way. A true counter had not been produced, and the options open to development were immense.

The Day Of The Tank — The Battle Of Cambrai

Reversion to the defensive by the Germans in 1917 brought about a previously unsurpassed feat of fortification construction. It involved the digging by hand and machine of several hundred miles of deep trenches, three lines deep, with shelter for the troops in pre-fabricated dug-outs, fronted by massive anti-tank ditches and dense belts of barbed wire. Vastly expensive in manpower and material to build, but designed to conserve life by holding the frontal areas thinly and stationing strong reserves in the rear for quick counter-attack, this, the Hindenburg Line, was a first essay in countering the threat of artillery and the tank as economically as possible.

The Allied plan for 1917 was an attack at Arras and in the Chemin des Dames in April, employing artillery tactics which were a vast augmentation of those of 1916, but without using unregistered fire. They failed, and at such enormous cost in material and lives that elements in the French Army mutinied. Again the British failed when their attack at Ypres, based solely on registered artillery tactics, led to a squalid slaughter. The assault sank in the quagmire of Flanders mud — created by the guns — without gaining even the initial objectives.

Desperate for some sort of victory before Christmas to offset the debacles of the year, General Sir Douglas Haig took up a suggestion of Brigadier General Hugh Elles, Commander of the Tank Corps, to breach the Hindenburg Line by surprise on firm, unbroken ground covering the city of Cambrai. The instigator of the idea, Lieutenant-Colonel J F C Fuller, proposed a revolutionary attack by 476 tanks and infantry to a depth of 4 miles, unheralded by a preparatory artillery programme, in which the assault forces would assemble secretly and 1003 guns would adopt unregistered fire, witholding their contribution until dawn when the first wave of tanks began to advance. It would be the infantry's task to follow the tanks after these had crushed the wire, filled the anti-tank ditches with large bundles of wood called fascines, and dominated the enemy machine-guns and artillery. Smoke fired by the artillery would screen off enemy observation positions; aircraft would bomb and machine-gun communication centres, gun pits and entrenchments. And when the three trench lines had been subdued, a mass of cavalry would debouch into the open ground beyond, seize Cambrai and enforce a wholesale German withdrawal in northern France.

To achieve a totally unrevealed concentration of forces, observation by enemy aircraft had to be prevented. Complete denial of the air to their reconnaissance machines was achieved by intensive fighter patrols, along with meticulous concealment of the tanks, guns and dumps of artillery ammunition. Because the new Mark IV tanks (their armour thickened to offset the armour-piercing bullet) were capable of running only 20 miles before their tracks and sprockets wore out, they had to be brought close to the front by rail at night with the least possible noise. Special railway tracks had to be constructed and the noise of tank movement, from the rail flats to the cover of woods, drowned by low-flying aircraft. Despite these precautions, the Germans did catch a hint from captured prisoners of something brewing, and a cautionary order was issued on 19th November to expect a local attack on a narrow sector near Flesquières. No hint of the vast dimensions of the impending blow leaked across the lines, however.

The night before the battle passed in relative quiet as the infantry and tanks eased forward systematically to the start line, while the surveyors completed their work (positioning the last gun only 40 minutes before zero hour), and the gunners received the latest weather reports for final adjustments to the planned aiming of their pieces. As dawn approached, troops in the line stood-to as of routine until, at 0610 hours, the throb of tank engines allied to the squeak and clank of tracks could be heard approaching. Ten minutes later, the sky to the west lit up with a thousand gun flashes, heralding vortexes of fire engulfing the German front line and battery positions, cloaking the higher ground in smoke to mix with the early morning haze.

Tactical surprise was achieved as the result of technical surprise, because it persuaded the enemy to dive into their dug-outs and remain there on the comfortable assumption that it might be days before an

The First Major Tank Battle — Cambrai, 20th November 1917

Cambrai, an important German communication hub at the centre of the Western Front, was protected by the Hindenburg Line with its thick belts of barbed wire, wide anti-tank ditch and complexes of deep dugouts and well-prepared fire trenches and gun positions.

The concealed secret assembly of British cavalry, infantry, artillery, machine-guns and tanks was effected without discovery by the Germans. Movement took place at night into wooded terrain and German reconnaissance aircraft were driven from the day skies by fighters.

Stage One

The pre-registered artillery bombardment was witheld until tanks and infantry began to advance. It then concentrated mainly on enemy artillery positions behind the crest and infantry strongpoints, leaving the tanks to flatten gaps in the wire for the infantry to follow. Aircraft bombed enemy gun positions and suspected headquarters as barrage-firing, medium machine-guns sprayed front-line trenches to suppress hostile infantry. Smoke laid by the artillery across the forward slopes and crest line was chiefly designed to blind enemy artillery observers and machine-gunners.

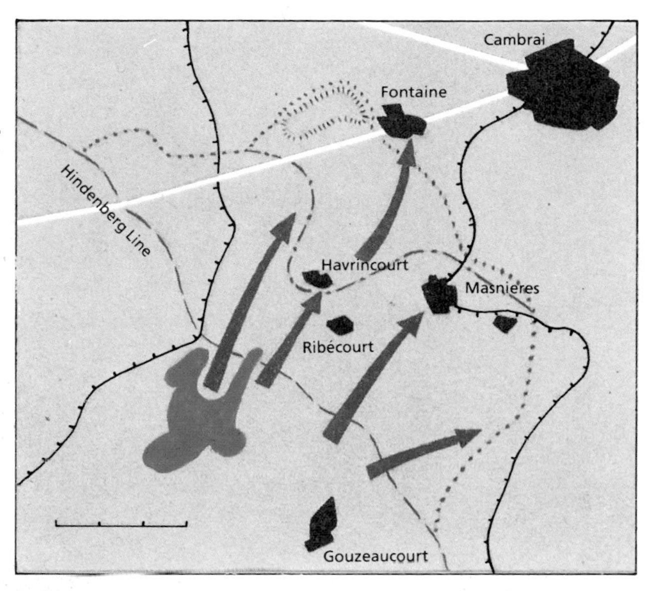

Stage Two

The tank infantry drill involved waiting for the leading wave of tanks to crush the wire, fill the anti-tank ditch with their fascines and then dominate the enemy trenches until the infantry arrived with the second wave of tanks to capture the positions.

Stage Three

Once across the trenches, tanks and infantry fanned out to occupy the entire enemy front line, while a third wave of tanks and infantry passed through to seize the objective in depth and overrun the German rearward lines and forward artillery positions. Thereafter the cavalry were meant to take over and, with artillery, infantry and air support, thrust deeply into the enemy rear to capture Cambrai and bring about a widespread enemy withdrawal in northern France and Belgium. In practice, the tanks tended to move too far ahead of the unarmoured infantry and, as a result, were unable entirely to overrun the guns. And, inevitably, the cavalry once more showed that, against even the slightest opposition, it was powerless.

assault was made. The German Higher Command also drew the same false conclusion and withheld counter moves, with the result that the successive lines of defence were overrun with little resistance.

On the right flank, where defences were weakest, the tanks crushed the wire, crossed the deep ditches and, in accordance with a pre-arranged drill, helped the infantry capture the forward lines with grenade and rifle fire before lunging, flat out at 4 mph, down the long slope to the vital bridges across the Escaut Canal at Masnières and Marcoing. To quote a German soldier, 'The riflemen were so totally surprised that there was almost no defence . . . the tanks looked terrifying . . . and many brave soldiers lost their heads and panicked.' Not that the British infantry was all that enthusiastic, stopped as it could all too easily be by the lightest opposition. So the tanks reached the canal an hour ahead of everybody else, and were then unable to go farther because a bridge collapsed under the weight of a tank and no portable bridging equipment was available, simply because it did not yet exist. Here the battle devolved into a dual between artillery and tanks to which the infantry and the cavalry, when the latter appeared at lunch-time, were mainly impotent spectators. German gunners who kept their heads soon discovered that at ranges beyond 300 metres, the chances of the relatively inaccurate 77-mm shell hitting the tanks were low. Tank gunners, when at last they could pick out the artillery through smoke and camouflage, were lucky to locate targets much above 200 metres. Dug-in German guns, drawn by horses, were at a disadvantage because they could not manoeuvre, but a 77-mm anti-aircraft gun, mounted on a truck which raced to Masnières, not only knocked out a tank at 500 metres after 25 rounds, but lived to tell the tale and established the value of the dual-purpose gun — a fact not lost on the Germans.

At Flesquières, on the vital left flank, the tanks raced ahead of the infantry. Nearly all the German guns emplaced just behind the crest had been destroyed by shellfire (a testimony to the accuracy of unregistered shooting) and those which survived were not only short of ammunition but hampered by a new type of shell for which the fuse-setting key was missing. As a result the gunners had to fire what, in due course, would be regarded as a superior penetrator of armour — solid shot. Tanks were destroyed at point-blank range whenever they attempted to enter the village or cross the crest.

But the attacks at Flesquières and on the right of the line came to a halt chiefly from technical and planning deficiencies which inhibited the co-ordination of tanks, artillery, infantry and cavalry. Once tanks and artillery had overcome the enemy machine-guns, the vehicles pulled back to rally and replenish, and the troops had to await orders by telephone to engage targets not included in the initial programme. The infantry, having caught up with the tanks, felt unable to progress without armour, as did the cavalry. And even if a plan had existed, its implementation would have been emasculated by inadequate communications in the absence of portable radio sets. Passage of information was governed by vulnerable telephone cables (and the time it took to lay them), or the time taken by a horseman or runner to move the distance from one concealed location to another when under fire — if he survived. A few tanks carrying radio, which were used for the first time here as report centres, barely contributed. It usually took two to three hours to arrange a fresh attack. In consequence the British, having captured over 10 000 prisoners, 123 guns, 179 mortars and 281 machine-guns for the loss of only 4000 of their own infantry and 65 tanks destroyed, were denied the strategic prizes of a tactical triumph. Next day the tanks, their inherent unreliability leading to a far greater decline in numbers than enemy fire, came up against much stiffer resistance. German troops reinforced the sector massively by means of a remarkable nose-to-tail traffic in railway trains from far afield — a movement that neither long-range guns nor bombing aircraft could prevent. Using several mobile anti-aircraft guns to check the tanks at Fontaine, the Germans stabilized the situation and rapidly built up a counter-stroke.

The Germans too had developed the technique of unregistered artillery fire but, having neglected to copy the tank idea, were compelled to exploit it with improved infantry tactics. When their storm of fire unexpectedly engulfed the British defences early on 30th November, their assault tactics were those already successfully experimented with against the Russians at Riga, on 1st September, and the Italians at Caporetto on 24th October. Specially trained and motivated 'storm troops' provided the spearhead for *ad hoc* Battle Groups composed of infantry machine-gunners, light assault artillery and engineers who attempted to by-pass opposition in order to infiltrate the enemy lines and undermine resistance through flank attacks. In what was only a limited, hastily improvised attack, the new methods worked admirably. Not only did the British lose much of the ground they had won on 20th November, but they were also ejected from nearby territory, and suffered a defeat which cancelled out the tank victory.

Nobody could ignore the multifarious lessons of Cambrai, which ended in stalemate amid winter

Above: **Attempts in 1918 to combat the problems of cold and anoxia above 10 000 ft are illustrated by the observer front-gunner of a twin-engine pusher Gotha G-IV biplane bomber. He is sucking on a pipeset mouthpiece of an Ahrebdt and Heylandt liquid-oxygen apparatus which can be seen inside the cockpit beyond his right hand.**

gloom. Few battles introduced such a rich variety of change to influence the art of war. For underlying the event was a lesson detected by very few at the time — that mechanized, armoured warfare created a saving in lives relative to the results achieved. At Cambrai it was noticeable that, after an advance of four miles in only seven hours, remarkably few bodies were in evidence.

Dawn Of A New Era — The 1918 Campaign

Blurred as the image of submarine blockade may have appeared, it did fail to cripple the Allies while they, by applying a surface blockade, benefited in the long run by starving the Central Powers of essential supplies. The neutralization of Russia, plunged into revolution and civil war, helped Germany only in the short run by enabling the transfer of enormous armies to the West, finally to crush France and Britain before the latent strength of America (whose technical contribution was meagre) could be applied. Their initial breakthrough, achieved most strikingly in March against the British before Amiens, broke down for reasons similar to those of 1914: a foot and hoof army possessed neither the stamina nor the logistic strength to advance beyond a restricted distance, and the Germans still had far too few motor vehicles to transport loads beyond rail heads through areas already devastated by battle.

Similarly, air power, considerably advanced in capability as it was by the introduction of bigger and better machines, could not impose a solution on its own. It remained ancillary to sea and land forces. The entry into service in 1917 of the twin-engine German Gotha IV bomber, with its range of 522 miles, operational ceiling of 21 000 ft and (on short flights) 1300 lb bomb load, made possible the bombing of London from Flanders by day, just as the introduction a few months later of the twin-engine British Handley Page 0/400 made it possible to retaliate against targets in western Germany, though from a much lower altitude. But no matter the height, the chances of doing much damage with the largest 250-lb bombs were poor and, once the initial shock had dissipated, the effect on people in the target area was as much to goad anger and hatred as cause paralysing fear. Moreover, bombing accuracy further declined when defending fighters and guns compelled night attacks only. The resulting crashes from navigational errors, bad weather and mechanical unreliability far exceeding losses from other causes.

Fighter performance had virtually doubled since 1914. The British Sopwith Camel could climb to 22 000 ft and fly at 120 mph; the French Spad XIII could reach 138 mph; and their principal adversary, the Fokker DVIII, was comparable with an excellent performance at high altitude. Engines up to 400 hp were coming into service, although power fell off seriously at altitude in the rarefied atmosphere. Air crew performance also declined above 10 000 ft, and the need to equip men with oxygen to combat anoxia had to be faced. Despite the introduction of wind tunnels for experimental purposes (by the Wrights in 1903), the basic construction of aircraft had made tardy progress as a result of rigid adherence to biplane designs that were often over-engineered and subject to excessive drag. A virtual ban on the far more aerodynamically 'clean' monoplane was in reaction to earlier structural failures, before the problem was thoroughly tackled by a genius, Professor Hugo Junkers. Commencing in 1915, Junkers designed and built a series of experimental, low-wing monoplanes of

The British Mark IV Tank

Trench warfare had resulted in deadlock at the mid-point of the Great War in 1916. The Mark IV tank was the technical solution to this battlefield stalemate. The type shown here is a Male (an emplacement destroyer, as distinct from the man-killing Females). At the Hatfield Park test ground of the firm of Fosters, efforts centred on developing a satisfactory drive mechanism, power plant and tracks that could withstand battlefield conditions.

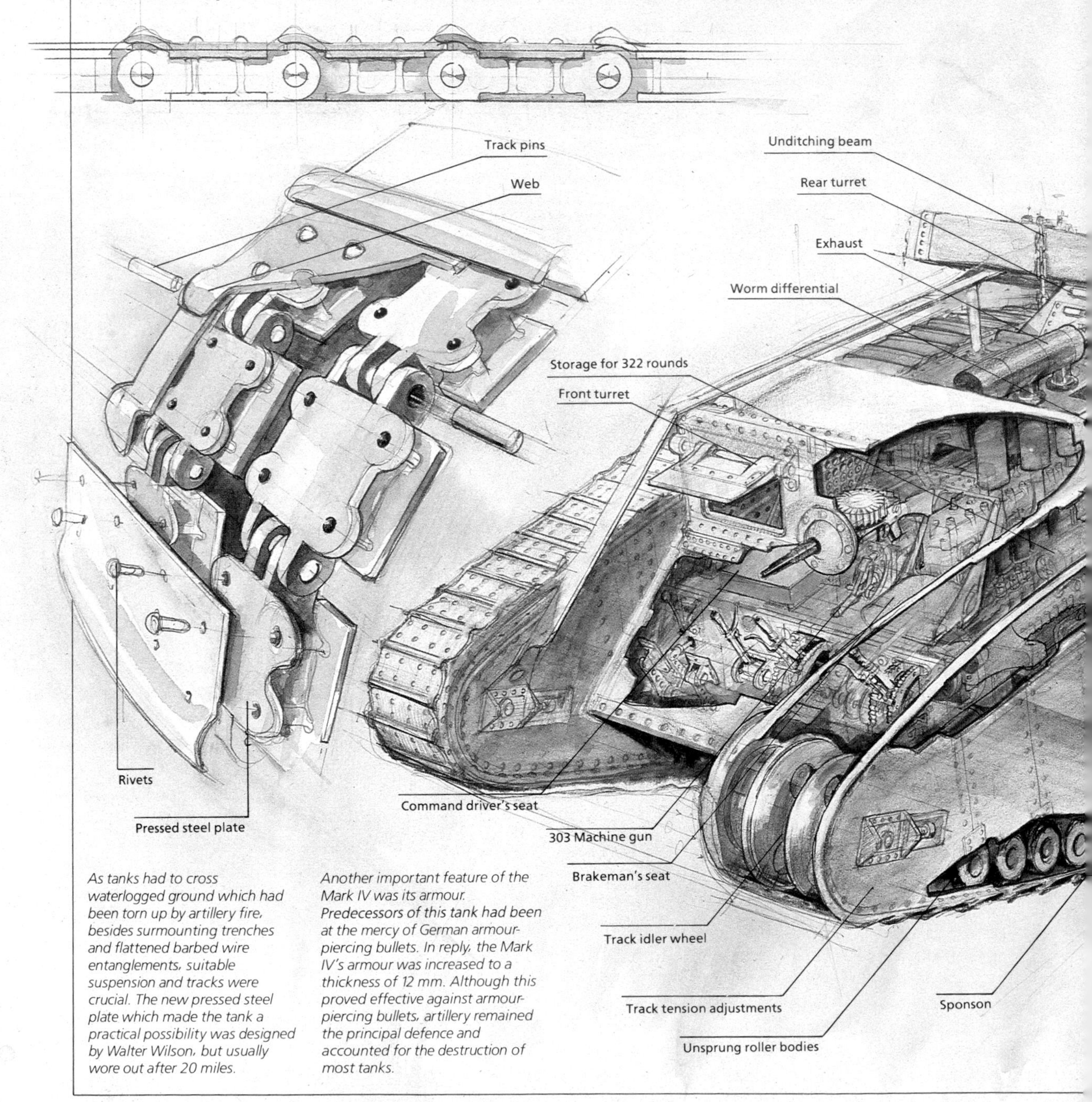

As tanks had to cross waterlogged ground which had been torn up by artillery fire, besides surmounting trenches and flattened barbed wire entanglements, suitable suspension and tracks were crucial. The new pressed steel plate which made the tank a practical possibility was designed by Walter Wilson, but usually wore out after 20 miles.

Another important feature of the Mark IV was its armour. Predecessors of this tank had been at the mercy of German armour-piercing bullets. In reply, the Mark IV's armour was increased to a thickness of 12 mm. Although this proved effective against armour-piercing bullets, artillery remained the principal defence and accounted for the destruction of most tanks.

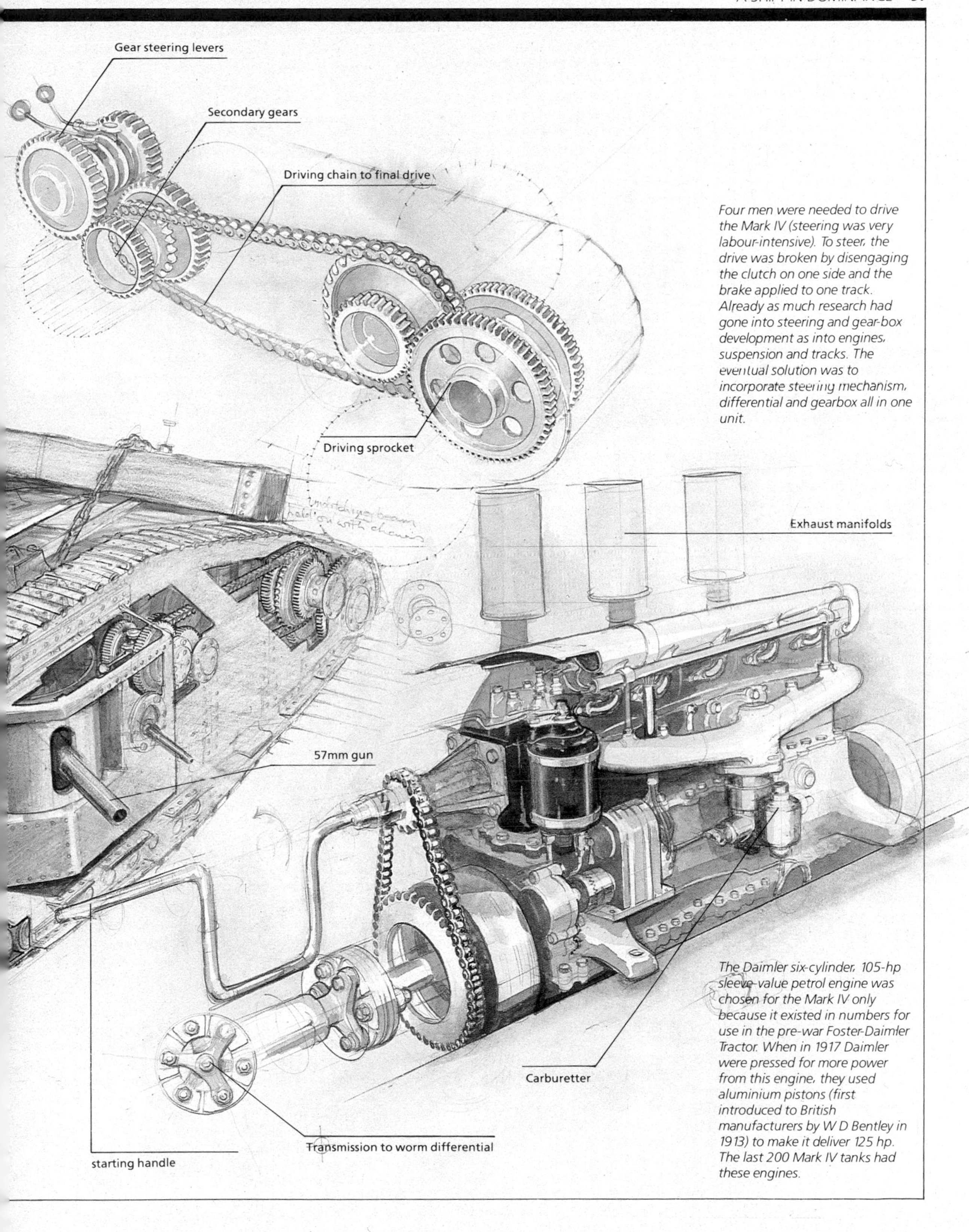

Four men were needed to drive the Mark IV (steering was very labour-intensive). To steer, the drive was broken by disengaging the clutch on one side and the brake applied to one track. Already as much research had gone into steering and gear-box development as into engines, suspension and tracks. The eventual solution was to incorporate steering mechanism, differential and gearbox all in one unit.

The Daimler six-cylinder, 105-hp sleeve-value petrol engine was chosen for the Mark IV only because it existed in numbers for use in the pre-war Foster-Daimler Tractor. When in 1917 Daimler were pressed for more power from this engine, they used aluminium pistons (first introduced to British manufacturers by W D Bentley in 1913) to make it deliver 125 hp. The last 200 Mark IV tanks had these engines.

which the J I, nicknamed Tin Donkey, was the first all-metal, cantilever-winged aeroplane. Breaking away from conventional methods, Junkers not only used welded steel and duralumin in his designs, but incorporated stressed metal panels (some corrugated) on wings and fuselage to improve load-bearing and help reduce weight. Side-tracked as his designs were throughout the war, Junkers' concepts were the true precursors of future design.

By 1918, massed production of complex aircraft from a burgeoning industry was in full swing. Germany built nearly 21 000 in 1918 and her enemies far more with their greater resources and free flow of supplies. Hence air combat intensified. The technique of endeavouring to attack from height, out of the eye of the sun, dictated that stalking by solo hunters should take second place to hunting in pairs or the manoeuvring of several dozen or more machines under skilled leaders prior to dog-fight clashes. Then it became every man for himself. Yet the concentration of effort on army co-operation was mandatory and led to various improvisations. Among the most far-reaching was the proposal by Colonel William Mitchell, an American advocate of strategic bombing, to give infantry a quick course in parachuting and drop them in the enemy rear on Metz from converted Handley Page bombers. Gradually, indeed, the latest aircraft were designed to be multi-purpose. The fitting of small bombs to fighters for low level attack against troops and guns assumed great importance — and caused such high losses to the attackers that a demand for the fitting of armour was heard and implemented on the latest machines.

A point was reached in 1918 when the conservation of life in order to preserve morale became critical. It was therefore all the more amazing that several air

Mechanized Cavalry And Machine-Gunners

The British Rolls Royce armoured car of 1914 (right), the British 14-ton, Medium A tank (Whippet) (above right) and the French Renault FT 7-ton light tank (above), both the latter of 1917, indicated the line AFV development would take in the future.

The British machines, with their relatively high speeds, would gradually assume the role of cavalry since they could move faster, further and with a greater chance of survival against fire than the horse. The French machine, on the other hand, enabled a machine-gunner to achieve superior protection and mobility while working in close co-operation with friendly infantry against hostile machine-gun posts. With a speed of 5 mph, the Renault was just able to fill this role.

forces, including the British, persistently rejected the parachute on the grounds of unacceptable weight and the possibility that it would undermine aircrews' aggression. Their men thus remained exposed to a fiery death in machines which were flying petrol tanks. The call to make AFVs safer fell on more sympathetic ears, even though each tank's life was reckoned to be short. Already, in July 1916 (before the tank's debut), the British designers had experimented with the idea of a 100-ton, shell-proof machine and later seen it rejected on the grounds of cost and manufacturing difficulty. But their proposal for an 8-mph, 14-tonner found favour, and the Whippet entered action against the Germans in March 1918 with marked success. Meanwhile the French, having independently built two unsatisfactory types of Holt tracked armoured assault guns, had at the instigation of Colonel Jean Estienne put the firm of Renault to work on designing a two-man tank. Little more than a slow armoured machine-gun carrier of dubious cross-country performance, it nevertheless had the advantage of production in mass, and presented only a very small target. Thus there came into use five distinct types of AFV: the heavy rhomboidal tank, much improved with a more powerful engine and capable of being driven by only one man; the medium tank; the light tank; the armoured assault gun; and the armoured car.

The rotating turrets on the Rolls Royces and Renaults merely copied naval practice. It was an obvious development but one which needed three men for maximum efficiency, rather than the one-man French turret which over-worked the machine-gunner who had also to act as loader and commander.

Quite apart from the physical effect that AFVs had on enemy weapons, and the terror they usually inspired, the most significant result of the tank was to make obsolete, after Cambrai, the entire concept of an invulnerable trench and wire system. Whether or not combined with unregistered artillery fire, AFVs restored surprise to the attack (and defence, too, if required) by the use of firepower, protection and mobility to which there was no sure counter. Revolutionary though this appeared, it was only a convenient modern way of applying the age-old principles of war. Indeed, it could be argued that it was not the AFV which was a freak (as some claimed) but the complete stalemate imposed by earthworks which was unprecedented. It was no longer worthwhile constructing an endless and costly barrier of trenches that could be penetrated even at its strongest point with relative ease. Throughout 1918 both sides totally revised their offensive and defensive tactics to cope with the new technical pressures. Defence had to be 'all around', established in greater depth and keyed to vital ground, so allowing the enemy to manoeuvre more freely and thus create wider spaces and loosen defensive fetters. Break down or be destroyed as AFVs inevitably would, the mere suggestion of their presence was enough to dilute defences and create a panic that undermined resistance, thereby restoring mobility to the attack.

Repeated counter-blows by tanks working in dense formation and by massed artillery, with infantry following up, smashed five successive German offensives which, by the middle of July, had cost nearly 500 000 men on each side. But it was a massed attack at Amiens by 604 Allied tanks on 8th August that irrevocably turned the tide. Here the Cambrai method tore a 20-mile hole in German defences which were much more lightly held than at Cambrai, and through which Whippet tanks and armoured cars, cavalry and infantry, were able to drive up to six miles in a few hours. The sheer magnitude, pace and shock of this blow brought home to those in Berlin and at the front that the war was lost. Soldiers, starved of food and despondent at the failure of their despairing efforts to win a decisive victory, began to surrender in droves as the massed *material* superiority of their opponent was inexorably revealed on the ground and in the air, day after day, blow by blow. The brute force, persuasion and coercion of technical supremacy was shown, as never before, to destroy will-power and courage.

Also demonstrated was the inbuilt stamina of mechanized forces, particularly those with the simpler, reliable kind of equipment. Not only in Europe, but dramatically in the Middle East against defeated Turkish forces in desert conditions, horsed cavalry were outclassed in long-distance pursuit. Claim as cavalrymen would a lion's share of the success in a 300-mile, 36-day pursuit from the battlefield of Megiddo to Aleppo in October 1918, it was armoured cars and lorries that outstayed the horses and, adroitly assisted by aircraft, frequently reached their final objectives first, when men and beasts were exhausted.

Wider Effects In The Aftermath Of The War

The subtlest of changes in military attitudes at the end of the war revolved around a more willing acceptance of technology and its psychological application against specific targets. Colonel Fuller's proposal to direct mechanized forces against enemy concentrations of command, control and logistics in deep penetration advances was one such approach in the tactical sense, yet bordering also on the strategical. More insidious was the attack upon enemy will and morale by means of propaganda introduced through neutral sources and by the dropping of leaflets, as well as bombs, from the air. Although in their infancy, the cumulative potential of these methods was recognized because the Germans admitted in the aftermath that they had made a positive contribution, and principally because the seeds of discontent fell on the fertile ground of a disillusioned though better educated populace. As the demand for female, as well as male, labour flourished to capacity in the war factories, revolutionary, emancipatory and political ideas were given the opportunity to breathe and prosper at the call of leaders imbued with a conscience. Retarded though the advancement of technology might be in the post-war decade, insistence upon progress towards a new social order, once shown to be attainable, was insatiable and beneficial.

Surprisingly, in the course of a war which cost nearly 15 million military and civilian lives, the effect on surgery was minimal despite the enormous number of cases demanding treatment. Rather it was changes in technique which carried surgery into a new era, the outcome of a host of younger surgeons being given the opportunity at last to use new skills resisted by their

Above: ***To meet serious manpower shortages, women were recruited in large numbers into service and manufacturing industries, such as this shell-making factory, and into the armed services (left), leading to a pronounced movement towards emancipation.***

elders and, as in any scientific experiment, to search for improvements that would overcome the fatal consequences of post-operational putrefaction and gangrene. Such new antiseptics as were developed during the war were of only marginal assistance. The important advances in reducing fatalities lay in thorough cleansing of wounds.

Reflection and Experiment in the 1920s: The Naval Race

For conflicting reasons, a reduction in technical research and discovery ensued in the aftermath of the war, to be replaced by studies in technique. Linked to a thorough-going revulsion of battle, and the restrictions imposed on Germany by the Treaty of Versailles (banning its possession of warships over 10 000 tons, U-boats, aircraft, heavy artillery tanks and gas) vast quantities of weapons were scrapped and orders for development of new ones largely halted. Fighting in various parts of the world continued, of course; there is never a moment in history entirely devoid of war. Russia was embroiled in civil war while trying to invade her neighbours. Britain slipped into war with Afghanistan; the Turks were fighting the Greeks; and Communism was stirring up widespread unrest. But existing weapons supplied these outbreaks plentifully from war surplus stocks. Nevertheless, latent demand existed, notably in Japan. That country's involvement in the war on the Allied side had been symbolic; its dream of expansion in the Far East was alive and very much attached to naval expansion.

A new naval race was triggered by Japan's programme of four 40 000-ton battleships in 1920,

Above: **HMS** *Furious*, **the first full-decked aircraft carrier with the Royal Navy, was used for many initial experiments with aircraft at sea. It had been developed from a fast battle cruiser.**

with eight more to follow. The USA responded with a programme of 14 of similar displacement and the British with a comparable quota which would include 48 000-ton ships armed with 18-in guns. Each ship symbolized the view that the battleship remained supreme because the latest designs, including more compartments and thicker armour, made them unsinkable. But the cost was enormous and brought second thoughts to the negotiation of the Washington Naval Treaty of 1922. The treaty placed proportional limits on numbers for the five leading naval powers, and a restriction on displacement at 35 000 tons. Like all such agreements, there were escape clauses and all participants (notably the Japanese) were intent upon exceeding restrictions and making technical improvements. For example, the Germans in 1929 set about building the armoured cruiser *Deutschland*, with six 11-in guns and 4-in side armour. By using diesel engines in a major warship for the first time, employing electro-welding in construction and fitting lighter components, they saved so much weight and space that they exceeded the 10 000-ton Versailles Treaty limit by only 2000 tons.

In any case, the future main challengers to the battleship were already in service. British experiments with aircraft carriers had been successful in 1917 with landings and take-offs from a ship with an unobstructed deck. In June 1918 they had been crowned by fighter operations from the carrier *Furious*, which downed a German seaplane at 21 000 ft over the North Sea. Another milestone might have been passed had the war not ended just in time to prevent an attack by carrier-borne torpedo aircraft on the High Seas Fleet in harbour. The terms of the Washington Naval Treaty actually stimulated the construction of aircraft carriers, with America and Japan both following the British example of converting existing, fast capital ships or laying down purpose-built carriers, such as the Japanese *Hosho*, completed in 1922, and the British *Hermes*, completed in 1923. The many problems connected with carrier operations were diligently tackled and propagated by the pioneers, if only to preserve their lives. For example, deck arrester gear, first tried on the US Navy's *Langley*, was an essential remedy to prevent overshooting on a ship where the landing area was too small. Other improvements included means to help aircraft find the carrier in mid-ocean and ways to clear funnel smoke to ease approaches for landing; and, as the Japanese threat loomed, competition between Britain, America and Japan hastened improvements to shipboard layout. Lifts were installed to carry aircraft to and from deck and hangar. Techniques to make fullest use of each new device went hand-in-hand with the production of well-trained aviators who formulated their dream of a future when aircraft, suitably designed and equipped with bomb and torpedo could dominate war at sea by sinking the unsinkable battleships.

Nothing like the same attention was paid to underwater weapons. The invention in 1918 by the British of a mine which detonated when influenced by the magnetic field of a ship overhead was immediately discounted. Counter-discoveries had shown that electrical sweeping was feasible and that it was possible for a ship's magnetic field to be reduced by 'degaussing' — that is, reducing the magnetic field out of sympathy with the mine. Similarly, improvements to submarines were delayed because it was believed that the convoy system, in conjunction with Asdic, provided an adequate counter. The introduction of double-hulled boats which could dive to 500 ft in an emergency still left them vulnerable to depth charges which, although currently exploded at 80 ft, could later be set to explode at 750 ft. The Achilles heel of defence was Asdic, its defects obscured by unrealistic British trials which were largely carried out in good weather against an assured target. (Even so, these produced only 50% detections, with many anomalies caused by such factors as water temperature layers and shoals of fish.) Not once was the system tried out in a single exercise on convoy protection.

Mechanization On Land

Resistance to controlled, realistic experiments was a military habit at this time, partly through shortage of funds, but too often through bigotry and a mystic belief in intuition. Among the world's armies, only that of Britain deliberately created an independent armoured force (in 1920) to investigate unexplored tactical theories and certain aspects of the wholesale mechanization inevitably occurring because civilian commerce was already replacing the horse with the motor vehicle. While the need was established for armoured vehicles to accompany, if not actually carry, infantry, the idea of long-range penetrations as proposed by Fuller demanded experiments which had to wait until 1927 before enough medium-weight tanks with cross-country speeds approaching 20 mph, reliable suspension and tracks and radius of action of 150 or more miles could be assembled. The key vehicle was the Vickers Medium tank of 12 tons. Its 8-mm armour was hardly bullet proof, but it incorporated a trend-setting rotating turret, efficiently housing a co-axially mounted 47-mm cannon and a machine-gun, and was manned by a commander, gunner and radio operator/loader.

Trials carried out by the British over a period of seven years showed beyond doubt that all-arms formations, in which artillery, vehicle-borne infantry and engineers, combined with AFVs, could out-manoeuvre and out-fight existing military formations — and would do so all the more effectively and rapidly if the vehicles could be made more combat-worthy; if aircraft were thoroughly integrated for reconnaissance, communication and strike purposes; and if radio and cable communications could be developed to enable the commander, as in the past, to control the battle from the front while retaining close contact with his subordinates. How technical air and communication problems were tackled is described later. Here it is necessary to mention the central problems surrounding vehicle and tactical development, such as were presented by the cheap 25-mm, high-velocity, armour-

***Below:** Vickers medium and light tanks of the 1920s and 30s, of the types used predominantly in experiments with mechanized armoured warfare. Upon designs of this kind were based, respectively, the world's future main battle and reconnaissance AFVs.*

piercing guns being offered by industry which in turned caused soldiers to focus their minds upon anti-tank tactics — the problem of the eternal weapon versus armour race, in fact.

By 1930 it was mooted that 12-mm armoured 6-ton AFVs were unacceptably vulnerable and that speed was no protection. Nevertheless, priority was given to their construction because they were required by the colonial powers to police their possessions — and they were cheap. Most agreed that the heavy tank in excess of 20 tons with 70-mm armour was a very powerful weapon — and then rejected it on grounds of expense and because, in the 1920s, major war was safely over the horizon. Medium tanks of 16 to 20 tons with 20- to 30-mm armour were also rated too expensive, but it was from prototypes of these and of a very few heavies that research funded by niggardly sums produced considerable information for thriving motor industries. The motor industry satisfied the requirements of mechanization in the lighter-weight range of cheap, simple wheeled vehicles. But due to an almost total lack of civil demand for expensive, complex, cross-country tracked, half-tracked and four-wheel drive vehicles, the industry had to be subsidized to experiment with these and with heavy-duty shock absorbers and brakes, long-life tracks (of which the Hadfield manganese steel type was a superior example), regenerative steering to reduce loss of power when turning a tracked vehicle, and special functions such as swimming, obstacle crossing and mine clearance.

An element of showmanship could raise interest and the leading tank showman, the American J Walter Christie, designed and made several high speed AFVs, some of which could travel at 60 mph on road wheels and 25 mph on tracks. But although some were impressed by demonstrations of such pace and by the ability of his original, big-wheel suspension to make it possible, experts detected the engineering weaknesses of Christie's work. Not until Russia and Britain had bought the rights to manufacture the Christie suspension and set their own (or, in the Russian case, imported American) engineers to redesigning did a breed of tanks emerge which was to last in service for the next 50 years.

Drama In The Air

Showmanship was a mainspring of progress in aviation because, after 1918, through lack of civil transport airlines and as a result of government policy, the market for aircraft dried up completely and nearly all aircraft factories closed down. The leading protagonists of the unity of air power, wielded by air forces independent of sea and land forces, were intent upon propagating the concept of bombing aircraft capable of winning a war by attacks on industry and centres of population. The chief apostle of this creed was the Italian General Giulio Douhet whose 1921 book *The Command of the Air* was read widely by a great many responsible people who grew fearful that bombers could successfully undermine a nation's will by terror. In fact, only the British RAF accepted this theory almost in its entirety and nowhere were scientific studies launched to discover if technically, let alone psychologically, such theories were feasible or viable. Trials, such as William Mitchell's fudged bombing and sinking of an anchored passive battleship from low level in 1921, and press liking for anything sensational (including Mitchell's court martial in 1925 for exceeding his authority) kept aircraft before the eyes of a public ever enamoured by the romantic deeds of top-scoring fighter aces, and by post-war stunt flying and record-breaking. Thus theory grew into delusion.

It was the record-breakers who thrust technology ahead in the 1920s. The world speed record of 200 mph in 1922 had been raised to 318 mph by 1928, mainly due to competition for the Schneider Trophy. The altitude record passed 30 000 ft in 1920 when an American biplane clawed its way to 33 115 ft in 1 hour 47 minutes before the pilot's oxygen supply ran out and the machine plunged 6 miles until he recovered consciousness. The Atlantic was flown non-stop in 1919 by a converted twin-engine British Vimy bomber in 15 hours 57 minutes. These records were mainly the result of improvements in aero engines. The aerodynamically inefficient biplane still predominated into the 1930s — despite clear evidence from Junkers and Fokker of the feasibility of building robust monoplanes, their tri-motor airliners setting high standards of safety in service.

Engines not only increased in power, from 600 hp in 1919 to 1000 hp by 1929, but improved in performance at higher altitudes by the introduction of the supercharger. A compressor geared to the engine or turbine, the supercharger was driven by exhaust gases to supply enough oxygen to the carburettor for efficient combustion. German experiments with a supercharged Gotha bomber in 1918 raised its operational ceiling from 12 500 to 19 357 ft, but it was several years before metallurgists achieved reliability by developing material for turbines which could withstand temperatures of 1500°F when revolving at 20 000 rpm. Much engine power would continue to have been wasted had not the British designer Hele Shaw managed to perfect in 1928 a variable pitched airscrew which, like the gears of a car, adjusted blades

THE NEW YORK HERALD.

PART TWO.

NEW YORK, MONDAY, JUNE 16, 1919.

PRICE TWO CENTS

TWO BRITISH FLYERS MAKE 1,900 MILE DASH ACROSS ATLANTIC TO IRELAND IN 16 HOURS; ALCOCK AND BROWN WIN $50,000 PRIZE

8,000 WORD SUMMARY OF GERMAN COUNTER PROPOSALS MADE PUBLIC; FOE TO GET NEW TREATY TODAY

Peace of Justice Forsaken for Peace of Might, Enemy Delegates Charge

Answer to Counter Proposals Contains Concessions Making Signing Easier.

GERMANS ALLOWED ARMY OF 200,000 MEN

Must Decide on Acceptance or Rejection Before Noon on Next Saturday.

GERMANS WILL GET NEW TREATY TODAY

VICKERS-VIMY BIPLANE LANDS AT CLIFDEN AFTER A PERFECT FLIGHT AT 120 MILES AN HOUR

Flies Upside Down.

WIRELESS PROPELLER LOST SOON AFTER START FROM NEWFOUNDLAND; RADIO NEVER USED

Intrepid Aviators Circle Marconi Station Near Galway and Then Land in Bog, Where Machine Ploughs Its Nose Deep Into the Earth.

KING SENDS HIS CONGRATULATIONS TO PILOT

Hawker Says It Was Amazing Achievement.

AMERICAN TROOPS CROSS RIO GRANDE AND ENTER JUAREZ

Twenty-Fourth Infantry, with 3,000 Men, Now on Mexican Soil.

TO PREVENT DAMAGE ON AMERICAN SIDE

Congress to Take Action This Week in Regard to Muzzling of Dr. Ellis

Representative Mott Has Resolution Directing State Department to Investigate Charges That British and American Officials in Egypt Are Hampering Herald Correspondent.

KING GEORGE SENDS WARMEST CONGRATULATIONS TO AIRMEN

The Extension Of Long-Ranged Flight And The Atlantic Crossings, 1919

Top: The American Curtiss NC-4 flying boat, one of three that set out on the first trans-Atlantic crossing between 8th and 31st May 1919. NC-4 was the only aircraft to complete the crossing, stopping seven times to refuel. *Above:* The Vickers Vimy bomber used by Captain John Alcock and Lieutenant Arthur Whitten Brown for the non-stop crossing of the Atlantic on the 14th/15th June, 1919. Their coast to coast flying time lasted 15 hours 57 minutes. Weight was saved for the crossing by removing the nose wheel.

Birth Of A Fighter

The Schneider Trophy provided the greatest single impetus to fighter aircraft design since the end of WWI, instigating technical advances that were ruthlessly copied and developed in the search for speed.

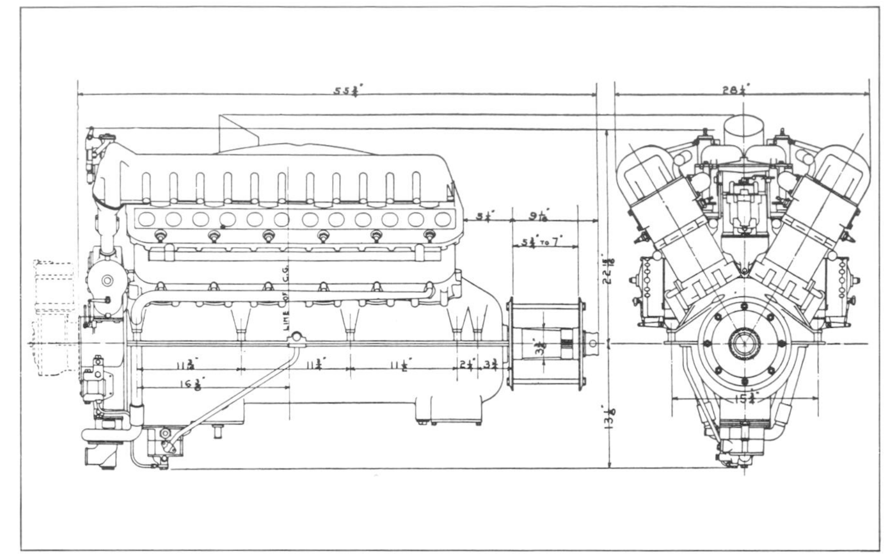

The Merlin engines of the Spitfire were descended from the revolutionary Curtiss D-12 engines (right) that powered the US to success in 1923. Rolls Royce, given an engine on loan by the Air Ministry were told to go one better and in 1926 produced the F-X. This was an entirely new departure for the Rolls Royce engineers, but within 3 years they had produced the world-beating 'R' (for racing) engine which powered the Supermarine S.6 and S.6B seaplanes to success in 1929 and 1931 respectively. Supermarine had been successful in 1927 with the Napier-powered S.5 (below) and it is popularly believed that the Spitfire developed as a continuation from the final S.6 version. However the designer, R J Mitchell, had returned to the lines of his earlier S.4 design as the basis for his classic aircraft.

(now to be made of metal) to suit the conditions of flight and altitude.

New materials lay at the heart of most aspects of aircraft development, in the struggle to reduce weight, increase strength, endure wide ranges of temperature and resist corrosion without raising costs prohibitively. Without aluminium alloys of an increasing and useful variety, progress would have been stifled. It was crucial when methods of casting duralumin were discovered, for this metal could be utilized to replace heavier steel among many engine components, including crank cases, pistons and steel-lined cylinders. Opinions remained equally divided as to the relative merits of liquid-cooled in-line and V-type engines and air-cooled rotary engines. But there was no disputing the enhancement to pay-load when an engine such as the rotary Wright Cyclone 9 could weigh in at under 1 lb per horse power in 1929, compared with a figure twice that in 1919, or with still greater economies made by the V-type Rolls-Royce Merlin in 1934.

Behind these increases of power lay the collaboration between engine designers and chemists who, in the 1920s, exploited higher quality petrol for engines of increased compression ratio by 'cracking' petroleum hydrocarbons; that is, by rearranging their molecular structure during refining process. This was advanced a stage further in the mid-1930s by using the discovery of the Frenchman Eugene Houdry that materials known as catalysts could produce even better spirit and obtain higher yields from the basic raw material.

Military aircraft developments by America, Britain and France were, however, retarded by a political desire for disarmament linked to curbs in expenditure and the conservative attitude of those to whom aeroplanes, especially (and with reason) monoplanes, were dangerous contraptions. They liked to concentrate effort and finance on a plethora of small step-by-step experiments; and, at the start of the Great Economic Depression of 1929, frowned on large-scale production of anything. They were largely uncomprehending of what was brewing in nations with predatory ambitions, such as Japan, which, with British advice, was on the eve of establishing an aircraft industry. There the Nakajima and Mitsubishi firms, which later built fighting aircraft of their own design, led all others to become the strong striking arm of an aircraft carrier force. People also overlooked activity in Russia, where American and British advisers were privately at work and where the Germans, deprived of their own aircraft industry, were collaborating in secret with a nation, isolated by its Bolshevism from official Capitalist co-operation, whose historic urge was expansionist.

At the start of the 1930s the war clouds again began to gather, fuelled by classic destabilizing factors such as economic chaos, nationalistic ambition, idealistic impulses and feelings of vengeance. Decisions now being taken to shape the military forces of the next decade had been extensively discussed in public in the 1920s. But while discussion tended to centre on hardware — ships, tanks and aeroplanes — the less discernible field of communication technology blossomed secretly with new devices and methods. Not only was this vital for the future of aviation, but it assured yet another all-embracing revolution as the world moved into the Electronic Age.

Communications, Electronics and Cyphers

The first transmission of speech by radio was a natural progression from cable telephone. It was made in 1904 by Valdemar Poulsen, one of many talented scientists already engaged in the exciting world of electronics — as the technology of electric vacuum tubes would become more generally known in the mid-1920s. This new light industry offered immense returns from relatively inexpensive research in clean, scientific working conditions; the prospects for the future were boundless. By 1914 voice radio signals were transmitted 1750 miles to two Italian warships at sea, and music had already been broadcast to interest, if not enthuse, the world of entertainment. The war and the need to develop better microphones, valves, circuits and speakers held back serious commercial application until the early 1920s. But as soon as a flood of inventions had been channelled by entrepreneurs into commercial broadcasting, those involved in the fight against crime, the sending of news and assistance to ships and aircraft, diplomacy and the military had little need to worry about shortage of funds or encouragement for their interests. Up to a point, civilian enterprise was doing it for them and, consequently, training strong cadres of experts in a world-wide, burgeoning industry at a time when every defence organization worth its salt was forming specialist signal corps. The subject was of such military operational and intelligence importance that it had to be given the highest priority — the equivalent of a weapon system — and more specialized equipment than civilian applications needed.

The use of improved radio in ships needs no further elaboration here, since it was well established before the war and a proven asset. But land forces were far from satisfied with 'unquenched' spark sets, which were sensitive to interference and remained extremely bulky and almost impossible to use in the forefront of battle. The signals pioneers of the 1920s, particularly the British and Germans, seized upon the latest 'quenched' method; adopted successively higher frequencies above 2 Megaherz (M/Hz) and built sets which were smaller, easier to tune (often by the use of crystals), more robust and capable of use in man-packs and in vehicles on the move. In so doing, they made practical a gigantic leap forward in communications for combat troops of all arms, at the very moment when the full exploitation of mechanization was being curbed by communications networks that simply could not cope with speeds in excess of 5 mph. Clearly, efficient radio sets were of even greater importance in aviation, since an aircrew out of touch with the ground could be at hazard in darkness, bad weather or over featureless sea or terrain. Bright neon light beacons suffered from obvious limitations and were superseded to some extent by radio beacons which repeated signals from locations along airline routes. But military aircraft might need to operate in an environment lacking such obliging navigation devices. In the late 1920s civil airlines had established ground stations which could speak to an aeroplane flying overhead in order to give its position. These could also be used by the military, but they found much more helpful a system of parallel radio signals directed along the intended flying course. The pilot could then hear whether or not he was off or on track.

Equally significant improvements were being made to field telephone and telegraph systems. Not only were the latest instruments more reliable, but their traffic capacity was also dramatically increased by multi-channel cable which could be laid from motor vehicles at a rate of about 100 miles a day. Simultaneously, the development of radio teleprinters in the 1920s in America, Britain and Germany made the sending and printing out of type a much speedier operation, particularly when at a later stage machine-fed punched tape was used to send a message. Line transmissions had one other vital military advantage; they were secure because, unless the enemy managed to tap in on the cable, they could not be overheard — and by 1919 every combatant nation was aware of the fruits to be garnered from listening in on the enemy while preventing him from doing the same.

'Scrambling' devices were of only marginal help to security because they could easily be unscrambled. But cypher machines were essential if vastly increased volumes of traffic, generated by complex networks, were to be securely and rapidly handled — and few nations needed this facility more than Germany as she deviously engaged in secret weapon manufacture, mostly in Russia and Sweden. A cyphering machine called Glow-Lamp, but better known as Enigma, was invented and marketed in 1923 by the German Arthur Scherbius to provide a superior way of 'keeping correspondence secret'. Using, to begin with, four wheels with lettered studs, interconnected with complicated electrical wiring and worked by a typewriter keyboard, it had the facility of encyphering and decyphering an almost infinite number of different wheel and plug settings which could be changed at will, and which, by their intricacy, were deemed unbreakable. In simplest terms, if the operator struck B on the keyboard, the machine might encypher it to X. And if X was struck for decyphering by the receiver it would print B. In 1926 the German Navy put Enigma into service. Soon all the German fighting services, the diplomatic corps, and many other agencies employed it to conceal a mass of secret traffic. At last radio communication was secure, or so the Germans and the Japanese, who introduced a similar machine in 1931, believed.

Of far greater importance than almost any other advance in communications technology was the idea of television, which had first been proposed as feasible by the British inventor A A Campbell Swinton in 1908 and described, moreover, in much the same way as the electron cathode ray tube system eventually evolved. Although the initial attempts to transmit a picture, successfully accomplished in 1926, employed different, mechanical methods, the true course was fixed in 1923 when the Russian-American Vladimir Zworykin patented the iconoscope cathode ray camera tube upon which television and much else besides would depend in the future.

The cathode ray tube (CRT) had the facility of 'firing' electrons onto a photo-electric luminescent screen to produce an image, not necessarily a picture but perhaps a displayed radio pulse or other electrical phenomenon. With the existence of the CRT it was but a matter of time before commercial electronic television was a reality and, once achieved, a number of other uses, both military and civil, were envisaged for it. Of those in the military and, particularly, the aviation world, none was of greater importance than the incorporation of the CRT in radar sets. First suggested in 1904 by the German Christian Hülsmeyer, radar was not considered seriously until the early 1930s (and then mostly in secret) by Germany, France, America and Britain — approximately in that order. Its vast impact upon the location of objects, range-finding, navigation and other functions inevitably became of central importance as the next major war grew closer.

New Horizons; Greater Depths

1931-1944

At the end of a period of tensile peace, alloyed with revolutionary outbreaks and economic crises, the 1930s got off the mark with an arms race feeding upon the many remarkable scientific discoveries of the 1920s. Circumstances were not so very dissimilar to those of the 1850s – except that technical scope was now of a breadth and depth beyond all imagination. No precise point can be assigned to the beginning of this race. The contestants had been limbering up since 1919, the front runners in aggression being the Japanese and Russians to begin with, while the Germans proved only too eager to enter once the shackles of Versailles had been removed. Probably it was the Japanese, with a government falling into the hands of bellicose militarists and industrialists hunting new markets in China and elsewhere, who fired the loudest starting gun by provoking the 'Mukden Incident' in Manchuria. It was September 1931, and China's internal struggles between Nationalists and Communists were splitting the nation in two.

The campaign which followed, though short lived, provided a pointer to the future. For one thing it indicated that Japan's army and air force were worthy of respect, for she gained her land objectives with commendable speed and instantly won air supremacy. It also displayed the formidable strength of her navy's air force which, between January and March 1932, used a seaplane carrier and two aircraft carriers to support landings at Shanghai. The provision of over 80 fighters and torpedo-bombers finally helped to break Chinese troops who held out near the coast for over a month. But chiefly the incident warned Russia and America of a long-recognized threat they could no longer ignore.

Soviet Russia, now under the ruthless leadership of Joseph Stalin, was in any case on the eve of embarking on a massive rearmament of her army and air force, employing British, French and German technicians privately to assist with tanks, aircraft, gas and much else besides. The Japanese occupation of Manchuria justified Russian rearmament. Similarly, the Americans, anxious about Japanese ambitions, were taking tardy steps to study and prepare for trouble in the Pacific with resources made scanty by severe financial restrictions. Like the Japanese, they focused attention on aircraft carriers and, specifically, on the best way to attack ships with bombs and torpedoes.

Torpedo-dropping techniques were fairly well developed by 1930 and had the advantage that, in addition to inflicting underwater damage, the shock of explosion tended to upset the host of sensitive electrical and mechanical systems installed throughout a ship. So, too, could the 500-lb bombs experimented with by the Americans in 1927 and, to an even greater extent, the 1000-lb type tried in 1930 — and which could be delivered with commendable accuracy by the technique of dive-bombing, first suggested by the US Marine Corps in 1919. Dive-bombing required little allowance for wind drift or target movement: the pilot simply aimed his aeroplane at the target in a near-vertical dive, thus, incidentally, increasing the velocity of delivery against relatively thinly-armoured areas of the target. Both systems required special skills and technology. But already in the early 1930s, the British were achieving something in the order of 75% torpedo hits in favourable conditions, despite the restrictions imposed by the weapon. To avert break-up on dropping, the torpedo had to be released within close margins of height and speed — and in this respect the Japanese Type 91 torpedo of 1931 was superior to the American because it could be dropped at 300 ft (against 100 ft) and 250 knots (against 125 knots).

The Americans did manage to retain their initial lead in dive-bombers, but in almost all else they fell behind, particularly with tactical techniques in which the Japanese skilfully learnt to combine torpedo attacks at sea level with diversionary dive-bombing from above. Moreover, the Japanese at this point began to divest themselves of reliance upon their Western mentors and to build superior aircraft of their own design, such as the Nakajima A2N fighter which, in 1930, could fly at 200 mph.

The Trends And Traumas Of Rearmament

Rearmament was hag-ridden by doubts in all nations as to the right equipment policies to adopt. Traditionalists fought to retain sacred cows such as battleships, cavalry and bombers in the face of philistines who aimed to divert funds and effort into extraordinary and often opposing channels of development. As ever in the past, arguments centred on the struggle between promoters to convince politicians and industrialists of the merits of competing weapon systems. This induced all manner of devious manoeuvres among factions striving to prevent or develop controversial projects, or even to obtain funds for them at a rival's expense. The trouble lay in the extreme complexity of the amazingly revolutionary options offered and the continuing shortage in most governments of a central defence organization staffed by leaders with enough technical awareness to understand the scenarios presented. As a result most changes came about after protracted and frequently ill-informed and bitter wrangling. Yet this wrangling was often highly productive and often the result of a steady rise in

Below: The Japanese aircraft carrier Kaga *was converted in 1928 from a battleship. It could operate 60 aircraft and had an armament of 10 × 8 in and 12 × 4.7 in guns. Note the smoke deflector intended to avoid interference with aircraft operation.*

government-sponsored and controlled research organizations working in conjunction with inventors, scientists, industrialists and the military. Official and enlarged civilian and service technical departments began to steer and grip weapons development more positively than before. This process by no means denied the individual inventor his part, but it did place even greater emphasis on the need for teamwork among originators to collaborate with projects of immense ramification. At the same time it introduced bureaucracy into areas of risky innovation where it was not always at home.

Each nation had its own individual theme and priority, befitting its special circumstances. But one requirement common to all was the need to stimulate industrial production, get national economies moving (even at the risk of inflation) and reduce the number of unemployed. Enlarging and equipping military forces was one way to do that, whatever might be the purpose or their future use.

Russia, like most others, would plead that her aims were defensive, to guard against a threat by the Capitalist nations to her Communist system. Geography persuaded her to place greatest emphasis on land forces with many fighter aircraft for protection and nothing more aggressive than medium and light bomber aircraft in support. Germany, having disavowed the Versailles Treaty and secretly started to re-arm in 1934, provided her army with new weapons, but gave priority to the air force with a smaller share allocated to the navy for a traditional balanced fleet of warships. France and Britain, with their widespread colonial responsibilities, the need to face the Japanese threat and Italian moves in Africa and the Mediterranean, plus a fast emerging German threat in Western Europe, found themselves with major commitments to all three services. With their greatest anxiety the fear of air attack on their cities, they suffered from an inability to fund all priorities.

'Incidents', colonial wars and prolonged struggles in Spain and in the Far East removed the veils of new technology and techniques without quite, by the end of the 1930s, revealing the full scenario of world war as it would unfold in the 1940s. Sometimes, indeed, the exposé was misleading. For example, the Italian invasion of Abyssinia, whose armed forces were mostly irregular and primitively armed, reinforced the widely-held dread that chemical agents (including persistent liquid, nitrogen mustard gas, which attacked internal organs by blistering the skin and into which considerable research and production was directed) would be used as a matter of course. Eventually, along with the latest Lewisite, liquid systemic gas (first used by the Japanese against the Chinese in 1938) it fell into disuse; primarily because it was easily defeated by reasonably well protected victims; and secondly due to fear of counter measures — an example of the nullifying effect of a deterrent technology. Less clearly understood, because of less sensational publicity by the newspapers, were the merits and demerits of mechanized land warfare. Poorly armed but courageous tribesmen could knock out the thinly-armoured Italian machine-gun carriers with relative ease, and yet those none-too agile machines were capable of penetrating the most difficult terrain, and, supported logistically by a mechanized road-building organization, influencing operations out of all proportion to their numbers and efficiency. Hidden, too, behind vivid descriptions of bombing the undefended, was the story of how photographic survey by aircraft made possible well-planned advances into undeveloped terrain, and how transport aircraft supplied troops in ordinarily inaccessible regions.

The Spanish Test Bed And Stimulus

Far more significant pointers were to be obtained from the Spanish Civil War, which broke out in 1936 between left-wing Republican forces and right-wing Fascists, because several nations used it as a proving ground for the latest theories and weapons. At a time when the major navies had embarked on construction of all kinds of vessel from battleships and aircraft carriers to minesweepers and submarines, the meagre Spanish navy, mostly sailing under Republican colours, had little to offer by way of innovation. The sinking of a pre-war dreadnought by Republican bombers taught less than the interesting experience of the British destroyer *Hunter* when mined on an international patrol intended to enforce non-involvement in the war. Her welded construction was shown to be far more damage-resistant than would have been a rivetted ship. Blockade and counter-blockade, to a large extent applied by aircraft from both sides, by Italian and German submarines on behalf of the Fascists and by minelaying, certainly achieved results although none too thoroughly. Most air attacks were on ships in port with far more hits than those attempted against moving targets at sea where, out of 18 warships attacked, none were sunk. Submarines had their successes and were but slightly deterred by air or depth charge attack. It was only now that the British navy at last came to realize the inherent shortcomings of sonic detection and the inadequacy of Asdic.

Threats from above or below the surface failed to

deflect the hierarchy of naval officers, most of whom were gunnery specialists, from according priority to battleships. They concentrated construction upon ships with improved guns, ammunition, protection, range-finding paraphernalia and gyro-stabilized searchlights which would have turned the tables at Jutland. In 1932 France laid down a 26 000-ton battleship; Italy responded with two 41 000-tonners in 1934. Then Germany laid down two 32 000-tonners (announced as 26 000 tons) followed in 1936 by two 42 000-tonners (announced as 35 000 tons). Britain tardily replied to this with two 37 000-tonners in 1937, the same year as the Japanese laid down the two biggest-ever battleships — the 64 000-ton *Yamato* and *Musashi* with their 18.1-in guns each firing a 3200-lb projectile. Naturally the USA had to match this threat with a programme of six 35 000-ton ships armed with 16-in guns. All were at vast cost and distractive of other industrial effort, but justified on the familiar grounds of indispensability.

Gunnery addicts might discount the air threat but they tacitly admitted its danger by specifying that secondary armament of battleships and main armament on smaller vessels were to be high-angled for air protection. They also added batteries of multiple 20-mm — 40-mm cannon and multiple machine-guns to deal with low level attacks. And while declaring that aircraft carriers would occupy only an ancillary role to battleships, they initiated extensive aircraft carrier construction, incorporating the experience gained with earlier types. All would mount batteries of anti-aircraft guns and be enlarged, both above and below the flight deck, to carry 70 or more aircraft. Some, like most cruisers and battleships, would be fitted with compressed air operated catapults (long after the American Theodore Ellyson first took off from one in 1912). To ensure the safe return of aircraft at the end of a mission by day or night, all carriers would be fitted with radio homing beacons (of which the British Type 72 DM, successfully tried out during 1931–33, was a prime example despite a tendency to distortion by high-tension electric cables) and, in the case of the British *Illustrious* class, vitally ahead of the time with a well-armoured hangar and flight deck.

War on land in Spain at first differed little from 1918 for the simple reason that the armies were equipped with the weapons of the First World War. The tanks sent by Italy and Germany were light types armed only with machine-guns because in 1936 neither nation had anything better. Those sent by Russia were superior because, although lightly armoured, both the slow Vickers-type T 26 and the fast Christie-type BT2 were armed with 37-mm guns. Their employment reflected the furious debates then raging between the old and new schools of thought concerning armoured warfare. The old school argued that AFVs should be subservient to the infantry and move at infantry pace, as at Cambrai; the new school claimed that if tanks with accompanying mechanized artillery, engineers and infantry were driven through enemy lines in deep penetration raids (as Fuller wanted) a strategic decision would be obtained. General Pavlov, the Russian in command of Republican armoured forces in Spain, attempted two deep penetration attacks but failed to reach his objectives because the tanks had insufficient infantry and artillery in attendance. General Franco, for his part, insisted on using German and Italian tanks for close infantry support only — and with reason, since light tanks were hardly a dominant weapon.

None of this boded well for the German General Heinz Guderian, who was on the verge of forming three so-called Panzer (Armoured) Divisions composed of all arms with the tank as dominant weapon. It was not only that the reports from Spain seemed to decry the deep penetration concept; they also denigrated the need for wireless to control these fast-moving formations. Guderian was not only a General Staff officer of outstanding drive and vision, but also a signals specialist. He took as his model the series of British experiments between 1927 and 1934 which had progressively and publicly demonstrated the kind of 'mix' suitable for modern, all-arms formations up to divisional level, and had shown how effectively one man could sit in a mobile command post with a short-wave, crystal-controlled radio set and control the movement of a considerable number of tanks by voice. He had seen, too, how on exercises (no matter how unrealistically arranged) these formations could out-manoeuvre and surprise much larger conventional formations. Guderian was completely convinced that radio was the key instrument in making armoured forces function to their full capacity, even if their armour and guns were inferior to those of the enemy. His concept also included an entirely new tactical method in which the tank and infantry element of a panzer division acted merely as the spearhead of infiltration to seize vital ground which the enemy could not afford to surrender. Having seized it, his tanks were intended to withdraw to allow anti-tank guns and infantry to beat off subsequent enemy tank attacks and thus destroy the enemy armour while preserving his own.

In Spain the latest anti-tank guns with 2000 + f/secs velocity demonstrated their ability to knock out all known tanks at ranges up to 600 yards with a 5 to 1 chance of scoring a hit (a much higher accuracy than lower-velocity field artillery). Of course, the existing 20-mm to 47-mm solid shot did not guarantee a 'kill' with each hit. That was reserved for a quite devastating weapon with a pedigree reaching back to the 77-mm anti-aircraft guns which had shot so well at Cambrai. Now it was the latest 88-mm anti-aircraft gun with a muzzle velocity of 2657 f/secs and an armour-piercing shot weighing 21 lb which could wreck any medium tank with 60-mm (let alone 30-mm) armour beyond 1200 yards, and very likely hit it at the second or third attempt. This gun was tried out in Spain.

The moment was near when the 88-mm gun and its counterparts in other armies were the only pieces capable of penetrating at 800 yards the latest heavy tanks. Once the French and British had concluded in 1931 that a mass of fast-moving, lightly-armoured AFVs were unacceptably vulnerable to the smallest anti-tank guns, they stepped up the gun/armour race by embarking on the construction of slower, heavily-armoured tanks which in the 1920s had seemed prohibitively expensive. In the lead were the French. They rejected 70- to 80-ton models in favour of the Char B weighing 32 tons, armed with a turret 47-mm gun and a short, 75-mm howitzer in the hull, and with 60-mm frontal armour which was proof in most conditions against the German 37-mm gun. This, on paper, was a useful machine with a 300-hp engine giving a speed a little under 20 mph. Unfortunately for the crews, the French remained wedded to the one-man turret, first adopted in 1917. This might have been satisfactory when the gunner, who was also commander, had only to load, aim and fire a machine-gun; but once he had to perform those duties with a cannon and also control his driver and, over the radio, other tanks, efficiency decreased. Far more formidable was the British heavy Matilda II tank, ordered in 1938, with its three-man turret, 80-mm frontal armour, 40-mm gun, 174-hp twin engine and 15 -mph speed .

As each nation began to expand its armoured forces it met the problems of inferior industrial knowledge and facilities. In 1932 only the British firm of Vickers and the French Renault company could be said to possess an adequate tank-building capacity — and even they were incapable of meeting the sudden demand for mass production employing many new techniques (such as casting of armour) and designs such as regenerative steering. A shortage of experts, technicians and factories delayed the immediate delivery of high-quality, combat-worthy AFVs. Several armies were compelled to buy inferior machines in order to begin training crews. All might have been well given wholehearted desire among service ministries and general staffs to rectify these shortcomings. As it was, armies, and notably tanks, were held quite low in estimation and order of priority for re-equipment. The navies took their traditional share; air forces were given high, sometimes top, priority; and land forces had to make do with what was left over.

Aircraft factories of 1932 were not, in any case, coiled springs primed for dynamic release. The wariness of air forces and civil operators, restrained by economic recession and confused imaginations, was

Below: **German 37-mm anti-tank gun hauled into position by troops in the Spanish Civil War.**

anathema to capital investment. Contradictory theories still abounded in 1936, and were being explored in the skies above Spain. The opinion that civil airliners were readily adaptable as bombers (thus making it easy for Germany to assemble an air force under the guise of her civil airline) died hard. Such adaptations were frequently unsatisfactory if only because a good passenger cabin does not readily convert to a bomb bay, and the mounting of special defensive armament is not efficiently improvised. Yet the Junkers Ju 52 tri-motor transport had its successes in this role. Franco benefited from it tremendously at the start of the civil war when 20 of these aircraft, later joined by nine Italian Savoia SM 81 bomber-transports, ferried 13 523 troops and 570 000 lb of war material from Morocco to Spain to establish the core of the Fascist army and make the revolution possible by over-flying the Republican naval blockade. And on 14th August 1936, before people had awoken to the peril, a Ju 52 planted two bombs on a Republican battleship from 1500 ft, putting it out of action.

The theory that fast, well-armed bombers would survive (particularly if flying in tight formations protected by interlocking fire from their machine-guns) held water at first — but only because the fighters of 1936 lacked the speed to reach them and the armament to do serious damage. For example, a Russian Tupolev SB-2 twin-engine monoplane with a speed of 255 mph was a difficult target to intercept by an Italian biplane Fiat CR32 with a top speed of 233 mph — although on occasion this feat was performed. Evidence of this sort underlined the widespread opinion of those, such as the British Prime Minister Stanley Baldwin, that 'the bomber will always get through'. It was assumed that this technical imbalance would persist and that, in any case, fighter actions would be impossible if their speeds increased much beyond the extant 220-mph mark. A natural reaction was to build bigger, faster and heavily-defended bombers in the pious hope that their existence would deter an aggressor from using his bombers — in much the same way as it was hoped that the possession of gas would deter its use.

In practice, only two nations wholeheartedly created strategic air forces mainly equipped with large, four-engined bombers: the USA and Britain. The former rather drifted into it, as the Boeing 299 of 1934 (which evolved into the B-17 Flying Fortress) was originally conceived as an anti-shipping machine. British specifications of 1936 called for long-range, four-engine bombers, defended by power-operated machine-gun turrets, which could survive in daylight. Russia and Germany also laid plans for a four-engine bomber strategic air force. But the former's Tupolev TB-3 machine was far too slow and vulnerable for this and mostly achieved fame as a transport for the men of the first-ever parachute assault division created in 1934. Germany abandoned its ambitions for political reasons associated with the Spanish Civil War and technical requirements related to the perennial struggle of naval and land forces to achieve closely co-ordinated support for their operations.

***Left:** The German 88-mm dual-purpose anti-aircraft and anti-tank gun first designed as an anti-aircraft gun in 1915. With a muzzle velocity of 2600 ft per sec and a 21-lb projectile, it could engage aircraft at 32 000 ft and, up to 2000 yards, knock out all tanks in service in 1940.*

Germany's movement towards abandoning a long-range bomber force began in June 1936 when its main advocate, the Chief of Air Staff designate, ex-infantryman General Walther Wever, was killed flying. His successor, ex-artilleryman General Albert Kesselring, not only leaned slightly towards naval, and above all army, support air forces, he was almost at once faced by the need to recast long-term German plans which aimed at being ready for war by 1943. Action in Spain and the aggressive utterings of his C-in-C, General Hermann Göring, compelled Kesselring to rationalize equipment under development and concentrate on machines in readiness for an outbreak of war in 1938 or 1939. The four-engine bomber fleet was incompatible with this time scale. Its viability was also in question as Kesselring reasoned that the low state of the navigational art precluded finding and hitting targets at long range.

So Germany's air force, like most others, concentrated on twin-engine medium bombers; single-engine fighter-bombers and single-engine reconnaissance machines, plus a few specialized flying boats, sea planes and four-engine, long-range naval reconnaissance aircraft salvaged from the strategic project.

The shape of the next generation of twin-engine medium bombers appeared in the USA in 1932. The Boeing B-9 was a cantilever wing monoplane, all-metal with a top speed of 188 mph and a retractable under-carriage — its contribution to improving aerodynamic designs. Almost simultaneously, the Martin company went a stage further with their B-10 which had two 775-hp radial engines, could fly at 207 mph, climb to 24 000 ft and had a range of 600 miles. With fully enclosed crew cabins and bomb bays, along with a rotatable gun turret in the nose, the B-10 allowed for operational efficiency in the cold of extreme altitudes, dependent as it was on skilled aircrew working in as comfortable an environment as possible. Within a few years, bomber speeds in the region of 260 mph were normal, momentarily keeping ahead in 1937 of the latest fighters in service and providing, incidentally, a useful vehicle for high-altitude reconnaissance with machines such as the German Dornier Do 17, equipped with the latest high-definition cameras. But it was the German Heinkel He 111 with a bomb load more than twice that of the Do 17 which made the biggest impression. It hit the headlines when, in conjunction with Ju 52s on 26th April 1937, it was used to pulverize the town of Guernica. A prime example of justifiable tactical bombing of a strongly defended locality this may well be — and successful, too, since Guernica fell without resistance two days later. But to the world it was presented as a callous experiment in the strategic and psychological process of undermining enemy morale through terror. Guernica became the symbol of air power's crushing influence, and a warning to weaker nations who dared to withstand Germany's predatory intentions.

In fact, the Germans undertook no deliberate major strategic attacks in Spain, their current doctrine eschewing it as they concentrated more upon the use of fighter-bombers to attack ground targets by low-flying and dive-bombing. The debut in this role of the latest biplanes (and the last in German service) was not entirely a happy one. Against the latest Russian Polikarpov I-15 biplanes (229 mph) and I-16 monoplanes (321 mph), the Heinkel He 51 biplane at a mere 210 mph was cannon fodder, as was the Henschel Hs 123 biplane dive-bomber (211 mph) which on test had exhibited a nasty habit of shedding its wing in the high-speed dive. But formations of nine He 51s, dropping six 20-lb bombs each on fortifications from 500 ft (each machine sometimes flying seven sorties a day) had a stunning effect on enemy troops and convinced the Germans of their vital importance to the army. This conviction was redoubled when they experimented with heavier bombs carried by the Hs 123s and, in 1937, the latest Junkers Ju 87 monoplane dive-bomber. But despite these robust aircraft proving their worth as heavy artillery, it was also realized that if they were to survive, they needed escort fighters of a calibre higher than those which were coming into service.

Although many air forces continued to specify biplane fighters (the RAF took delivery of the Gloster Gladiator, the last of its line, in 1937 and the Italians their Fiat CR42 in 1939) the writing was on the wall for biplanes, even those with retractable undercarriages. It was clear that these aircraft had almost reached the limits of their development and that

Trends In The Use Of Multi-Engined Aircraft

Above left: *An early Junkers Ju-52 airliner adapted as a bomber. Note the corrugated duralumin skin for strength and the lower, retractable dustbin gun turret to fend off attacks from behind and below.*

Above: *Russian parachute troops, the first in the world, emplaning in their TB-3 aircraft which was converted from an inadequate strategic bomber design. They would jump out through the side doors, a method which became standard practice worldwide.*

Left: *The trend-setting 207 mph Martin B-10 medium bomber. Its gun turret in the nose represented the triumph of designers who saw that the day of the biplane was over. Note the radio direction - finding loop on top of the fuselage.*

another revolution was impending. Charles Lindbergh's solo crossing of the Atlantic in a monoplane in 1927 did more than anything else to overcome prejudice, which was further eroded when nearly all the seaplanes entered for the Schneider Trophy in the late 1920s were monoplanes. From 1927 onwards, monoplanes monopolized the world air-speed record, 300 mph being exceeded for the first time by an Italian Macchi M52 that year. Even so, doubts about fighting monoplanes were not removed until the air war in Spain dispelled the objection that gravitational forces (radial acceleration, expressed as g) precluded violent manoeuvres at high speeds such as were to be expected in the traditional dog-fight of the First World War. High-speed racing fliers of the 1920s had sometimes lost consciousness momentarily when applying more than 4g in tight turns, and their aircraft often suffered structural failures. Experience in Spain indicated that these forces were tolerable up to a point and that, in any case, the tail-chasing dog-fight was but one transient phase in air combat. Most attacks consisted of a fairly direct line of approach prior to opening fire, followed only subsequently by the mad scramble of tail-chasing which seldom ended in a 'kill' because it was impossible to bring guns to bear in a turn. As an alternative, deflection-shooting in head-on or beam attacks improved somewhat with the introduction of optical, reflector sights. These were better than the existing ring and bead sights, but still left the pilot as the sole arbiter of the correct moment to open fire, a decision requiring much practice and fine judgement.

Armament Improvements To Aircraft

Guns and gunnery, crucial as ever to aircraft and particularly fighters, were also entering a new phase with specifications for the latest machines paying great heed to the armament offered. The complexity of available choices was neatly displayed by the demanding specifications for the 300-mph-plus German Messerschmitt Bf 109 and the rival British Hawker Hurricane and Supermarine Spitfire as they evolved to the dictates of new technology between 1932 and 1933. In configuration the designs were similar, their slim fuselages made possible by adopting in-line engines which traced their ancestry to similar power plants fitted to Schneider Trophy racers. Wing design and aspect ratio varied according to operational requirements, such as the variable emphasis on manoeuvrability or on rapid rate of climb needed to intercept bombers before they could enter defended air space. But when it came to armament there was a striking contrast: the consensus that the French Hispano automatic 20-mm cannon had highly desirable destructive power was tempered with concern about providing an adequate ammunition supply, about its reliability and its mounting which, for accuracy, had to be extremely rigid. The German solution was like that of several other nations let into the gun's secret — to mount the gun in the V between the engine's cylinders and to point the barrel through the propeller boss. The British preferred to mount the gun in the wings but deferred a decision until satisfactory mountings could be devised. Meanwhile, they adopted batteries of four of the notably reliable Browning (Maxim type) .303-in machine-guns in each wing, and these could spray bullets. The removal of guns from the pilot's immediate care and the elimination of interrupter gear were blessings in disguise, despite the complaints of some pilots. But the sensing (if not the sure knowledge) that armour was certain to be incorporated at some later stage in fighting machines raised doubts about the potential of machine-guns as main armament.

The entry into universal use of the monoplane and the unavoidable interdependence of several essential systems within aircraft all contributed to far greater complexity and a resulting demand for advanced specialization in maintenance, handling and tactical use. The overcoming of natural phenomena lay within the province of scientists and engineers. In few aspects was this more crucial than at high altitudes, particularly once the clear evidence of the early 1930s showed that simple oxygen equipment would not support life above 33 000 ft, when a sharp decrease in arterial oxygen saturation occurred. A pressurized oxygen supply to the individual was one acceptable, if clumsy, solution. A rubberized pressure suit, rather like that of a deep-sea diver, also proved successful when worn by the pilot of a Bristol 138A monoplane which was flown to 49 967 ft in 1939. But most attractive in its utility was the pressurized cabin, first shown to be feasible by German experiments between 1928 and 1932 and exhibited in commercial form in 1937 by the American Lockheed XC-35 twin supercharged engine transport.

The raising of normal combat ceilings from about 14 000 ft in 1918 to 20 000 ft and above by 1938 did more than challenge the technicians. Such an immensely expanded void enforced changes to the age-old struggle for command of vital heights which centred on how to find, let alone tackle, the enemy. Sound locaters and visual detection to help guide aircraft or direct gunfire no longer sufficed.

Technologists were called upon once more to satisfy the needs of tacticians with even more striking solutions at the frontier of knowledge.

Enter Radar

In the 1930s the ingrained habit of disposing of revolutionary notions with the words 'not in our lifetime' began to fade. A slightly wider-based military, technical community began dramatically to prove their worth and so influenced policy makers to adopt, almost with aplomb, innovations which previously would have strained credulity and been summarily dismissed. In no area of innovation was this new surrender to science and technology more pronounced or better rewarded than that of radar.

In 1885 the German scientist Heinrich Hertz had been the first to demonstrate in his laboratory the reflection of radio waves from a metallic object. In years to come investigation of this phenomenon,

The Apparatus Of Radar

Above: *Aerials at Dover: far left, Chain Home Low and, far right, Chain Home 'floodlight' type. The relic of a previous age of warfare (left), the castle is still used as a vital navy and army command centre.*

Left: *Interior of a radar station. In the background the operator watches 'blips' on the CRT. Telephone terminals for communication with the control and plotting centres are visible on the right.*

although publicized, took second place to sound radio. Indeed, it was through experiments with ways to improve long-range radio signals by day and night that led in the early 1920s to the discovery of the ionosphere, an ionized layer some 60 to 500 miles above the Earth. This pointed the way to study the applications of very high frequency microwaves which, when linked with the invention of the CRT, provided the essentials of practical radar — quite apart from 90% of all radio and television systems.

In 1933 in Kiel harbour the Germans managed in secret to detect a ship by radar and went on to develop range-finding equipment which, by 1939, was ready for installation in their larger ships. Shortly afterwards the French announced that they would fit radar equipment to their new liner, the *Normandie*, for the detection of icebergs. By this time, with war clouds gathering, every nation with an electronics industry was taking notice and grading the subject 'secret' once their governments had understood the vital importance of devices which might counter the bomber menace and give an edge in all kinds of combat.

Slightly late in examining radically new approaches to air defence, and inspired by Elihu Thomson's discovery in 1896 that ultra high frequency radiations (such as X rays) were destructive of matter, the British Air Ministry in 1934 asked Robert Watson-Watt of the National Physical Laboratory to examine the feasibility of 'cooking' aviators with a high-energy 'death ray'. He said this was pure fantasy but that a way of detecting aircraft at long range was possible by two methods: by a pulse beam or by the simpler 'floodlighting', using a metre beam. By either method, the position of an aircraft could be displayed in three dimensions on a CRT making it possible to plot position, altitude and course. Moreover, by fitting friendly aircraft with a discrete pulse repeater it would be possible for an operator to identify friend and foe (IFF). Intercepting aircraft could then be directed by radio to a tactically advantageous position for attacking bombers. Successfully tried out in February 1935 at a range of eight miles, this was but the first step towards an almost boundless leap into the future, exceeding the dreams of the most irrepressible optimists.

To begin with, British and German developments followed somewhat similar lines: that is, with ship-watching and ranging sets such as the German *Seetakt* 80-cm wavelength; the British Chain Home early-warning 'Floodlight'-type sets working between 8 and 13 metres; and 'beam-type' sets, such as the British Chain Home Low (CHL) at 1.5 metres and the German *Freya* at 2.4 metres. Both 'beam-type' sets could scan at a much lower level than those of higher

Stalwarts Of The Offensive Air War

Left: *Engineers examine a crash-landed Ju88 for vital intelligence on the enemy's technological status. Probably the most versatile aircraft of the war, the Ju88 served in both fighter and bomber variants and underwent more modification than any other aircraft.*

Above: *The most enduring aircraft of the war, the Me 109 was built in larger numbers than any other fighter aircraft.*

Top: *British Wellington bomber of 1937 with its 4000-lb bomb ready for loading. Built of the 'basket-work' geodetic system, this extremely robust machine's derivatives remain in service 50 years after the initial design which saw operational service throughout the Second World War*

frequencies and thus were useful for finding the height of aircraft. The *Freya* worked in conjunction with the *Würzburg* beam type on 53 cm to provide direction for anti-aircraft guns and searchlights.

In 1937 the British moved ahead with the first rudimentary airborne short wave-length radar set. This promised not only to enable a fighter to detect and fly within visual distance of a bomber at night, but also to locate ships at sea. It also completed the first phase of the radar revolution by making practical the long-range detection of aircraft. By 1938 the British Chain Home Stations set up to scan the eastern and southern skies were reaching out with 60% reliability to 70 miles at 20 000 ft. They were able to plot the subsequent movements of aircraft; to control their more accurate engagement by surface guns; and to organize their interception in most weather conditions by fighters guided to the vicinity by ground controllers studying the radar plots. The final attack then depended upon airborne radar or visual detection.

This may have been but a child's step at the foot of a climbing frame — but in Britain the infant was particularly lusty because the dread of bombing was there most acute. For example, when work on gun laying (GL) 6-metre radar for anti-aircraft guns commenced early in 1937, the initial purpose was to guide the existing optical instruments in the general direction of the enemy, with a parallel project aimed at range-finding. In May it was felt that the problem 'would not be solved in our time'. A few weeks later a ranging accuracy of 300 ft at a distance of 42 000 ft had been achieved and was, of course, regardless of weather conditions such as inhibited the existing optical instruments. Three years later a set was in production which took over all the functions of previous instruments and improved the 'kill' rate from 20 000 shots per aircraft in 1940 to 4000 per aircraft in the spring of 1941. This was but one of a number of successful projects then in course of development for use on sea, land and in the air; their appearance in action will be described in their appropriate place.

The Bureaucratic Weapon

Developments in modern communications tended to fasten an ever-stronger grip on home fronts, so that a relatively few men could, through a highly centralized bureaucracy, condition large blocks of people to their will. The dominance of the printed word for mass dissemination of news and propaganda was strongly challenged by sound radio as the number of homes without a radio set rapidly diminished. For the first time, politicians and propagandists could speak to the 'man in the street' directly, without the distortion of an editorial system. Oratory, which once infected

Below: ***The Russian Polikarpov I-16 monoplane fighter which, despite its speed in the region of 300 mph in 1938, could nevertheless be outclassed by the skill of Japenese pilots and, later, by those of Germany.***

relatively small audiences, could now be heard by all those who chose to switch on. Speakers such as Franco, Benito Mussolini, Adolf Hitler, Franklin Roosevelt and Winston Churchill who possessed or studied the technique to seduce the radio listener were handed a tool of persuasion over the nations they ruled. Furthermore, they were aided by skilled broadcasters, themselves controlled by information organizations given the task of concealment, distortion and exhortation. In some instances, they were also backed by a biased judiciary and police security ordered ruthlessly to suppress dissent. By a cynical twist of fate, the communication revolution which idealists hoped would enable nation to speak to nation was converted into an instrument of unscrupulous dictatorships such as enmeshed Germany, Spain, Russia and Japan. The exploitation of lies, coercion and suppression made possible by centralized organizations, could lead an entire population supinely towards the horrors and privations of war, indoctrinated with sincere belief in their cause (however misguided) and the determination to persevere to destruction (however unnecessary or ridiculous). Thus the battle of words assumed a more influential place than ever in the prosecution of war and therefore, in the nature of the classic response, compelled opponents to reply with propaganda offensives of their own. These, through radio, had the unique quality of being able to overcome the barriers of censorship in reaching the enemy ear. Alongside the battles for material acquisition of territory now loomed a pernicious struggle of words for men's minds.

Nor were words the only medium available. The cinema was beginning in the 1930s to expand beyond pure entertainment to the dissemination of news and propaganda to a mass audience. And, first shown to a few privileged viewers in Britain in 1936, there was high-definition television transmitted from CRT to CRT by means of high-frequency signals.

The Sino-Japanese Incident And The Second World War

Contrary to the commonly-held view that the Second World War started on 1st September 1939, the date is more realistically fixed at 7th July 1937, when Japanese troops ambushed Chinese troops at night at Lukouchiao. Accidental it may have been, but the subsequent rapid capture of Pekin on 28th July was on too large a scale to be dismissed as a punitive expedition. This was war in which technique rather than technology had the principal effect on land operations. Japanese infantrymen, well trained in infiltration tactics, concealment and fieldcraft since they had suffered so badly during massed attacks in 1904, managed prodigious advances from Manchuria and from the sea. Despite a few costly set-backs against the fiercely resisting Chinese, they seized the important cities of Tientsin, Nanking, Hanchow and Canton, but did so with few of the latest weapons, using their handful of lightly-armoured and armed tanks exclusively in the infantry support role.

It was at sea and in the air that new technology was to the fore. And here the other nations, to some extent, learnt about the relative merits of their own designs — and could have learnt even more about advanced Japanese technology had they taken real interest. But it is unlikely that they could yet have learnt about the new liquid-oxygen-propelled Long Lance torpedo with its range of 11 miles, speed of 49 knots and warhead of 1210 lbs. This promised a minor tactical revolution by far out-ranging American and British torpedoes, enabling launching to take place outside the range of ships' secondary armament. What they did see, and hardly reacted to, was a special amphibious warfare ship which carried mechanized landing craft and which most efficiently launched them in the lightning seaborne assault that took place on the city of Tientsin and later on Canton.

Above all, foreign observers should have studied the lessons of air warfare: for example, the manner in which the entire offensive on central China was supported by 264 aircraft from three aircraft carriers lying off shore when no land bases were at once available. Likewise, the vulnerability of unescorted medium Mitsubishi G3M bombers flying at 258 mph when pounced upon over Hanchow by American-built Hawk III biplane fighters of lower speed. Or the manner in which the balance was redressed when 27 escorting monoplane Mitsubishi A5M monoplane fighters, fitted with long-range fuel tanks, fought off Chinese fighters over Nanking, claiming 11 'kills' out of 16. Still more could be learnt when, as an immediate result of the Russians entering a pact with the Chinese, large numbers of Polikarpov I-15 (biplane) and I-16 (monoplane) fighters joined the battle and with their significant advantage in speed were able to inflict losses on the more highly manoeuvrable A5Ms. Regardless of the performance differential and, incidentally, disposal of the theory that high-speed fighters could not indulge in dog-fighting, here was clear proof that the essential weapon of air supremacy remained the fighter. The side which seized and maintained just a local superiority managed, as of old, seriously to hamper all the other reconnaissance and attack tasks so important to surface forces.

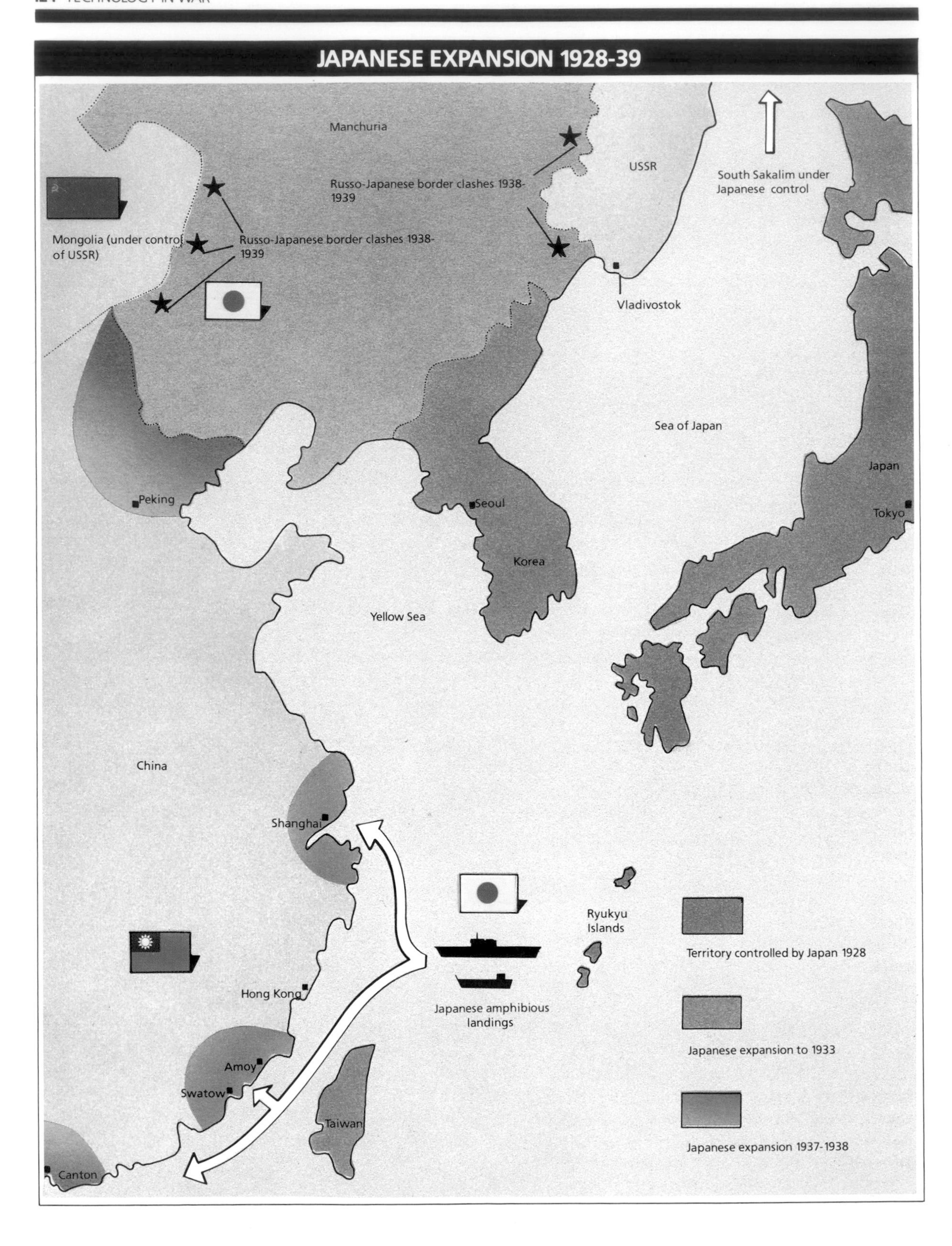
JAPANESE EXPANSION 1928-39
Manchuria
USSR
South Sakalim under Japanese control
Russo-Japanese border clashes 1938-1939
Mongolia (under control of USSR)
Russo-Japanese border clashes 1938-1939
Vladivostok
Sea of Japan
Japan
Peking
Seoul
Tokyo
Korea
Yellow Sea
China
Shanghai
Ryukyu Islands
Territory controlled by Japan 1928
Hong Kong
Japanese amphibious landings
Japanese expansion to 1933
Amoy
Swatow
Taiwan
Japanese expansion 1937-1938
Canton

Sheer technical merit was not, of course, the sole reason for Japanese superiority in air combat. They also outnumbered the Chinese and were far better trained. But although this meant that they had a free hand in the air and that their land forces benefited significantly from strong air support, it slowly became evident that air attacks in isolation did not win wars — not even when directed with impunity and ghastly effect upon cities. Initial shock among soldiers and civilians would begin to die away if a dose of bombing was not repeated: and if too often repeated it began to lose effect. In fact, symptoms of defeatism were often replaced by an angry determination to 'carry on' — providing it could be shown (if only through propaganda) that defensive measures of a belligerent as well as a passive type were being prosecuted, plus reprisals upon the enemy.

This pattern was repeated when the Japanese, true to their practice of striking without warning before war had been declared, started bombing Russian positions in Mongolia as the latest of a series of incidents along the ill-defined border line. Heavily outnumbered as the Japanese were by 1500 Russian aircraft, they plunged into attacks which triggered clashes of unprecedented dimensions. From the start it was plain that, although the Russian machines were not inferior to the Japanese, their crews were. Setting aside exaggerated claims for losses inflicted on each other (an inevitable phenomenon in the circumstances of high-speed flying and fleeting engagements when only occasionally is a 'kill' positively confirmed), it was made evident in adversity to the Russians, but not in victory to the Japanese, that fighters needed protective armour plus armament heavier than the machine-gun. Until this was implemented the Russian pilots tried to avoid dogfights, preferring to stalk and dive on unescorted bombers. And while bombers frequently got through, it was apparent that, if harassed, their effectiveness rapidly declined. Attacks on airfields embarrassed the enemy, without preventing an air force from operating. Against ground troops the destruction inflicted none too accurately by small bombs was of marginal effect only, whether against front-line troops or logistic installations.

On the ground, however, new technology was far more in evidence than it had been in China. Russian industrialization, concentrating more on military equipment than anything else, had produced in addition to a large air force a great many AFVs. It was 498 of these vehicles that the Japanese infantry met after entering disputed territory on the far side of the Khalkin River. Biding his time, the Russian commander, General Georgi Zhukov, slowed the Japanese advance with artillery and infantry accompanied by tanks, and then struck back hard. Regardless of bombing, his infantry and tanks managed to make progress against stiff opposition which included 180 Japanese tanks. But the decisive stroke came from a concentrated brigade of Russian tanks sweeping round the enemy flank to fall upon the rear of a Japanese division, prior to interposing a strategic barrier of tanks, guns and men across the Japanese rear. In places the Japanese were panic-stricken by tanks in mass. Overall they were defeated, with losses varying between 18 000 and 40 000 depending upon who made the claims and on what basis they were established. Far more important than the outcome was the clear refutation of experience in Spain which had decried independent tank operations. But nobody in the West had time to notice or digest this lesson for the future: operations in the East were cloaked in secrecy on the eve of the Second World War spreading to Europe.

*Left: **Japanese expansion in the Far East from 1937 until 1941. Various aircraft carrier-supported amphibious landings are shown along the Chinese coast from Shanghai to Swatow and Canton, past Hong Kong. Also plotted are advances inland and the areas of friction with Russia in 1938 and 1939.***

Mechanized Warfare In Europe From 1939 To 1941

Germany's almost unique feat in historic terms when she went to war against the combined strength of Poland, France and Britain was to be in possession of land forces almost ideally designed, equipped and trained to win swift and devastating victories. Hopelessly outnumbered in all departments at sea and only marginally superior in the air, she nevertheless employed her forces with strategic and tactical techniques which made outstanding use of the latest technology. Not that such techniques infallibly paid off. For example, the initial attacks on Polish airfields misfired since the Poles, anticipating trouble after a year-long war of nerves through propaganda, had dispersed their aircraft to secret landing fields to escape bombing. Nor was the air war very intense. It should be recalled that only some 1300 German and 400 Polish aircraft were involved and that of 600 lost in three weeks of combat, only about 20% were from air combat. One reason was the difficulty of finding enemy aircraft once visual early-warning systems had broken down in a war almost devoid of radar.

Although bombing caused immense destruction and disruption of Polish land communications and mobilization, it failed of itself to impose a crushing victory. That was credited to the army, spearheaded by six German panzer divisions, and a few mechanized infantry formations which struck too hard and moved too fast for the largely horse-and-foot Polish army. It handicapped the Germans little that of their meagre stock of 3200 tanks, only 300 were thoroughly battle-worthy medium types. Concentrated in close co-operation with artillery, infantry and dive-bombers they could crush opposition, filter through the gaps created and spread alarm and utter confusion as they drove deeply into the Polish rear — in one instance 125 miles in five days. It was emblematic that when Hitler expressed the assumption that a battery of Polish guns had been destroyed by bombing, General Guderian was able to inform him, more as a matter of myth-disposal than simple pride in tank achievement, that it was accomplished by his panzers.

Post-campaign impressions of events in Poland caused as much confusion as clarification of new technology's impact on battle. Air force propaganda focused on the pulverizing bombardment of Warsaw by medium- and dive-bombers and concealed the artillery's distinctive part against a broken opponent fighting mainly out of nationalistic pride. The sweeping advances by panzer divisions concealed concern about both the infantry which, lacking support from tanks, had performed poorly and the logistic system, which still depended extensively on horses. The remarkable achievement of modern communications in enabling command and control to be imposed firmly from the front, even at the tip of the farthest-flung spearhead, scarcely attracted attention. When the German forces turned westwards, their prospective victims continued to harbour delusions of the omnipotence of anti-tank forces but were terrified of the air weapon. Self-induced terror by the enemy admirably suited German leaders who well knew that surprise and technical excellence could overcome superior numbers, even though old-fashioned concepts still deemed them decisive.

Technical Pointers In Phoney War

The extraordinary lull which ensued between the end of September 1939 and the beginning of April 1940 served to provide pointers to the technical struggle ahead. The massacre in daylight of British bombers with power-operated gun turrets by cannon-armed German fighters on the fringes of German air space

Above: ***Bundles of propaganda leaflets await despatch through the chute inside a British bomber.***

compelled the RAF to concentrate on night flights over Germany. These in turn indicated how hopelessly inaccurate was astral navigation. A good interception rate of German aircraft encroaching on British air space gave proof of radar's utility — and alerted the Germans to the existence of the Chain Home system which, prior to the war, they had been prevented from discovering. Yet the Germans, in the correct belief that navigation fallibility made night-bombing unprofitable and the incorrect notion that guns and searchlights were an adequate defence, did little to improve their night defences.

At sea, the Germans staked much on the introduction of magnetic ground mines laid in shallow waters, hoping in the autumn of 1939 to paralyse British ports. Unaware of any possible counter-measures, they were disappointed when the British recovered a mine specimen, unveiled its secrets and introduced effective measures immediately. And when in August 1940 the Germans introduced an accoustic mine which reacted to a ship's presence, 'noisy' counter-measures rapidly put paid to their intentions.

The Germans achieved more with their *Seetakt* radar. It was of crucial benefit in enabling their armoured cruiser *Graf Spee* to inflict severe damage on three British cruisers on 13th December and its nature, though not its existence, was successfuly concealed from British investigators when *Graf Spee* was duped by false reports into scuttling herself at Montevideo. But in the initial stages of their submarine campaign the Germans were far from happy. By April they had lost 18 U-boats for the sinking of 764 000 tons of Allied shipping — a record which might have been bettered if their torpedo pistols had not been suffering from almost total malfunction. The old contact type was affected by diving too deep and the latest magnetic type (which functioned in response to the target's magnetic field) failed to the extent that in the critical April period, only one out of 38 exploded.

However, this miserable U-boat failure was hardly the result of the Allies' own prowess or their use of Asdic. It must be stressed that the other related cause was poor German tactics which prevented them from co-operating with each other and assuming sound attack positions.

Far-reaching as were the technical revelations of the Phoney War, their war-winning potential was as nothing compared with two deadly secrets which the Allies were concocting in the winter of 1940. One was centimetric radar: the concept of a pulsed, narrow beam of short wavelength capable of detecting its target with fine discrimination yet remaining free of random echoes and unsusceptible to jamming; it also had the advantage of requiring only a small, rotating aerial. Dependent upon vacuum valves of previously unattained power, centimetric radar was made possible by the single-minded work of Dr J T Randall and H A Boot in producing the powerful, resonant cavity magnetron valve. With this step forward in miniaturization, small radar sets, suitable for carriage in aircraft and with a wide potential for other uses, became assured and put the British years ahead of the Germans and still further ahead of the Japanese who had as yet hardly studied radar.

The second Allied secret was an electro-mechanical computer based upon a so-called 'Bombe' which the Poles had invented in the early 1930s to crack the cyphers of the original German Enigma machine. Matched to the electrical wiring of the Enigma wheels, the original Bombe could only slowly solve changes of key and these were being made progressively more complex and being changed more frequently by the Germans. Seeking to decrypt keys within hours of their entry into use, the British speeded up the original Bombe by programming it with punched-hole sheets

The Instruments Of Secrecy

Above: American version of the Anglo-Polish Bombe computer, used to decrypt messages encoded by the German Enigma machine. Lacking a memory, it was merely a tool programmed by people and not always fast enough to satisfy operational requirements. But as a war-winning weapon it, and the organization it served, was vital.

Right: The Enigma machine in service during the German advance through France, May 1940. General Guderian stands in his armoured command half-track, surrounded by the apparatus of control. In the foreground is the Enigma with three wheels and typewriter keyboard. Next, wearing headphones, is the operator of the command radio link sitting before his HF set. Behind wait despatch riders ready to carry messages by hand. Not far away is another signals vehicle containing teleprinters and telephones served by land line.

The German Conquest Of Western Europe

1940: The German invasion of Norway in April and its evacuation by the Allies in June; the three-pronged invasion of Holland, Belgium and France culminating in the evacuation from Dunkirk; and the final conquest of the remainder of France by fast-moving forces.

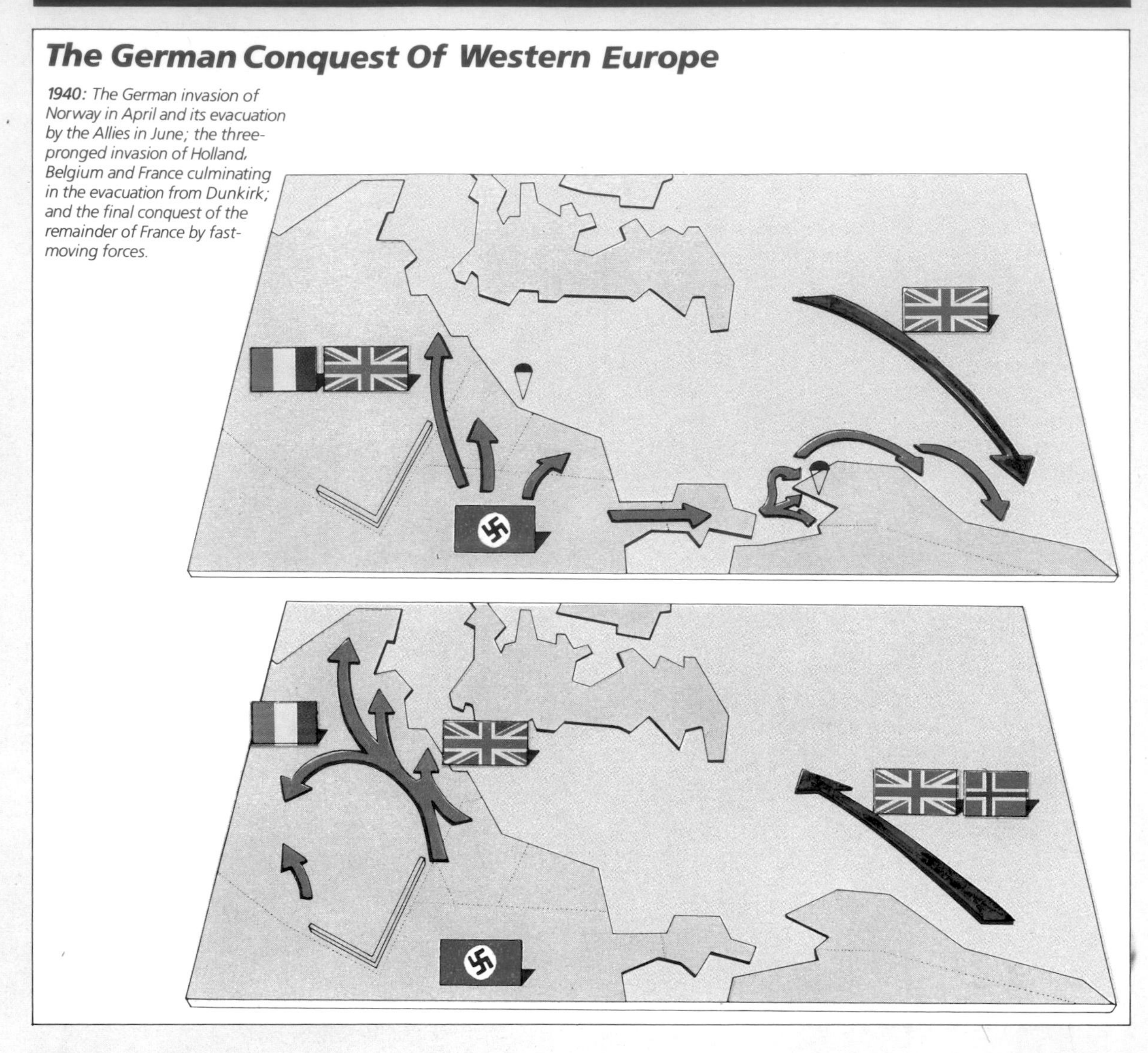

of paper. These were related to the punch card, invented by Joseph Jacquard in 1801 for semi-automatic weaving machines, which since 1930 had been used for analytical calculating machines — one of which was the Bombe.

Western Europe

Yet another veil was stripped from Germany's secret visage when her parachutists spearheaded the unannounced seizure of Danish and Norwegian airfields on 9th April 1940. The classic employment of surprise, deception and concentration at vital localities assured a stunning success in a campaign which, otherwise, witnessed combat with gun, mine, bomb and torpedo as of old. The decisive factor was intelligence. While German losses at sea were heavy, they might have been even heavier had they not been decyphering by traditional means 30% of British naval codes, and if their air reconnaissance had not kept track of Allied ships. But these examples of intelligence-gathering were unimportant by comparison with an event on 15th April when, to the delight and surprise of British technologists, their first high-speed Bombe came into operation and within hours — not days as previously — broke the latest Enigma keys used for Norway by the German army and air force. From this moment a voluminous supply of enemy operational information, intentions and deployment was available to the British Ultra organization. The handling of

Enigma decrypts would be at the very heart of a struggle to maintain the facility of reading the enemy mind, inextricably entwined in the contest between opposing scientists and technologists for weapons superiority.

On 10th May the struggle intensified when Germany launched an all-out, but by no means unexpected, invasion of Holland, Belgium and France. Faced by the threat of flooding in Holland, fortifications barring defiles into Belgium and the formidable but shallow Maginot Line of forts built by France to guard her eastern frontier, the Germans opted to go over and round these obstacles by capitalizing on technical surprise. They spearheaded the invasion of Holland by dropping airborne troops on key locations deep in the hinterland, disrupting the plan of defence and so undermining will-power that resistance collapsed in four days. Similarly, the capture of the Belgian Eban Emael forts by glider-borne troops not only engineered the precise landing of an élite on top of the objective, but capitalized on the so-called 'Munroe effect' as a highly effective way of penetrating thick steel and concrete to blast the garrisons out of their bunkers. In the 1880s an American chemist called Charles Munroe claimed what many others since 1800 had known, namely that explosives in a conical shape, placed at a critical distance from metal or concrete, would concentrate a jet of gas at 28 000 ft per second with a pressure of as much as 2000 tons per square inch. The discovery was welcomed by mining engineers and safe-crackers alike for its economy and, in the 1930s, was proposed as a good way of attacking tanks if the charge was incorporated with a metal liner. Within a few hours at Eban Emael small, but precisely-placed, lined hollow charge devices did the job which in 1914 had taken imprecise heavy artillery several days.

It is, however, the celebrated drive by German mechanized formations through the Ardennes, outflanking the Maginot Line and moving rapidly onwards to the Channel coast which is enshrined in history as the epitome of Blitzkrieg. As an example of the destruction of numerically superior forces by a superbly-trained, excellently-led and appropriately-equipped élite it stands supreme. Yet it represented no more than Fuller and Guderian (the latter of whom controlled the spearhead through exemplary use of encoded radio signals, cable teleprinter, telephone and personal contacts) had foretold and should not have come as the surprise it did. Indeed, it is the failure of mere mortals to comprehend fully the offerings of technology which lies at the heart of the ironies of the advance to Dunkirk. Initially, for example, there was scepticism among the orthodox military of the feasibility of passing massed mechanized forces rapidly through the enclosed Ardennes terrain and then executing, without pause, an assault-crossing of the well-defended River Meuse. Only Guderian's faith

Below: ***German parachute troops leap from a Junkers Ju 52. The drop had to be made from a very low level, in order to least expose the parachutists to fire, making the aircraft extremely vulnerable to light anti-aircraft fire.***

Mechanized Warfare In 1940

Top: German PZKW III tanks during the advance into France, May 1940, in company with a motor cycle combination used by reconnaissance troops and mechanized infantry. The horses seen in the background were used in large numbers to draw guns and transport— epitomising the semi-mechanized state of the German army throughout the war.
***Left:** French Hotchkiss H 35 light tank with its comparatively thick armour, none too powerful 37 mm gun and extremely inefficient one-man turret.*

carried the project through — but he was abetted by Allied commanders who dallied in reinforcing the threatened area since they, too, underestimated the latent threat, never coming to terms with the momentum of the German advance. Allied communication and logistical systems were broken and overrun; they were unable to concentrate mobile forces at the critical points; and their scattered armoured units were destroyed piecemeal on vital ground. All this was achieved by an opponent whose defensive offensive tactics were devastating; whose communications were swift and sure; and whose stamina was hardly impaired by an advance of 150 miles in seven days.

Yet, at this moment of triumph, German mis-evaluation of technology robbed them of the full fruits of victory. Fears that a 50% reduction of tank strength would fatally weaken subsequent operations led to the famous order by General Gerd von Rundstedt to halt, on the very eve of the panzer forces seizing Dunkirk and so sealing the fate of the main Anglo-French armies in Belgium. In fact, not only was German tank strength still perfectly adequate for the task ahead, but von Rundstedt had also overlooked the fact that a high proportion of unserviceable vehicles could rapidly be replaced by local repair and maintenance. Worst of all was Hitler's acceptance of the claim by Marshal Göring that escape of the Allied armies by sea could be prevented by the air force alone — a boast which the existing state of the art and technology of air warfare made quite impossible to fulfil. It was not simply the telling effect of gunfire from ships and attacks by RAF fighters which prevented the German bombers at extreme range from pressing their attacks accurately. The true cause of failure lay in air-power's inability to maintain a constant presence in all conditions. The vast majority of over 330 000 men taken off the beaches left under cover of night and bad visibility.

The Air War Develops

The failure of air power at Dunkirk did not affect German opinion which, on the fall of France, continued to believe that defeat of the RAF would ensure the collapse of the British nation under unrestrained bombing followed by invasion. Their failure to eliminate an enemy air force through bombing or air combat was set aside. Additionally, the Germans were in some ignorance of the technical attributes of British air defence based on radar, radio control and a strong fighter force operating from airfields many of which lay beyond the range of German fighters. Over-confidence ruled the Germans, above all in their belief in the inviolability of Enigma. The German air force, in particular, sent a high proportion of its operational messages by radio, eschewing the far more secure land links (including the captured civilian systems which were utilized at once). These had provided the army with an almost uninterrupted service even at the extremes of its advance — and had thus concealed its intentions. Starting on 22nd May 1940, Ultra's high-speed Bombe broke the latest Enigma key and began to decrypt messages at the rate of 1000 a day, providing a service to the end of the war without once being suspected by the Germans or seriously hampered by any radical, precautionary modification of the system. With intelligence of this quality, the British and their allies had a service of war-winning potential.

While Ultra's contribution to the defence of Britain was limited to a sporadic outline of intended air attacks (and never disclosed the manner and place of an invasion) the part played by radar was fundamental. Its range and efficient interpretation by fighter-controllers were improved by practice over a gradual increase in the intensity and scale of attacks. But radar might have been crippled had the Germans only persevered with initial attacks upon transmitters, instead of drawing the false conclusion that they were impossible to put out of action. As it was, the only serious threat to RAF communications came from bombing of inland fighter airfields where, unknown to the Germans, the apparatus of fighter sector control was lodged — and, sporadically, damaged.

Since the British began the Battle of Britain by over-estimating German air strength by a factor of three, and the Germans persistently under-estimated that of the British by a factor of more than three, the struggle may be termed a conflict of delusions. German miscalculation was fatal. In prematurely assuming the RAF fighters to be exhausted by losses, the Germans felt safe to relax their punishing attacks on airfields in order to indulge in the bombing of London. This gave their enemy a chance to recover and, crucially, to shoot down bombers which in order to reach London had to fly partly unescorted at the extremes of their fighters' range. Failure to win air superiority put an end to Germany's hopes of invading Britain by air and sea — although it is arguable that, despite her technical unpreparedness for such a venture, it could have been achieved if attempted within five weeks of Dunkirk, while Britain's defences were at their nadir. As it was, the delay gave Britain time to improve her radar screen and build more fighters to defeat the day bombers. This led the Germans in turn to the substitution of inaccurate night bombing — and with that the disclosure of yet another German secret, the directional radio beam.

Air Defence 1940

Early warning of enemy aircraft approaching at high altitude is given by the Chain Home radar stations, while those flying in at lower level are picked up by Chain Home Low, and transmitted by land line to HQ Fighter Command. Warnings and orders are sent by HQ to Fighter Groups, Fighter Sector airfields, Observer Corps posts, anti-aircraft gun positions, balloon barrage sites and the Civil Defence authorities. Fighters 'scrambled' under Group orders are then controlled by through radio from controllers at the Sector Stations who plot the enemy's position, altitude and course from information received from radar, the Observer Corps posts and gun positions. The controller's aim is to locate his fighters in a tactically advantageous position and minimise the dangers of being fired on by friendly guns. Radar inadequacies and possession of the initiative by the enemy frequently foiled these attempts.

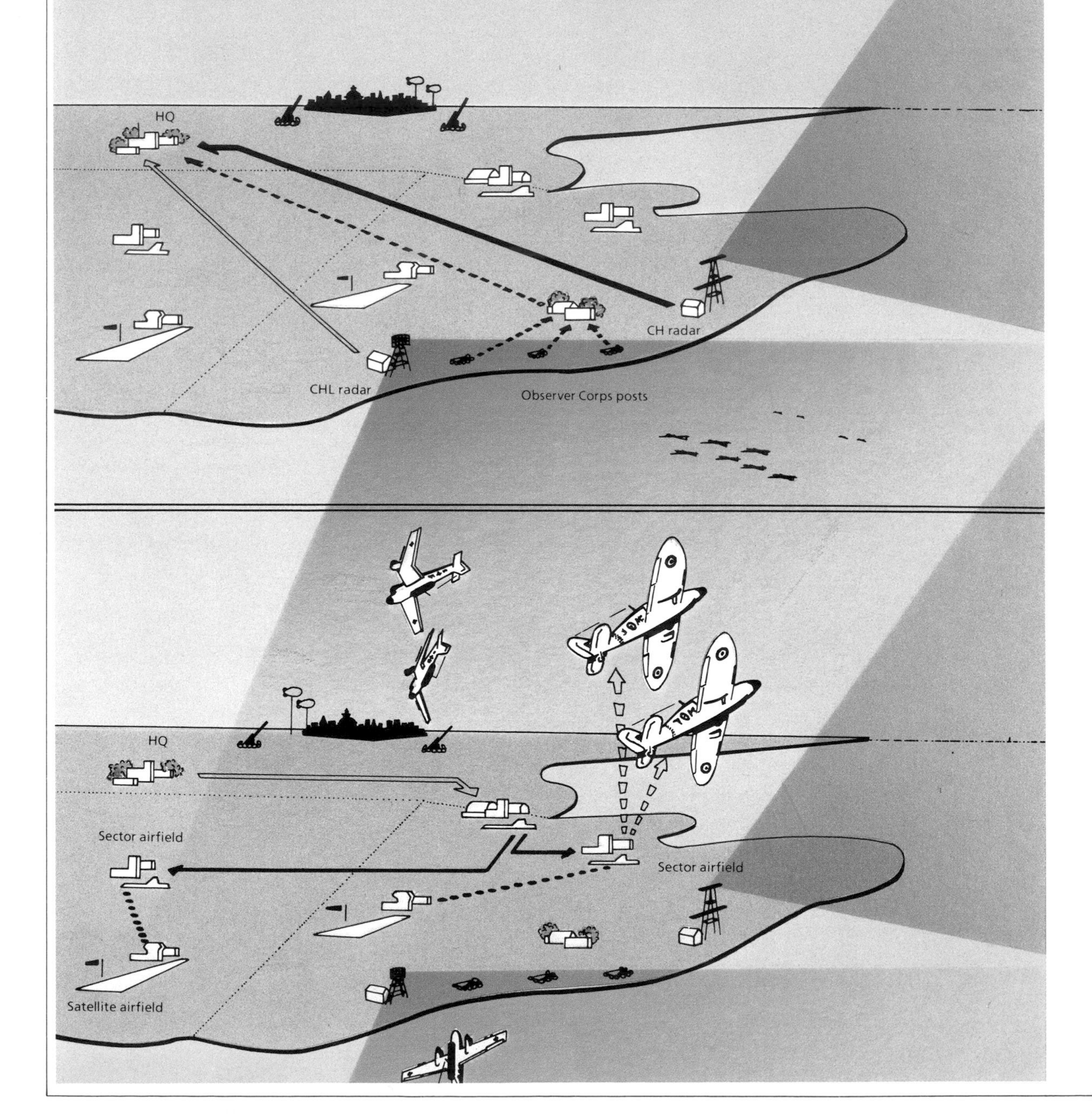

Approaching German bombers flew to a pre-arranged course and plan, the fighters in tiers as close escort to the bombers at medium or low altitude, or as top cover. The British controllers, warned by radar of the German approach, would endeavour to order fighter take-off in sufficient time to acquire tactical advantage. As a rule (although this did not always happen in practice) the fighters would try to climb above enemy formations and attack from out of the sun. The higher performance Spitfires aimed for the single-engine Bf 109s, Hurricanes for the lower performance Bf 110s, with the bombers fair game for anybody and rated a prime target.

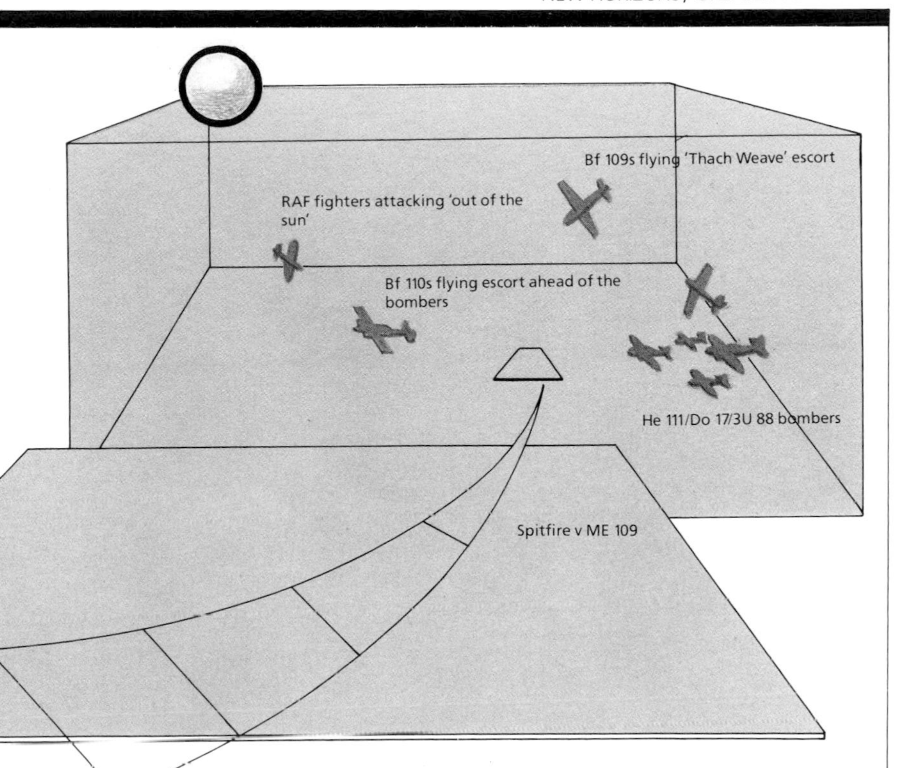

Spitfire and Bf 109E. The Spitfire, with its speed of 356 mph at 19 000 ft, was at a disadvantage against the Bf 109E at high altitude due to inferior engine performance: it could also be out-dived by the German machine. In manoeuvrability and at lower altitude it had the edge, however. The balance of armament, between the British 8-gun machine-gun battery and the German single-cannon, twin machine-gun arrangement, was slightly in the German favour. Therefore there was little to choose between the two machines and it was the more flexible, loose German tactical formations which, initially, gave the Germans superiority over the more rigid, close British formations. Besides which, too often, the British were unable to reach maximum altitude before they were 'jumped' from above by an opponent with the initiative.

The classic British fighters of World War II. Nearest is a Hurricane, flying alongside two Spitfires.

Directional Radio Beams

Starting in 1937, the Germans had begun to convert existing civil aviation radio beacons and Lorenz beams to military purposes. These proved double-edged weapons, as nearly all electronic devices are prone to be. When British monitors passed information concerning frequencies to the RAF, its bombers made use of them over Germany. Then, in August 1940, it introduced 'meacons' to render useless the beacons or actually mislead German aircrew into landing or crashing in Britain while in the belief that they were near their home bases.

More deadly were the directional Lorenz beams, the existence of which was disclosed to the British by captured air crew, documents and equipment, decrypted Enigma signals and from trials with the beams over England. Through the perseverance of Dr Reginald Jones in presenting the evidence and overcoming scepticism among fellow scientists, the beams were detected with time enough to prepare counter-measures. The German system, called *Knickebein*, directed parallel beams for a bomber to follow until an intersecting beam announced it was over the objective. *Knickebein* was jammed and later diverted by synchronizing bogus signals with the original, thus confusing aircrew until they lost faith in the device. Another system called *X-Gerät* also consisted of combined and intersecting radio beams which enabled aircrew to bomb blind, and to illuminate the target area with flares and fires from incendiary bombs as a beacon for following bomber waves. Jamming was again effective and soon noticed by the Germans who, without success, tried procedural counter-measures and sudden changes of radio frequency — many of which were discovered by Ultra — before abandoning the system.

By winning this phase of what was to be a ceaseless radio war, the British minimized, without preventing, accurate bombing. They also eased the task of their fighters and guns, increasingly assisted by radar, to get to grips with bombers compelled to operate in bright moonlight in order to find targets.

This German acceptance of technical defeat was symptomatic of a fatal distraction in 1940 from wholeheartedly pursuing new technology when uplifted by conquests. Believing the war in the West was won and that a projected invasion of the most dreaded enemy, Russia, would attain equal and rapid success in 1941, they economized on several weapons (such as advanced radar) opting instead for improvements to those in existence. For example:

A. *Tank up-gunning* was delayed in order to equip the infantry with the powerful, long-barrelled 50-mm anti-tank gun to replace the 37-mm type which had failed to penetrate the best French and British tanks. There was logic in this since in 1940 far more enemy tanks had been knocked out by field anti-tank guns than by tanks whose gunners rarely did better than score one hit in ten. But the down-grading of tank priority, implied in diverting scarce production to guns of limited traverse mounted on tracked hulls, was risky. In 1941 German tank strength of 3400 was little greater than in 1940, even if, at 2000, a higher proportion was of the medium types.

B. *Aircraft propulsion improvement* was also deferred by vacillation over developing types within their grasp such as would have out-classed anything Britain, Russia or America had in preparation. Since the 1920s the Germans had experimented with rocket missiles as a substitute for the heavy artillery they were forbidden. In 1938, Professor Hellmuth Walter had devised a practical controllable unit offering thrust which did not vary with atmospheric conditions and therefore could be used in space. Of equal importance was the flight in August 1939 of a Heinkel He 178, powered by a turbine engine which produced jet thrust by compressing and expelling heated air at high velocity. This was the newest stage in the ceaseless development of gas turbines which had intensified for commercial purposes in the 1920s. It spurred on metallurgists, who were asked to devise turbine blades (often nickel

Above: The 540-mph jet-propelled Messerschmitt Me 262. Designed in 1939, its entry into operational service was delayed until autumn 1944. Of short range and erratic in handling, its combat potential was never realised, and it was preceded into action by the British Gloster Meteor twin-jet fighter.

chromium steel alloys) which could withstand temperatures up to 1000°C, high centrifugal tensions at 30 000 rpm, as well as bending, erosive and corrosive effects at high temperature in order to achieve maximum thermal efficiency and power. Several axial-flow twin-engine fighters (which in due course would be capable of a maximum speed of 540 mph at 20 000 ft with an operational ceiling of 37 500 ft while carrying four 30-mm cannon) were attainable by 1943. The project was not pressed vigorously because the Air Ministry opted instead to concentrate its efforts on improvements to piston-engine types with top speeds which would barely exceed 400 mph at 20 000 ft and which were more lightly armed.

C. *The submarines* which went into action in 1939 were little better than those of 1918. Of 57 German U-boats, only 22 were ocean-going. So important was it felt to raise production of proven types (which were equal to their task) that the development of higher-performance boats was set aside, leaving the true revolution to the operational techniques of Admiral Karl Dönitz, adapted from the rejected 'wolf-pack' tactics of 1918. Such techniques were assisted by intercepts of enemy convoy radio transmissions and the decoding of signals. Guided to their targets by reports from reconnaissance aircraft (which also attacked with bombs), groups of U-boats supplied by 'milch cow' submarines (introduced in 1942) could remain at sea for prolonged periods and co-ordinate concentrated attacks on convoys, often overwhelming the escorts by sheer numbers. These tactics were made far easier since the capture of the entire European seaboard from Norway to southern France gave freer access to the Atlantic Ocean.

D. *A brain drain* exacerbated German shortcomings in advancing science and technology, following the loss of over 10% of university teachers since 1932 through anti-Semitism and anti-intellectualism. Brilliant people such as the physicist Lise Meitner and hundreds more who were compelled to leave the country contributed their services to Germany's enemies. The accelerating pace of discoveries is caused by the doubling and redoubling in numbers of inventors and technologists. Subtracting from or obstructing their efforts (as Germany did, for example, by outlawing the theory of relativity because Albert Einstein was a Jew) puts a brake on development.

To surrender a technical lead is easy: to catch up against opposition, as the Allies now strove to do with uninhibited vigour, is much more difficult. They had to tackle the full spectrum of existing weapon technology plus several projects which were at the frontiers of knowledge. For example, in addition to building sufficient conventional ships and aircraft to beat the blockade, the Allies had, almost from scratch, to design, develop and construct an armada of specialized landing craft, eventually to carry men and equipment on to hostile beaches as the first stage of invasions which would bring about the downfall of Germany and her latest ally, Italy. They had also to work out techniques for amphibious warfare in which the Japanese, already pushing towards South-East Asia, were well ahead. To do so the British took the lead in establishing a Combined Operations organization which, activated by Admiral Sir Roger Keyes at the behest of Prime Minister Winston Churchill, began to assume the role of a Ministry of Defence in an attempt to co-ordinate the operational, technical and weapon procurement activities of the established sea, land and air forces. Resented as this was by traditionalists, the system, already implemented in the Wehrmacht, was considered essential by those who saw the need for a central agency to exploit all assets.

The Battle Of The Atlantic

A short pause in land warfare after the fall of France conspired to concentrate attention upon crucial aspects of sea and air warfare. While battleships lurked in the background, kept at a distance from foreseeable places of action by fear of underwater and air attack, aircraft carriers entered the limelight on 11th November 1940 when 11 torpedoes from a score of slow-flying British Swordfish biplanes, launched at night from an aircraft carrier, scored six hits on three Italian battleships in harbour at Taranto. Although this event put aircraft carriers on the map and foreshadowed the demise of the battleship, the struggle against submarines and commerce raiders was more important since Britain's survival depended upon the outcome. Technology held the key. It was not simply the inadequacies for detection and defence of Asdic and depth charges which gave the well-organized wolf-packs an advantage. Surface attacks also did much harm since German submarines (in particular) as well as surface warships were extremely difficult to detect at night. Radar provided the answer, but only by stages. Sets in the 1.5-metre band fitted in ships in 1940 could only detect a submarine at short range; they did little better in aircraft in 1941. Only a few U-boats were sunk with the help of this set, but it enjoyed a triumph in an Anti-Surface Vessel (ASV) role when it enabled Swordfish aircraft to find the battleship *Bismarck* in mid-Atlantic in May 1941 and score the torpedo hits which slowed

Arbiters Of Defence In The Battle Of The Atlantic

From 1939 to its turning point in mid-1943, the Battle of the Atlantic was not simply a conflict between convoys of merchantmen and roving U-boat wolfpacks. The battle was much more complex and was fought over a wide area of the Atlantic, involving the air, sea and intelligence forces on both sides.

Far left: *The heavy loss of merchant vessels resulted in the application of new techniques and technology to the ship-building industry. Prefabrication and assembly-line production, which had been pioneered by the American automobile industry, was practised on a much larger scale in the construction of Liberty ships. The ability, as celebrated in the photograph, to produce a ship in 4 days 15 hours and 29 minutes was as important to keeping the supply lines open as the ability to track and destroy the enemy.*

Top: *Depth charges ready for launch as an Asdic contact is made with a submarine.*

Above: *Swordfish aircraft, with wings folded,on a carrier being prepared for an anti-submarine strike.*

Left: *Flight deck of a Sunderland flying boat escorting convoys.*

her down, enabling her to be caught and destroyed.

Centimetric radar was to prove the decisive instrument, as demonstrated in April 1941 when a surfaced submarine was picked up at 10 miles range and a periscope detected at 1300 yards. After that it was a complicated matter of choosing sets which would fit into aircraft without creating radio interference; putting production in hand; training crews to use it; and launching it into action — always in the hope that it was not disclosed prematurely to the enemy in time to take technical counter-measures. Meanwhile, the Germans enjoyed two years of superiority in the Battle of the Atlantic because — even after the USA with her immense reserves entered the war in December 1941 — anti-submarine resources were totally insufficient in a vast watery waste against a hidden opponent who held the initiative.

Sound technique practised by well-trained and determined men could compensate for smaller numbers and slightly inferior technology — a dictum which could also hold true on land as on sea — as Germany made clear when she invaded Russia on 22nd June 1941.

The Invasion Of Russia, 1941

The collection of comprehensive intelligence from a variety of sources (not least among them the myriad dissenting peoples of the Soviet Union who were only too happy to betray their oppressors) and the selection of the day upon which to launch their attack, were important contributions to the success of the numerically inferior German air forces' initial strike against Russian airfields. Despite ample warning from sources including Ultra, the Russians permitted their aircraft to be caught undispersed, too close to the frontier and with their crews sleeping-off Saturday night junketting. In a single day 2770 German aircraft claimed to have destroyed 1489 aircraft out of 12 000 on the ground and 322 in the air for minimal losses. Certainly they established an ascendancy over the Russians in the air which was never entirely lost. The victory might have been even greater had not the Germans already diluted their effort, with 1600 aircraft committed to operations in support of the Italians in North Africa (where they had been badly worsted by the British); to the invasion of Yugoslavia, Greece and Crete (which had swallowed up the entire airborne force and several panzer divisions) and to the defence of Western Europe.

Technique *and* technology, of the proven pattern, gave Germany a victory in the air which, to no small degree, was credited to the superiority of improved German Messerschmitt 109F fighters over modified Polikarpov I-16 fighters. It was a victory entirely beneficial to the ground forces, although not in terms of direct destruction wreaked upon the enemy. A consensus of German generals welcomed any assistance from bombing of enemy troops, but they were prepared to sacrifice it in favour of logistic assistance from transport aircraft, and most of all, reconnaissance. In a country so vast that a continuous front was impossible, yawning gaps could only be covered by aircraft in conjunction with mechanized patrols, which detected enemy concentrations, reported details of defences and gave warning of enemy counter-strokes.

With first-rate information, much of it garnered from monitoring of insecure Russian radio links, the German army made sweeping advances to trap hordes of out-manoeuvred and out-fought Russian forma-

Above: ***The 30-ton Russian T34/85 with its 85-mm gun which so upset German composure. Extremely basic in construction, with a minimum of frills, it was distinguished by its sloped armour which enhanced protection.***

tions. Yet German technology was flawed and seriously challenged. At root, they had failed to appreciate the logistic difficulties imposed by weather and by terrain which was by turns a dust bowl, a quagmire or a frozen waste, and which was served by primitive roads and a railway of different gauge from Germany's. Two-wheel drive trucks whose engines were not proof against thick dust, unable to cope with severe cold and bogged down easily, placed a severe strain upon logistics. Replacements from an industry still not mobilized for total war fell short of demand. This was the first campaign in which the mechanized forces were asked to operate for longer than six weeks before withdrawal to the home base for refurbishment. Lack of adequate maintenance and repair services in the field exacerbated breakdowns which were far in excess of combat losses.

By incredible feats of endurance and improvisation, the concentrated panzer spearheads kept rolling, recording advances of 413 miles in 25 days against dispersed enemy mechanized forces. The 20 000 Russian AFVs were by no means technically inferior to the 3400 German machines. But communications and generalship were woefully bad due to poor training and leadership by an officer corps which had been decimated in Stalin's political purges. The only real shock — but a shattering one for the Germans — was the early appearance of two enemy tanks of unrivalled power and potential. Improving upon the original

GERMANY'S THRUST TO THE EAST

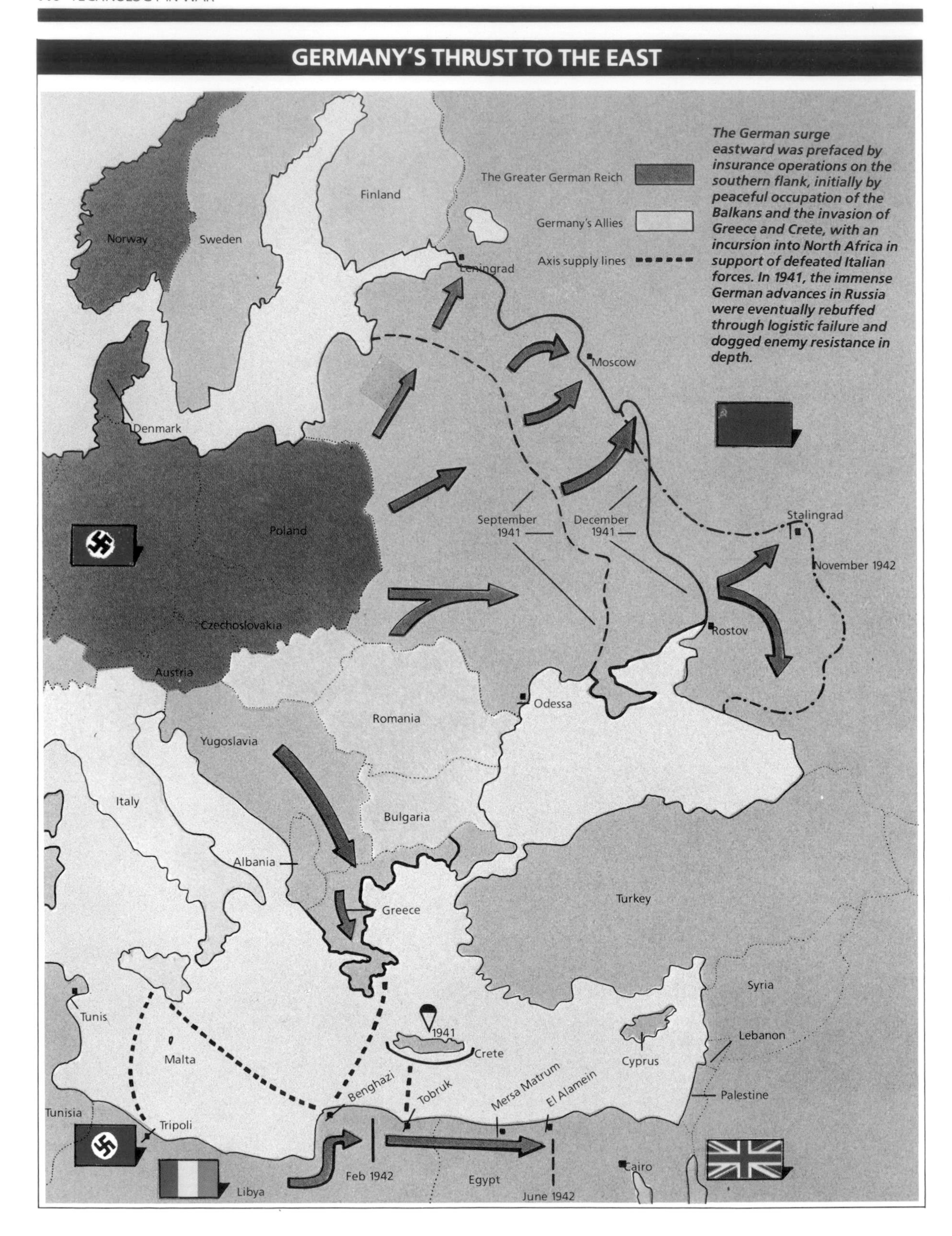

The German surge eastward was prefaced by insurance operations on the southern flank, initially by peaceful occupation of the Balkans and the invasion of Greece and Crete, with an incursion into North Africa in support of defeated Italian forces. In 1941, the immense German advances in Russia were eventually rebuffed through logistic failure and dogged enemy resistance in depth.

products of foreign companies and simultaneously training their own designers (of whom Michael Koshkin happens to be the most famous because he developed the T 34 tank from the Christie machine), the Russians produced in the T 34 and KV 1 two AFVs which were capable, with the 76-mm gun, of penetrating all German tanks at 800 yards and were proof frontally against even the latest 50-mm anti-tank gun at 300 yards. Unsophisticated and rugged in construction, they could withstand much of the punishment which the Germans, the terrain and their own ill-trained crews could mete out.

The initial, naïve German reaction was to propose copying the T 34 — until it was pointed out that this would take longer than putting into production the prototypes of their own next-generation tank. It was essential to improve in order to match the new designs of Soviet tanks, which were already on the way. Without option, the Germans were thrust into the vicious gun-versus-armour race which they had over-optimistically ignored. But it was nearly a year before parity had been attained through up-gunning and up-armouring existing models.

Meanwhile, technique had to compensate, and it was the sophisticated signal communications devices developed since the 1920s that made this possible. Across Soviet territory, which was largely denuded of high-capacity land links, a new network of wire-transmitted telephone and teletype facilities had to be laid. These incorporated the latest multi-channel, 'carrier frequency' systems; the sending along bare wires of simultaneous, electronically-separated modulations in enormous quantity to cope with intensive traffic. Reasonably secure, although prone to cutting by saboteurs, this network formed the foundation of military communications without supplanting the encyphered (but insecure) radio morse still used at longer ranges and across difficult or guerrilla-infested country, besides acting as a stand-by in periods of congestion. Narrow-beam microwave transmitters would be added to this in 1942. Working on line-of-sight (developed by A Clavier in 1934), they dispensed with wire while improving security over broadcast radio.

Partisan Warfare

By subjecting neighbouring countries to an oppressive rule, Hitler sowed seeds of discontent which naturally grew into revolution and primitive partisan warfare. But the guerrilla fighter of the 1940s was far better served by technology than his predecessors. For the first time partisans were provided with radio contact through miniaturized sets with distant bases in friendly territory. Resistance fighters could be rapidly inserted, extracted and supplied by air. Just one of several penalties paid by Germany for leaving Britain intact in 1940 was the island's importance as a base for

Above: ***Russian partisans armed with sub-machine guns preparing demolition charges on a railway track. Their accumulated efforts proved effective and were enhanced a moment later in this attack when they blew up the telephone lines beyond.***

resistance forces in Western Europe and elsewhere. The so-called 'Fourth Arm' of the services would be slow in making its presence felt through ambush, sabotage, hit-and-run raids by sea and air, assassination and, perhaps most fruitful of all, through the acquisition of intelligence. But its activities were deemed essential to distract enemy effort and in preparation and rehearsal for the day when massed Allied formations would attempt to re-conquer Europe.

The technology of subversion, like the operations involved, was something apart, virtually criminal in approach. Anything which had to be smuggled in by air, or slipped past sentries guarding targets earmarked for demolition, had to be cost-effective in size, power and simplicity of use. Lethal tablets to poison enemy agents or save friendly ones from torture; sub-machine guns; knives; limpet mines for attaching to ships' hulls; chemical time-pencil detonators; and the celebrated Cyclonite, a stable, mouldable plastic explosive sometimes called RDX, which was invented by the German Hans Henning in 1899 (but not manufactured until 1941) were all partisan weapons.

Carrier pigeons now gave way to radio for passing messages and a new twist was introduced to monitoring. Each side endeavoured to recognize clandestine radio transmissions, to locate the sender by direction-finders and capture the operators and their codes before they could hide the special, compact transmitter — or, better still, take the operators alive and 'turn' them against the enemy. This modernized form of underground warfare was raised to a high pitch between 1939 and 1945. It demonstrated the feasibility of diverting considerable enemy effort by means of small outlay, and concealed a potential value for the future that only experts could evaluate.

The Pacific War

It was a paradox of Japan's assault upon American, British and Dutch possessions on 7th December 1941, that her victims had the most advanced technology in the shape of radar and communications facilities, but made inept use of it, while she won great victories despite fundamental technical weaknesses. The American Magic organization (equivalent of Ultra)

Right: ***Destruction of ground and air forces was but a small part of the devastation inflicted by the Japanese at Pearl Harbor. Battleship losses were far heavier, but still more important was the proof of Japanese expertise in technique and technology, since it announced a shift in power to the East.***

having previously broken the principal Japanese code system (which was similar to the German Enigma but based on an electro-mechanical Type 97 machine, known to the Allies as Purple) had access to enemy plans to an extent totally denied the Japanese. And the Japanese, embarking on a campaign governed by sea and air power, lacked that vital instrument for those spheres — radar. Extent of experience lay at the heart of the paradox. The Americans at Pearl Harbor had not come to terms with the meaning of war — and certainly not on a peaceful Sunday morning. So their organization failed to disseminate in time the vital information from decrypts which told all; and the operators of a radar set which picked up the raiders on their way in had their warning rejected.

The Japanese, on the other hand, made up for their shortage of intelligence and lack of radar by outstanding use of the techniques and machines with which they had long practised against China and Russia. Moreover, with a foresight which Westerners chose to decry, they equipped themselves with the key weapons of maritime warfare — aircraft carriers (having nine to the Americans' six) and submarines — controlled by leaders and crews to whom innovation was welcome. This was soon apparent. It took a little longer to lay bare Japan's deficiencies in raw materials, including oil, without which modern war was impossible.

It should have been no surprise that the Japanese struck without formal declaration of war; it was not their habit to do so. Nor should the excellence of their equipment, such as the Mitsubishi A6M Zero fighter, have been a cause for astonishment; ample evidence was available for those who, unlike the Allies, took careful note. As for the sinking or damage in harbour of six American battleships, two cruisers and three destroyers by 360 aircraft (and three days later the sinking at sea of two British battleships by torpedo and bombing attacks) the feasibility of that had long ago been predicted by Admiral Fisher and affirmed at Taranto in 1940 and in mid-Atlantic in 1941 by the British. Yet it was not so much the deadly accuracy of the Japanese aircraft which was impressive at Pearl Harbor (where surprise and nugatory defences contributed) as the deft combination of high-level with

Amphibious War In The Pacific

The key to Japanese expansion across thousands of miles of water lay in their development of specialized craft to land assault troops rapidly on beaches, and thereafter to maintain them in position. But these forces depended to a large extent on air power to provide fire support for protection against enemy air attack. Thus the advent and eclipse of the Zero, the best Japanese fighter, was a critical symbol in the equation of victory and defeat.

Above: The 330-mph Mitsubishi A6M Zero fighter out-classed Allied fighters when it first appeared operationally in 1941, but its vulnerability to fire led to its eclipse in 1943.

Right: Japanese expansion in the Pacific Basin, showing the extent and rapidity of their gains by the use of a dominant fleet, amphibious forces and expertly co-ordinated air attacks.

low-level attacks to confuse the defenders — a technique which eased the sinking of the two British battleships in open waters off Malaya.

The five Japanese midget submarines carried to Pearl Harbor by ocean-going boats proved ineffective; one was sunk by depth charges before the air attack began. But submarines and aircraft provided a wealth of information to the Japanese commanders as they struck at the Philippines and towards Australia, and these also caused most damage in the naval and land encounters. Moreover, the mere threat of long-range air and submarine attack, besides administering shock to fearful inhabitants of America's west coast, did tie down resources to the defence of vital points which, in the past, would have been thought safe. How justified were the fears of American west-coasters would be seen when a midget submarine damaged a British battleship in port in Madagascar (after reconnaissance by a seaplane from an ocean-going boat); and four other midgets managed to penetrate Sydney Harbour, although without success, prior to their liquidation.

In the aftermath of Pearl Harbor and the capture of the Philippines and Malaya, it was their light naval forces backed up by distant battleships and aircraft carriers which made possible the rampant Japanese amphibious invasion of Indonesia, the South-West Pacific and the penetration of the Indian Ocean and Burma. In January 1942, four cruisers put paid to five Allied cruisers in the Java Sea. The victory owed something to the surprise use of the Long Lance torpedo, but was decided, basically, by superior Japanese command and control assisted by good air and submarine reconnaissance. Deprived of these facilities and hampered by the defects of a hastily improvised communication system between the ships of three navies, the Anglo-Dutch-American task force was always at a disadvantage. It was a slightly different story, however, when the Japanese fleet entered the Bay of Bengal towards the end of March. Here the greatly outnumbered British were forewarned by decrypts of Japanese movements, enabling them to take fairly successful evasive action, which the Japanese, who lacked intelligence, could not follow. As it was they managed to sink an aircraft carrier, two cruisers, a

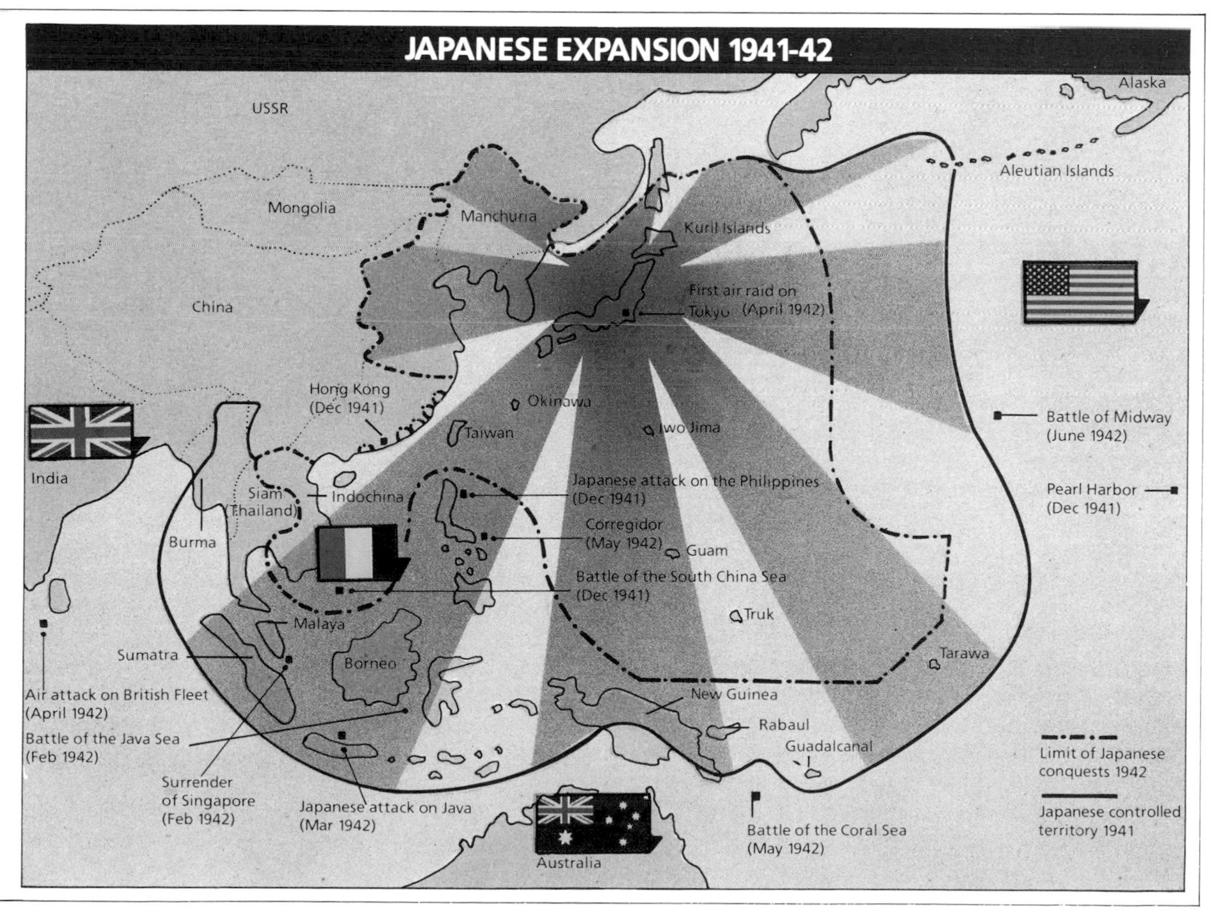

destroyer and 150 000 tons of shipping in a ten-day rampage which totally disrupted shipping in the Bay of Bengal. These were the products of matured new technology executed by seasoned warriors against an outnumbered opponent. Yet, as in Europe, there was a fine balance between élitism and technique on one side, and the brute force of mass production supplied by superior technology on the other. Administer a shock to the élitist system by the removal of its prime element and the scales could shift.

Tipping The Balance: Might Versus Élitism

The shift in balance appeared at Coral Sea when, early in May 1942, a strong Japanese carrier force supporting an amphibious invasion of Port Moresby was intercepted by an inferior American carrier force. Profiting from decrypts, the Americans blundered into position and then made as many errors as the Japanese made themselves. This was the first naval battle decided by aircraft carriers, one of which was sunk on either side by air attack in an engagement carried out at a range of 120 miles. On paper the Japanese won, but it was a Pyrrhic victory since they could ill afford the loss of so many irreplaceable, expert aircrews. Moreover, the total of 10 torpedo and 21 bomb hits shared between the two Japanese carriers indicated all too menacingly how fast American naval pilots were learning.

The extent of that learning was demonstrated a month later at the Battle of Midway in which carrier losses were four to one in the American favour; it was a victory which smashed for ever the Japanese élite and threw that nation on the defensive. Again, detailed American intelligence through decrypted messages exposed the Japanese to a concentrated blow by inferior forces of whose presence they were totally unaware until, far too late, ineffectual reconnaissance disclosed the danger. The Japanese might still have won, but their fate was sealed by faulty judgement on the part of their commanders; by the skill and bravery of American airmen, who were the equal of their own; and by the revelation of hitherto undisclosed defects in their carriers and aircraft. It was misleading reports about the presence of American aircraft carriers which led Vice Admiral Nagumo to vacillate and allow his own carriers to be caught unprotected — that and the inherent weakness of those carriers, with wooden flight decks that were easily penetrated by bombs which on detonation below ignited aviation fuel. Likewise, it was lack of protection for fuel which exposed the otherwise excellent Mitsubishi A6M Zero to sudden death when caught by a burst of fire aimed by Americans with sturdier aircraft, better flown.

It would be many months before the Japanese could muster the semblance of a force to replace that lost at Midway, and by then the first step on the way back had been made at Guadalcanal when an American amphibious force demonstrated techniques and technology to match its opponents. Not that it appeared so during the five months that the Japanese fought furiously on land, sea and in the air to repel the Americans. Operating by night without radar, the Japanese scored repeated successes against an opponent who was well-equipped with radar, generously supplied by intelligence from decrypts and air reconnaissance, but was also frequently too ill-organized to capitalize on these assets. The loss of four Allied cruisers on 8th/9th August in the Savo Sea stemmed from such incompetence. The loss to submarines of an aircraft carrier, with yet another

As the Second World War progressed, the need to provide air escort to convoys across the broad oceans became paramount. It was essential to construct, in addition to the large fleet aircraft carriers (such as the* Essex *type shown overleaf) classes of light, cheaper, easier-to-build escort carriers.
Above: ***The USS*** *Monterey* ***carried 31 aircraft and was of a class built in considerable numbers for the Allied fleets. Aircraft carriers of this sort played a vital role in overcoming the U-boats operating in the Atlantic.***

damaged later, along with a brand-new battleship, suggested the laxity of inexperience. But the sinking of a Japanese cruiser by radar-assisted gunfire at 5000 yards, without illumination, on the night of 11th/12th October told another tale; as did the destruction on 14th/15th November of the battleship *Kirishima* after eight minutes of intensive, radar-directed fire from the battleship *Washington* at 8400 yards (with nine hits out of 75 shots). Although the Japanese continued to win local victories by sinking or damaging many important American ships (throughout the Guadalcanal campaign, major warship losses amounted to one aircraft carrier apiece, two Japanese battleships, and seven American cruisers to three Japanese, in addition to many more cruisers damaged and destroyers sunk on both sides), the writing was on the wall. Having lost the cream of her aircrew, Japan was next in the process of losing the best of her surface fleet without inflicting sufficient damage to check the American build-up. It was a fatal attrition, as Allied technique, technology, sheer weight of industrial production and ample manpower were mustered for inexorable transformation into an irresistible force.

The *Essex* Class Aircraft Carrier

Conceived in mid-1939, this class of American carrier was an improvement on the smaller Yorktown class which preceded it, carrying 90, instead of 72, aircraft.

With a displacement of 27 200 tons and a length of 820 ft, the carrier's four Westinghouse geared turbines, producing 150 000 hp, drove the ship at 32 knots. Twelve 5-in guns controlled by two Mark 37 directors and 40 × 40 mm AA guns by Mark 51 directors gave the carrier a formidable defensive firepower, allied to close in-shooting by 55 × 20 mm guns.

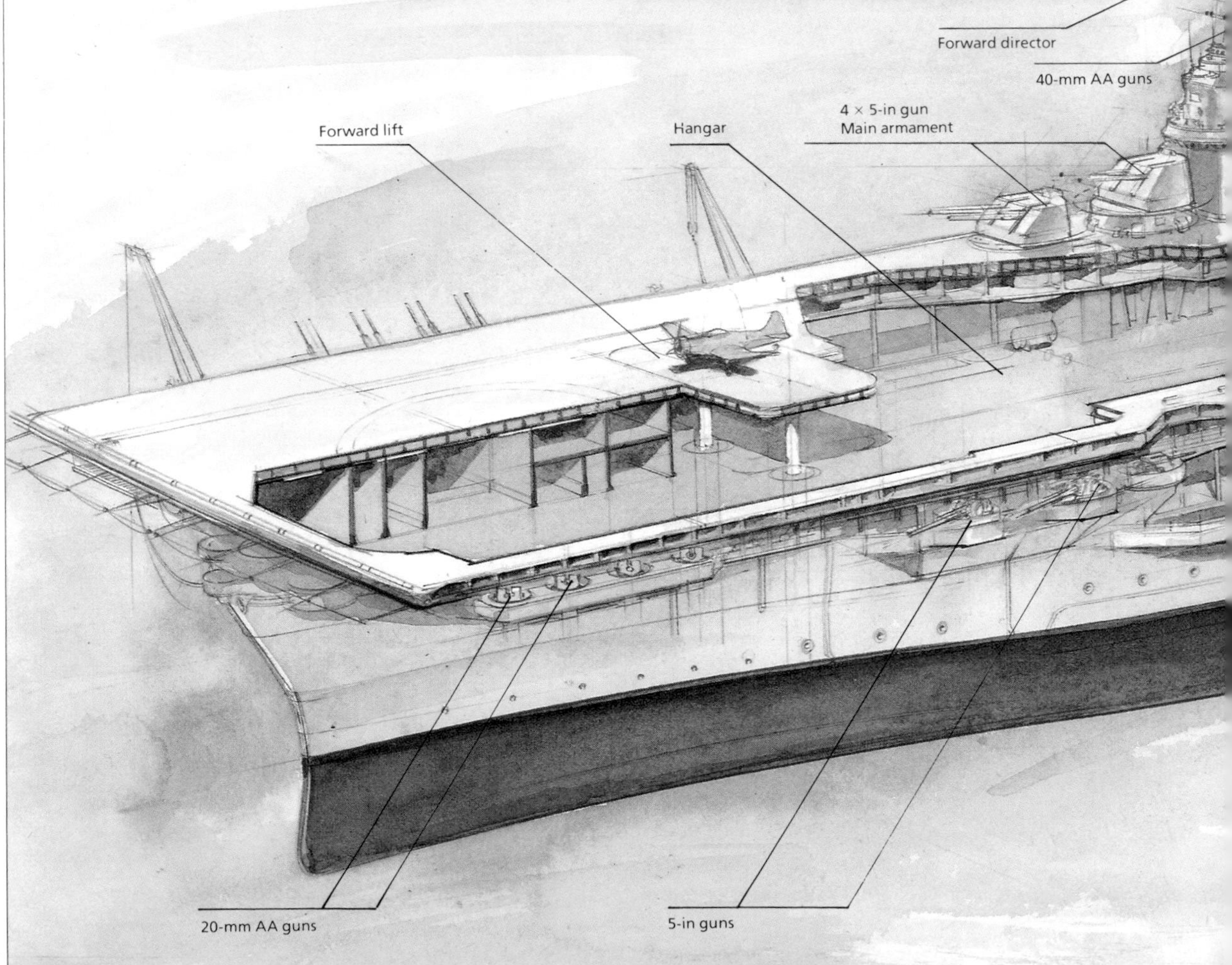

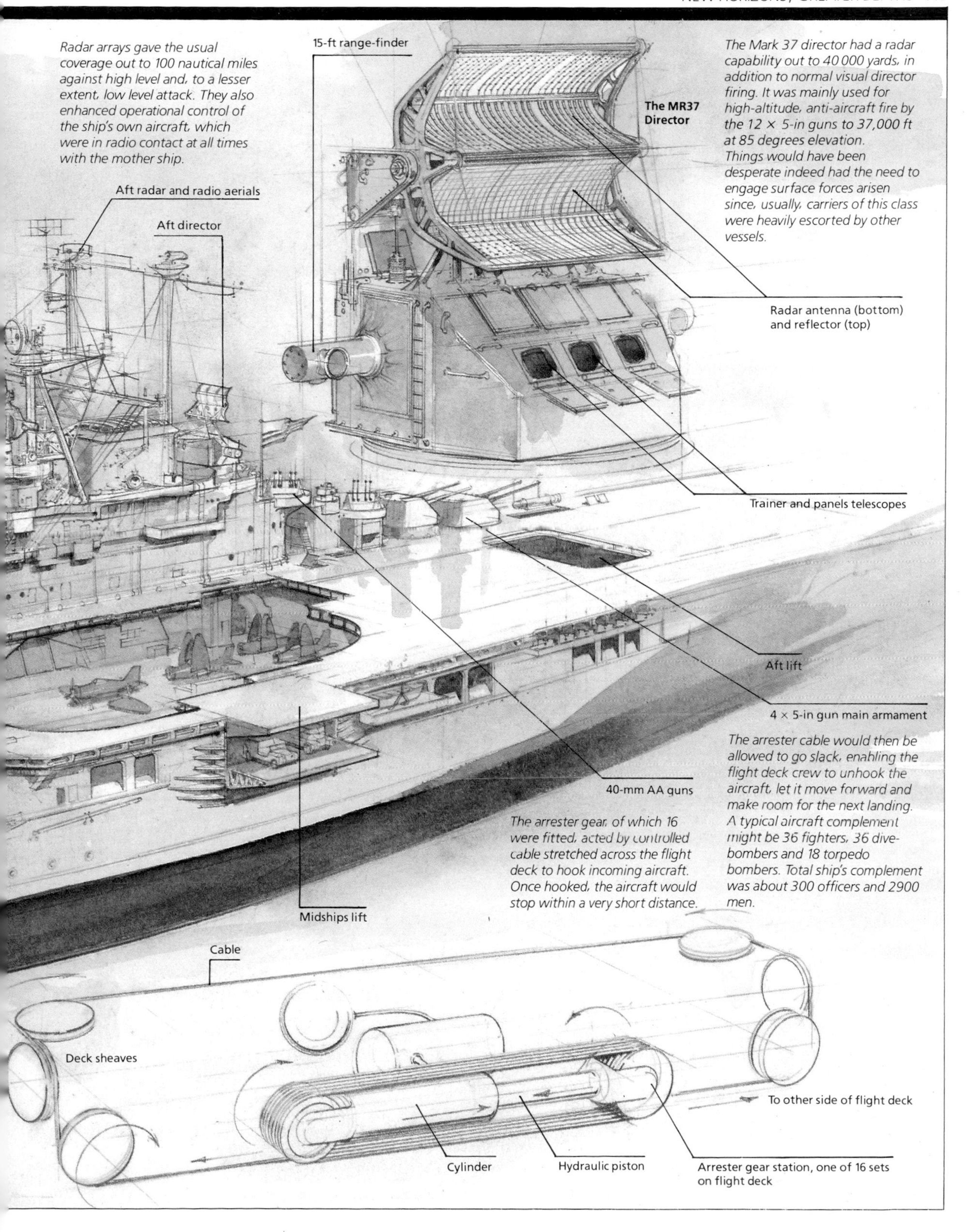

Radar arrays gave the usual coverage out to 100 nautical miles against high level and, to a lesser extent, low level attack. They also enhanced operational control of the ship's own aircraft, which were in radio contact at all times with the mother ship.

The Mark 37 director had a radar capability out to 40 000 yards, in addition to normal visual director firing. It was mainly used for high-altitude, anti-aircraft fire by the 12 × 5-in guns to 37,000 ft at 85 degrees elevation. Things would have been desperate indeed had the need to engage surface forces arisen since, usually, carriers of this class were heavily escorted by other vessels.

The arrester gear, of which 16 were fitted, acted by controlled cable stretched across the flight deck to hook incoming aircraft. Once hooked, the aircraft would stop within a very short distance.

The arrester cable would then be allowed to go slack, enabling the flight deck crew to unhook the aircraft, let it move forward and make room for the next landing. A typical aircraft complement might be 36 fighters, 36 dive-bombers and 18 torpedo bombers. Total ship's complement was about 300 officers and 2900 men.

The Battle For Europe: 1942 – 1945

By October 1942, territorially speaking, a state of check had been reached. The Axis partners had reached the fullest extent of their conquests; the Allies were poised to riposte. Technically, although as yet unrevealed, the balance had swung the Allied way; mainly because they had put so much effort into technology, but partly because the Germans had permitted a lapse in development work. But as evidence began to indicate that Japan was over-stretched and as Germany's forces began their retreat from El Alamein, the Caucasus and Stalingrad until within less than two years they had fallen back almost to their starting points of 1940 (see pages 128, 140), the realization that Axis survival depended upon defensive tactics and a revived technology drove their scientists, industrialists and military leaders to incredible feats of innovation and production.

In the aftermath of Germany's failure to defeat Russia in 1941, together with America's entry into the war and Hitler's consequent appreciation that the struggle was to be a long one, the policy of making do with marginally improved weapons and low production had to be revised. When Hitler's architect, Albert Speer, was appointed Minister for Armaments in February 1942 he did more than take in hand full mobilization of industrial resources through centralization and rationalization of supply and production; he also gave priority to the production of the revolutionary weapon systems which had been held back since 1939, at the same time encouraging projects beyond the frontier of feasibility. Time was short. The combat services were living close to the threshold of collapse on three overseas fronts while the home front was under attack from increasingly deadly bombing. Inter-service rivalries and objections to loss of traditional rights in industry were rife and never wholly suppressed. Shortage of manpower, along with certain essential materials, was prevalent. But such was Speer's organizing genius and ability to understand and relate a vast spectrum of technical subjects to essential demands, that he presided over a sustained

Above: ***The 56-ton German Tiger tank. Based on a pre-war concept and produced hurriedly in 1942, it was unreliable and relatively immobile. But its 88-mm gun and 110-mm frontal armour made it a dominant weapon.***

rise in production of new weapons while phasing out the obsolescent ones, and managed it against a background of air attack on factories, wrangling among factions and a progressive combing out of industrial labour to meet the requirements of the front.

At the heart of Speer's success lay the response of private enterprise to incentives of profit and influence. Inventors and entrepreneurs vied to share in the latest armaments bonanza, to catch Hitler's or Speer's eye with dramatic proposals — no matter how impractical or unrelated to necessity. For example, by 1943 no less than 40 different types of aircraft were under consideration — an incontinent expenditure of talent and effort. Only those which seriously impinged upon the war will be mentioned here, but these are merely a sample of an inventory of imaginative projects.

Machine-Guns, Artillery And Mines In The Land Battle

It must not be supposed that the time-honoured importance of infantry was abolished by new weapons. Men still had to occupy and hold ground which others had helped them seize, despite greatly increased volumes of shell fire heaped upon them. Whereas the German belt-fed MG 34 machine-gun of 1936 could fire 800 to 900 rounds per minute, its successor, the MG 42, could shoot 1200 to 1300 rounds with commendable reliability. It was a scourge of opponents such as the Russians, who favoured massed assault.

Rates of artillery fire, along with range of engagement and bursting power of shells, also improved in line with mortars which the Germans, in particular, used dexterously in defensive fighting. To a man crouching in a trench, it was immaterial how or with what he was fired upon when all manner of missiles tried to search him out. Technology's main contribution to soldiers continued to be the detection of targets by flash-spotting and sound-ranging, but enhanced by marginally improved instruments. Flexibility of response benefited from good radio communications between forward observers, command posts and gun positions, but chiefly through capitalization of those facilities. It was, for example, slick, simple procedures pioneered by the British which eventually made it possible for the concentrated fire of over 600 guns, deployed over a wide area, to be concentrated within 33 minutes of a request.

It had long been the practice for defensive minefields to be laid in order to hamper tank attacks, but in German hands, when mixed with anti-personnel mines and booby traps, minefields began in 1942 to impose immense problems of delay for attackers in a hurry. Mines could be laid faster than they could be detected and swept. Manual detection (by prodding) and use of electro-magnetic devices were both perilous and time-consuming — and neutralized when metallic mines were replaced by cardboard, wooden or synthetic plastic ones which were not susceptible to detectors. Faster and fairly reliable were devices such as heavy rollers, ploughs and, best of all, powered chains flailing the ground from rotating drums fitted to specially-adapted AFVs. But these proved successful only until delayed-action detonators were introduced which made the mines go off beneath the minesweeping vehicle. Mines remained predominant in defensive positions, being a lot cheaper and quicker to install than the elaborate concrete and steel fortifications which the Germans chose to construct along several thousand miles of west European coastline in their endeavours to prevent amphibious landings.

Although artillery continued to have limitations and the tank no longer reigned supreme (if indeed it ever had) guns and AFVs continued to play a dominant part in providing mobility and surprise against every kind of defensive wall — even those bordering the sea. They waxed mightily in size and power as the Germans reacted to the superior Russian T 34 and KV 1 tanks, as well as to their improved versions, and as the British and Americans sought to keep pace with the Germans. The mounting of guns above 75-mm calibre and up to 128-mm went hand-in-hand with sloped-frontal armour thicknesses of between 80 mm and 150 mm, and with engines of 600 hp, to create vehicle giants weighing over 60 tons. Against these infantry quailed and looked to mines, artillery and their own tanks for protection. True, they were assisted in some degree by their own small, shoulder-held, short-range, rocket-propelled missiles such as the American bazooka which had a hollow-charge warhead capable of penetrating armour up to two or three times the diameter of the warhead cone. But these cheap, light weapons which in the hands of determined men might destroy a large and costly machine also had their limitations. Sometimes the jet failed to do any damage. Also the enemy took stern measures to deter those who set bazooka ambushes within 100 yards of tanks.

In fact, concealment and stealth, particularly by use of natural cover or earthworks, contributed as much to the survival and striking power of the fighting man in his natural environment as did machines. No one weapon or combination of weapons dominated on land — which was more than could be said at sea and in the air where there were few places to hide and where technology ruled.

Electronics At Sea

At sea hardly a move could be made without reference to electronics. If surface vessels were not spotted in harbour by aircraft, located by direction-finding of their radio transmissions, or given away by decrypts of their messages, they all (particularly if German and not so well endowed electronically as their opponents) were liable to be located at sea by radar. This became an even more likely contingency as centimetric radar, with its greater power and immunity to jamming, replaced lower wavelength sets. Of course, camouflage in harbour and radio deception measures might provide concealment, just as it was possible to jam some radar — as the Germans so successfully did in February 1942 when sailing two battleships in daylight from Brest to Germany. But jamming was conditional upon recognition of the enemy capability and the introduction of a suitable jamming device in the right place. It amounted to a disaster for the Germans when, having in March 1943 at last captured a magnetron valve, they refused to credit the British with being able to make an ASV centimetric radar. The U-boat holocaust of spring 1943 was made possible by centimetric radar in ships and aircraft — above all the latter which, in May, put down 22 out of 36 U-boats sunk, several by a combined use of radar with a powerful searchlight to illuminate and attack

Above: ***The 105-ton German Schnellboot (S-boat, sometimes known as E-boat) with its speed of 42 knots, good sea-keeping qualities and 2 × 21-in torpedo tubes was a formidable adversary. Vulnerable to fighter aircraft in daytime, it tended to attack mainly by night.***

the boat on the surface. Not only were the Germans unable to detect or jam centimetric radar, they persisted in believing the trouble came from signal insecurity (leading to some tightening of their code procedures) or infra-red, photo-electric devices in which they too were interested. Not until 1944 did Dönitz sadly report to Hitler that 'it was 10-cm radar . . . which had caused all the German losses' — indicating that the complementary major part played by Ultra decrypts was still quite unsuspected.

Just as surface radar compensated for the relative failure of sonar to find U-boats beneath the waves, so acoustic torpedoes which homed onto the noise of the prey's propellers made up for the limitations of depth charges which, even when fired in salvoes, rarely scored direct hits. Meanwhile, sonar was being improved to the point at which a submerged British

Above: ***The essence of tight formation flying. Boeing B-17 Flying Fortresses in daytime flights over Europe practised pin-point bombing of vital targets. Without escorting fighter cover, however (identified here by the contrails above), they were extremely vulnerable to enemy fighters.***

submarine sank a submerged U-boat by sonar tracking of the target and predicted firing of a straight-running torpedo. But results were always hostage to many anomalies as well as deception measures.

The Germans conceded defeat for U-boats in the summer of 1943 and began modifications to outmoded U-boats while initiating production of radically new and faster vessels with deeper-diving characteristics. But these changes would take over a year to implement. Realizing that the majority of losses occurred on the surface, they adopted, as an interim measure, the schnorkel. Invented by I Wichers, a Dutchman, the schnorkel was a breathing tube which could be raised like a mast above sea level to allow prolonged diving under diesel instead of electric power.

Most urgently needed were increases in speed to assist evasion and tactical agility. And increased speed depended upon streamlined hulls and more powerful engines in the Type XXI boats — an impossible revolution at short notice: improvement had to be restricted to increases in battery power sufficient to raise underwater speed from 10 to 17 knots, but for

bursts of one hour only. Had the Germans foreseen centimetric radar they might have started earlier the brilliant innovations of fast boats which were built within six weeks by new, mass-production methods.

The pause of 1940 and 1941 in German radar development had wide ramifications. Their longer wavelength sets were not only more easily jammed than the British, but frequently were unable to detect surface objects. Few naval craft benefited more from this than the Allied motor boats and assault craft used for raiding enemy coast lines and supplying resistance movements across the beaches. When it came to combat, the German S-boats were every bit as good as their opponents, but were at an immediate tactical disadvantage in conditions where they were blind but could be seen. Moreover, on the narrow seas and shallow waters to which these mosquito craft were committed, precise navigation was essential but almost impossible by traditional methods. A technology devised initially for aircraft proved invaluable: the Gee navigation system worked on the interaction of pulses transmitted by a master ground station and one or more slave stations, enabling a receiver to plot its exact position be it in the air or on the surface. When fleets of small craft were being built against the day when the invasion of Europe would take place, this relatively simple instrument eased the intractable problem of how to train sufficient expert navigators.

The technique of handling armadas rarely bothered the Germans, whose main aim it was to sink

War of the bombers' black boxes

To enable bombers to find and strike targets more accurately and, at the same time, gain some immunity from enemy counter measures, a number of electronic systems were introduced. In this Lancaster bomber the principal devices are shown, each either passive and vulnerable to jamming or active and vulnerable to detection. H_2S (active), GEE and OBOE (both passive) are explained in more detail. BOOZER (passive) warned the crew it was being scanned by enemy radar. IFF (active) enabled radar operators to identify an aircraft's nationality (friend or foe). MONICA (passive) gave warning of an approaching enemy fighter.

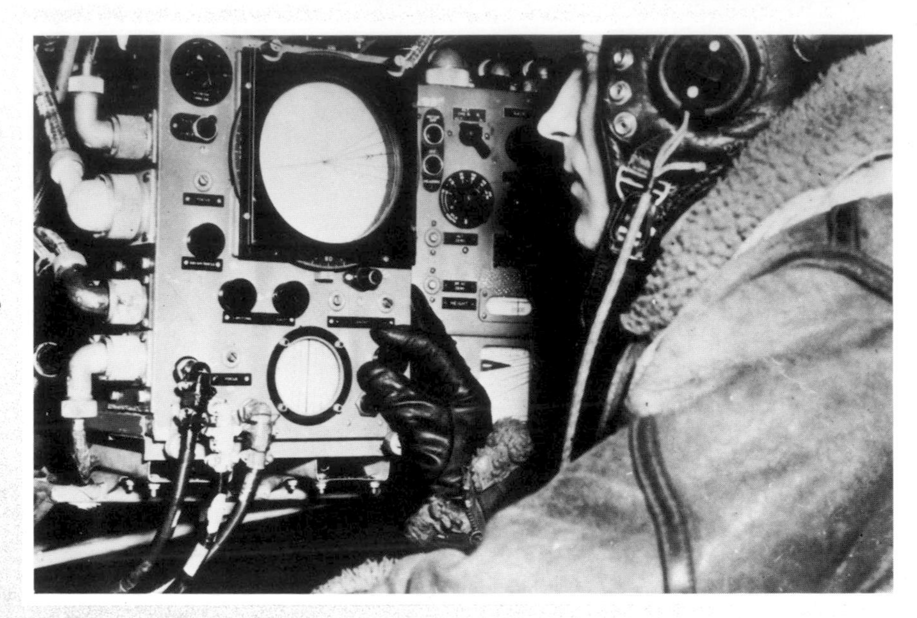

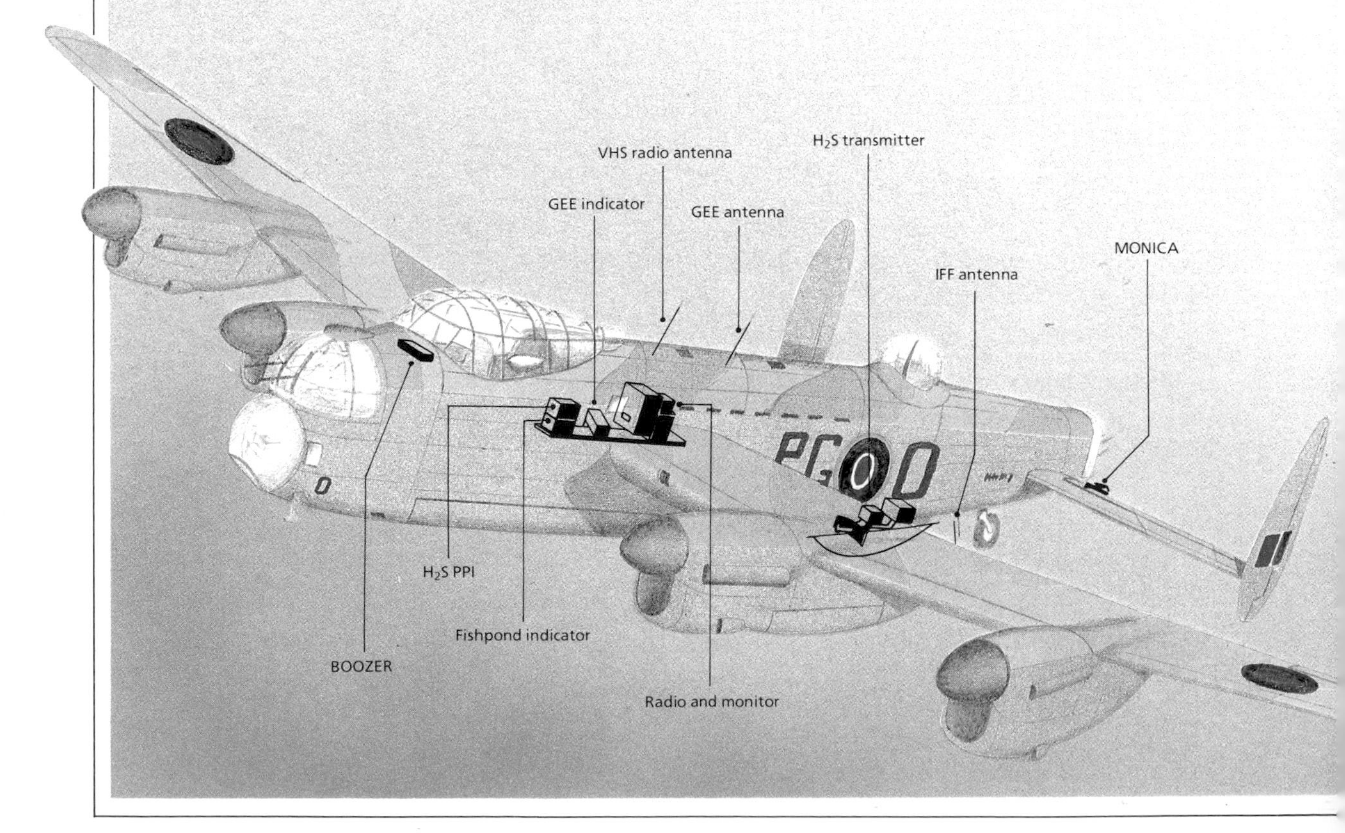

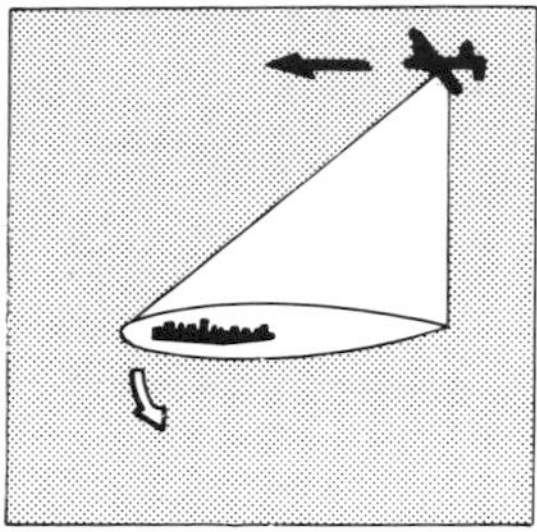

H_2S

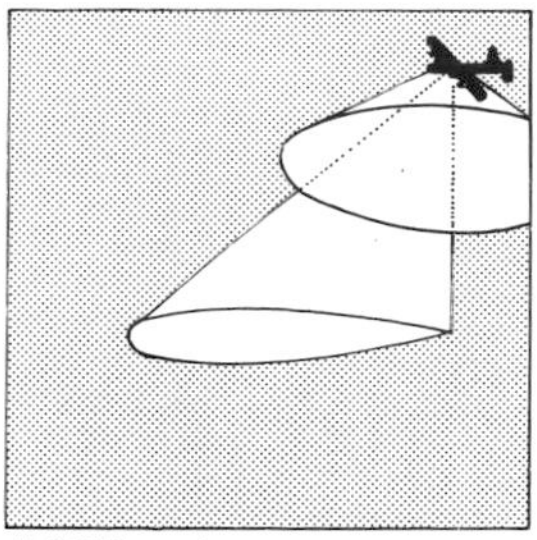

H_2S/Fishpond

H_2S was a centimetric radar set, tilted downwards, which was capable of picking up ground echoes and displaying them on a CRT – the Plan Position Indicator (PPI) – thus creating a radar map to help the crew find their objective. Because it was active, enemy aircraft were able to home in. FISHPOND was simply a more advanced, directly-pointing type of H_2S. But its effectiveness over large conurbations was limited. Over the ocean a similar device called ASV Mk III was invaluable in locating surfaced U-boats.

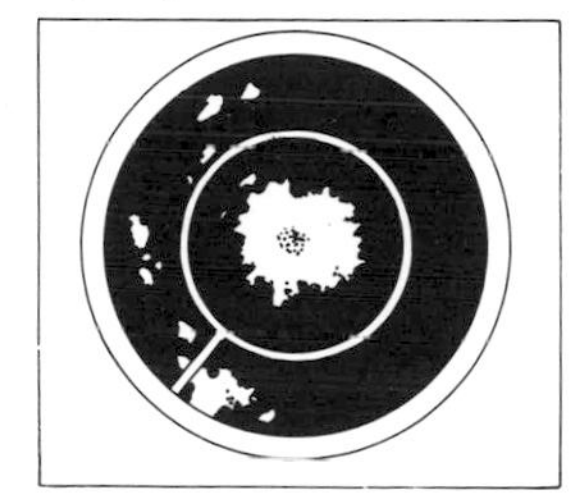

H_2S PPI map display

OBOE was a blind bombing system controlled by two ground stations, one of which (CAT) transmitted a beam along which the aircraft could fly; the other signal (MOUSE) told it when it was over the bomb dropping position. Very accurate, OBOE was in the centimetric frequency range and therefore difficult to jam.

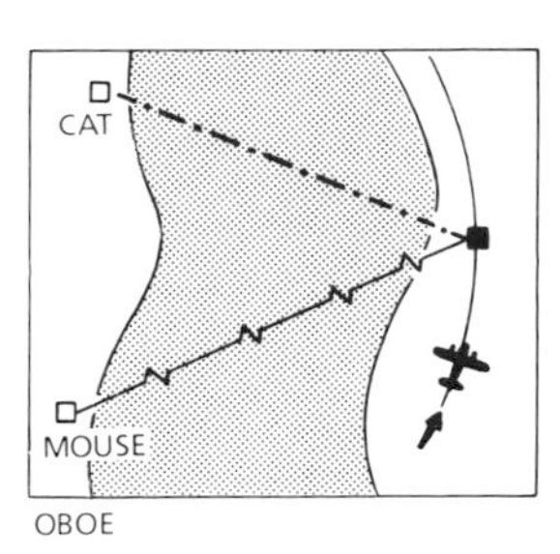

OBOE

GEE was a navigation system in which three master stations – A,B and C – radiated signals which, when picked up by the receiver in the aircraft, made it possible to fix positions with considerable accuracy, although not enough to allow accurate bombing. The active ground transmitters, were, of course, subject to jamming.

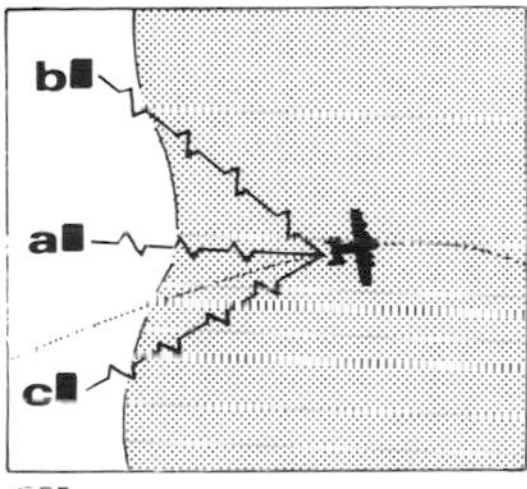

GEE

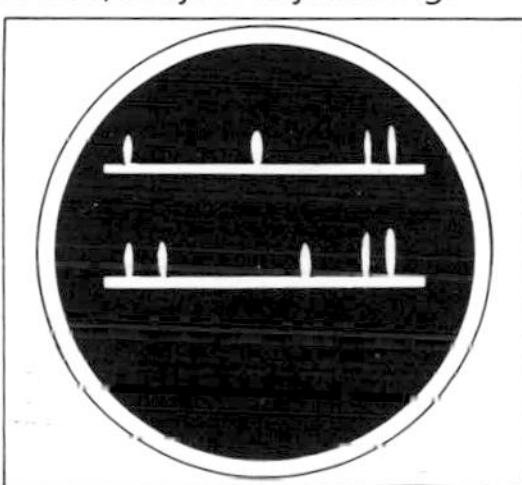

GEE indicator

Above: Counter measures adopted by the Germans in the electronic air war included the mounting of NAXOS devices, enabling crews to home on to British radiations, and various types of air-to-air radar. Shown here is a Lichtenstein SN2 antenna with NAXOS mounted in Me 110 fighters.

Left: Upward-pointing guns called Schrage Musik enabled radar-guided German fighters to attack bombers by surprise from their blind spot below.

ships by any means available, but preferably airborne to save imperilling their surviving few vessels. One idea was the *Fat* torpedo, launched by air among enemy flotillas to run in loops of decreasing circles until it hit something or ran out of fuel. Far more effective was *Gnat*, an acoustic torpedo fired from U-boats, but which could be lured astray by a noisy underwater device that, as a bonus to the enemy, scared stiff some U-boat crews.

Other new weapons were the radio-guided rocket missile and gliding bomb which, in August 1943, began to sink British ships in the Bay of Biscay and in September scored a notable first by sinking the Italian battleship *Roma* as she was on her way to honour the Italian terms of surrender to the Allies. These events should not have surprised the Allies as they did, since the British had experimented with radio-controlled aircraft before the war and the Americans were currently engaged in a project of that kind. Yet deadly as the rocket-powered He 293 and the gliding SD1400 bombs were when they struck, their Achilles heel proved to be the guidance systems which, once the frequencies had been discovered from captured missiles, were easily jammed and deflected.

By the middle of 1943 it hardly mattered what new weapons the Axis might have in action a year or more hence. In July, for the first time, Allied production of merchant ships exceeded losses as mass-production yards in the USA poured out ships and craft of all kinds at a stupendous rate. The main problem facing the Allies was simply to be ready with counter-measures conceived, if possible (and as was several times the case) before or immediately after an innovation's advent. To fail might expose men to severe losses affecting morale — as German morale was undermined by failure to cope with new Allied technology.

Give And Take In The Air

Pursuing their pre-war policies of attempting to subjugate an enemy through bombing, the British and Americans were by 1942 in possession of a rapidly-expanding fleet of several makes of four-engine bombers capable of carrying, in the case of the British Lancaster B1 for example, 12 000 lb a distance of 1730 miles, and in the case of the American Boeing B 17F, 6000 lb for 1300 miles. The major performance difference between the respective types lay in their ceilings of 24 500 ft and 37 500 ft, and their operating mode. The British intended mainly to operate in a continuous stream under cover of darkness; the Americans to fight their way through by day in packed formations protected by batteries of heavy machine-guns escorted part of the way by short-range fighters. Each method posed the German defenders with conflicting defensive problems, while their attackers faced the perennial difficulties of finding and hitting targets. Remarkable in technology as the contending aircraft and anti-aircraft guns were; well as their crews might be trained; skilful as the directors of operations would become in deploying several hundred machines at once over strategically important industrial targets, centres of communication and population, success in attack and defence depended in the final analysis upon a few electronic devices. Without them the bombers would rarely hit their targets; the fighters could not find the bombers; and the bombers could not counter a growing array of defensive measures.

It is possible to outline only a few complex moves here, starting with the night in March 1942 when 80 Gee-fitted bombers started fires in the Ruhr, to which another 270 bombers homed with devastating effect. This was more impressive than all previous raids, for a study of air photographs, linked to operational analysis, had indicated that only 30% of night-flying crews were dropping their bombs within five miles of the intended target. It was all the more successful because the Germans, who used Lorenz beams, allowed themselves to be deluded by subsidiary Lorenz transmissions. This shielded Gee from jamming for nearly a year, by which time mitigating measures (such as changing the Gee frequency) had been prepared.

Complementary to Gee was Oboe, a bomb-aiming radio aid which emitted a tone to the aircraft and instructed it exactly when to release its bombs to within 20 yards of the target. Fitted into another triumph of technology (the 400-mph, twin-engine Mosquito light bomber of wooden construction), Oboe spearheaded the technique of dropping special pyrotechnic candle 'marker bombs' as a guide for fire-raisers which, in turn, led the stream of bombers to the target. But because of screening caused by the Earth's curvature, Oboe was limited in range by the maximum altitude of the Mosquito at 37 000 ft. It was a downwards-scanning radar device called H_2S (because initially its sceptical designers thought its prospects stank) which gave accurate navigation to unlimited distances by displaying on the CRT a picture of the terrain below which could be related to a chart.

Oboe came into service in December 1942; H_2S

***Right:* The effects of carpet bombing on a Paris railway yard in 1944. RAF heavy bombers had been led to their target by the Pathfinder Force.**

PARIS M/yd
K2708.

Defence of Western Europe in 1944

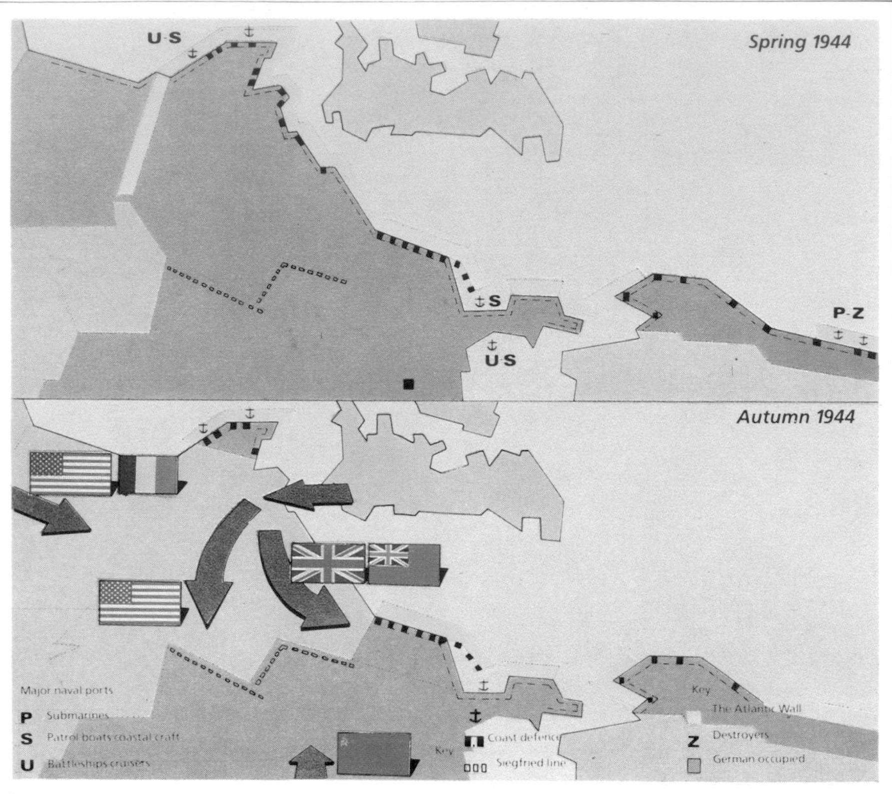

By losing the initiative in 1942, Germany was compelled to construct a costly and complex defensive system to cope with the developing enemy air and amphibious raids on its interior and periphery. While Russian land forces pressed in from the east and pressure developed throughout southern Europe, Germany's air defences were based on an intricate radar system linked to fighters and guns which, by 1943, were having to deal with heavy attacks by night and day. At the same time immensely strong, but linear, coastal fortifications were built, screened at sea by totally inadequate naval forces, and garrisoned by static infantry formations backed up by central mobile reserves, tasked to counter-attack landings.
These surface defences failed before the Allied invasion of Normandy in June 1944 and the advancing Allied armies moved inland. The German air early warning system was progressively overrun, exposing the heartlands and industry to round-the-clock bombing from aircraft able to take full advantage of their closer proximity to targets. German coastal fortifications were either assaulted or by-passed. The old Siegfried Line provided a temporary barrier of respite. But the fabric and communication systems of Germany were wrecked before the culminating invasion took place in 1945.

two months later. Along with Gee they enabled Allied bombers to find their targets unerringly. Moreover, Oboe escaped jamming because German scientists did not relate its signal to a bombing aid, and H_2S could not be jammed. The outcome was the delivery of vast tonnages of high explosive in bombs which, in due course, weighed as much as 10 tons. Their purpose was to pulverize factories, penetrate 15-ft-thick concrete U-boat shelters and sink pin-point targets, such as the 42 000-ton battleship *Tirpitz*, which was achieved with two 12 000-lb bomb hits and several near-misses. In addition, fire-bombs could engulf a city in flames to create a new phenomenon, the fire-storm, caused by air sucked through burning buildings with blast-furnace force. But just as this vast new technology of electronics, aerodynamics and metallurgy was invented for massive forces of destruction, defences were strengthened by parallel efforts every bit as impressive.

Improved ground radar sought to establish the position of the attackers prior to engagement by radar-directed guns and interception by radar-equipped fighters. Commanders and controllers connived with scientists to devise better techniques with existing equipment while calling for still more sophisticated devices. The battle swayed to and fro as one side or the other obtained some transitory advantage. For example, the employment of 'Window' — clouds of metallic chaff cut to a precise length to give false echoes on German radar — created chaos among the German air defence systems when first used on the night of 24th July 1942 during a major attack on Hamburg. Only 12 out of 741 bombers were lost. But the Germans had successes too. When 367 US Air Force B 17s attacked the Messerschmitt factory at Regensburg and the ball-bearing factory at Schweinfurt in daylight on 17th August 1943, they lost 59 aircraft and had 55 bombers damaged beyond repair as the result of superbly-directed fighter attacks. And that same night when 597 RAF bombers attacked the rocket weapons experimental establishment at Peenemünde, the night fighters, despite interference from Window, brought down or severely damaged another 72 machines. In terms of technical warfare, however, Peenemünde had a very special significance not simply as the first instance in which a Master Bomber aircraft directed the bombers by markers and radio instructions against pin-point objects within the target area; nor because of specific aiming, with some success, against the living quarters of the scientists and technologists; but chiefly because this was the first shot in the anti-rocket missile war.

Devastating attacks on cities, such as those on Hamburg in the summer of 1943 and on Berlin a few months later, seemed to Speer to have German morale reeling. Speer feared too that persistent attacks on key industrial targets might cause economic collapse. But invariably the Allies, from lack of intelligence or failure to hit the target or from severe losses, called off campaigns when Speer thought they were on the verge of success. On the other hand, there were periods in which the German defences inflicted terrible execution on Allied bombers, such as the end of 1943 and early 1944. The classic debacles were over Schweinfurt on 14th October, when 60 B 17s were lost and 138 severely damaged; or the night of 30th March 1944 when 294 German fighters shot down 94 RAF bombers out of 794 during an abortive attack on Nürnberg. Such reverses led to deep-penetration raids being called off until new methods could be devised. Both by night and by day the answer lay with the introduction of long-range fighter escorts fitted with drop fuel tanks, and by still more sophisticated Electronic Counter Measures (ECM) allied to deception measures and the jamming of enemy airborne intercept radar. It was struggle with no reprieve for errors and omissions. For example, the habit of British bomber crews to switch on their H_2S sets throughout the flight presented German fighter controllers with ECM-free plots until, after a year of fatal consequences, the Allies realized what was happening. Or the German failure to concentrate on jet fighters which might well have won air superiority and put an end to daylight bombing.

Persistent heavy losses weighed heavily against air forces. There had to come a point at which replacement of machines and crew became impossible or morale was severely shaken. Had punitive losses been inflicted upon the 10 000 or more aircraft committed to support of the invasion of Europe in June 1944, airborne troops might not have survived; and air support for the rest of the amphibious force might have been so impaired that the Allied cause would have suffered even stiffer resistance from the Germans than was actually the case. As it was, Allied air superiority gave their surface forces a free hand to bear down overwhelmingly upon the Germans at any time.

Codes, Colossus And Computers

It simply was not enough that a few German aircraft were technically ahead of the mass of Allied machines; that one or two submarines could out-manoeuvre whole flotillas; or that a handful of German tanks could out-shoot serried ranks of Russian, British and Americans AFVs. In 1944, sheer weight of numbers, machines and firepower was about to impose a closure upon a nation with technology and prowess equal, and in sectors superior, to that of its enemies. But such a classic theme, noted before in this book, was in the Axis case distorted by an almost unique variation. Never before had one belligerent nation been so well supplied with information about another's plans, internal state and weapons technology. For, quite apart from the traditional gleaning of intelligence from espionage and visual surveillance, the Allies were never prevented from reading a myriad low-grade radio communication networks plus no less than 72 high-grade encyphered German nets, in addition to diplomatic messages between the Japanese staff in Berlin with Tokyo. It is likely that during the Crimean war only a few hundred people were engaged in the gathering and synthesis of intelligence and, in the First World War, they were numbered in thousands. The staff so employed in the Second World War amounted to tens and tens of thousands, served and stimulated by a galaxy of complex machines. And of those, Colossus was mightiest of all with a tiny brain of its own.

Colossus was built in response to the emergence in 1941 of regular German transmissions of encoded, non-morse teleprinter messages such as had been experimented with by Britain, America and Germany for ten years or more. The *Geheimschreiber* (known to the Allies as 'Fish') had the facility to encypher and decypher messages and transmit them with the aid of perforated tapes at the rate of 25 letters ('marks') per second. It also incorporated more security safeguards than Enigma and was beyond the capacity of a Bombe to decypher rapidly.

In May 1943, after two Fish machines had been captured in North Africa, a semi-electronic machine called Robinson (after Heath Robinson because it was such a lash-up) and incorporating fewer than 100 valves, was tried out and found promising. Thereupon a far more powerful machine was commissioned, invented by T H Flowers of the British General Post Office Research Station. It contained 1500 valves in

Above: ***Colossus, the first programmable electronic digital computer, was designed and built within a few months by a British Post Office team.***

place of electro-magnetic relays, and in February 1944 proved capable of decyphering Fish messages within hours instead of the days taken by Robinson. Encouraged by such results, the GPO team was then told to produce a much more powerful version within three months; this they did by building it on site. Colossus II was to be epoch-making. Equipped with 2400 valves, binary adders, binary and decade counters and several more advanced elements, it also included a loop of friction-driven perforated tape, which was read photo-electrically and provided a limited memory. With a basic speed of 5000 tape positions per second, Colossus II was the first programmable electronic digital computer, and perfectly timed in its availability for work on 1st June 1944, within six days of the Allied invasion of Europe.

Instantly it would decrypt a mass of invaluable information passing along the vital *Geheimschreiber* link between Berlin and Paris, revealing a host of German intentions, strengths and weaknesses. One of the weaknesses — worsening German communication difficulties — lay at the root of increased *Geheimschreiber* use and an accelerating chain reaction of decay. For as wire links and vehicular despatch services were progressively destroyed within the shrinking Reich, Fish (which was originally regarded only as a complementary tool) had to carry more traffic, thereby yielding a still more bountiful harvest of prime intelligence to Ultra via Colossus. Supplied so well, the Allies could close in easier and faster, eliminating Germany's outer defences and, at last, exposing the heartlands to an air bombardment which was irresistible and utterly destructive of surface communications and production.

Chapter 6

Deterrence and Constraint

1945-1986

When Allied intelligence first got wind of the threat of attack by German rocket-propelled missiles, their suspicions were dismissed with incredulity by scientists who declined to believe the Germans capable of making such weapons. And if the German chemical industry had not found out how to concentrate, handle and store hydrogen peroxide (discovered in 1818 by the Frenchman Louis Thénard) and liquid oxygen, that would have been the case. These unstable fuels generated performances far in excess of any achieved by solids (such as cordite) with which the British and Americans were familiar. They made possible the Hs 293 guided bomb, the Messerschmitt Me 163 rocket fighter, the FZG 76 (V 1) pilotless flying bomb and the A-4 (V 2) long range rocket — unreliable as all were bound to be in the early stages of development.

The Hs 293, as already noted, was susceptible to radio jamming and difficult to control by inexperienced air crew. In spite of its 40 000-ft ceiling and speed of 600 mph made possible by a Walter hydrogen peroxide motor of 3750 lb thrust, the Me 163 was refractory. It was aerodynamically remarkable with a delta-shaped wing (an aid to overcoming compressibility at speeds above those of sound), but its duration of only eight minutes in the air posed serious tactical limitations, while vagaries in handling which caused the deaths of even well-trained pilots led to its withdrawal from service.

V 1 possessed the virtues of doing without radio control or pilot. It was guided by a gyroscopic automatic pilot monitored by magnetic compass and its dive on the target at a measured distance was determined by the revolutions of a small propeller. Driven by a pulse jet (patented in 1907 by the Frenchman Victor de Karavodine and proposed in 1919 as the power plant for a pilotless aircraft by a compatriot, René Lorin), V 1 was relatively cheap, had a speed of 350 mph, a range of 160 miles, flew at 2000 — 3000 ft and was mobile to the extent that it could be launched from simple catapult ramps or from aircraft. Ironically, the original elaborate concrete and steel launching structures were wrecked by bombing but thereafter did good service in decoying further Allied attacks and diverting attention from the secondary sites. Built in underground factories, V 1 was fairly safe from destruction until launched. It then posed awkward problems to a defence, stretching intercepting fighters to their maximum speed and

Above: The 600-mph Messerschmitt Me 163 Komet. The operational debut of this rocket-propelled, delta-winged fighter in 1944 was marred and finally cancelled owing to its tendency to catch fire and incinerate pilots. The brainchild of Alexander Lippisch, it was also handicapped by its short duration of 8 minutes powered flight.

flying just a little too high for light guns and a little too low for heavier pieces. Of 10 492 V 1s sent against England between 12th June 1944 and 29th March 1945, only about 7500 tested the defences, the rest defaulting. Of some 3531 which got through, the majority did so only in the early days, before the defenders had their measure, and very few hit military targets. Of nearly 4000 brought down, 231 hit balloon cables. Fighters and guns claimed an almost equal share of the rest, with only a handful falling to the British Meteor I twin-engine jet which was hastened into service in August 1944 just in time for the end of the battle. It was the guns which benefited chiefly from new technology by the employment of a proximity fuse which detonated within optimum distance of the target. Conceived in Britain in April 1940 as a doppler radar device, it was developed primarily against low-flying aircraft but also to render redundant the ranging and fuse setting required to achieve air bursts by field artillery. Mass-produced in the USA, its introduction in August 1944 proved the best antidote to the V 1, besides effecting vast savings in ammunition. Guns had previously been fired more in hope than with any certainty of hitting an elusive target — until the day when out of 97 V 1s sent against London, only four got through. Two had been downed by balloons, 23 by fighters and 65 by the guns. Nevertheless, a re-deployment of the guns had contributed to the success; it was considered desirable that they should fire out to sea rather than from inland sites owing to a reasonable fear that the new fuse might not self-destruct and so cause widespread harm to the people it was intended to protect.

V 2 enjoyed the same characteristics of mobility and freedom from attack and jamming as its partner and rival V 1, and could deliver about the same weight of explosive but at the longer range of 200 to 220 miles. Unlike V 1, however, it was impossible to intercept after launch, ascending as it did to 50 or 60 miles and descending with a velocity of 2200 to 2500 mph, three to four times the speed of sound as the double 'boom' which accompanied its arrival indicated. It was also much more difficult and twenty times as expensive to produce than V 1, as a launching figure of only 3195 against targets in Britain and Europe between September 1944 and March 1945 shows. Results did not fully justify the German effort. True, the fears instilled in the minds of bemused Allied leaders and the diversion of 117 000 tons of bombs against V weapon sites were a bonus. But less than one person killed per V 1 fired and slightly under two per V 2 could hardly be rated a holocaust such as Hitler yearned for. Indeed, the rocket campaign was a supreme example of premature committal of an unproven system at the behest of a leader whose political instincts favoured the dramatic and the bizarre and upon whom unscrupulous underlings and scientists were able to foist numerous dubious projects.

The prospect of an 'intermittent drizzle' of V 1s and V 2s cracking civilian morale in Britain or on the Continent was infinitesimally less than the chance of a 'drenching' by bombs achieving the same object against the Germans, who, in determined apathy, stood up to their ordeal. More intensive production of jet fighters, instead of rocket Me 163s, might have considerably mitigated that bombing. That the rocket weapons were highly significant was another matter. Both V 1s, launched from out-moded bombers which could not penetrate (without heavy loss) the radar-controlled defences of Britain, and Hs 293s, which were successfully guided against Russian bridges over the River Oder, were ancestors of the future stand-off bomb. The designer of the V 2, Wernher von Braun, invited trouble (and went to gaol) for protesting that his rockets had been designed for peaceful applications and the exploration of space, not as missiles. He might also have criticized the Reich for failing to concentrate its research and development on essentials and for indulging in fantasies which could never be related to a war in its final stages. Despite the fecundity of ideas, the Germans had no one weapon capable of winning the war in its own right.

A 'Pandora's Box': The German Inheritance

When the victorious Allied armies moved into Germany in 1945 it was to discover a stupendous collection of weapons, some of which paralleled Allied inventions, but several of which were originals. There was, for example, the Walter closed combustion U-boat engine, using hydrogen peroxide instead of air,

Guided Missiles

Left: Wernher von Braun (in plain clothes), genius of German rocket development.
Above left: The Wasserfall surface-to-air visual, radio-controlled missile designed by von Braun as a 'beam rider'.
Above: The Ruhrstahl X-4 air-to-air missile designed by Dr Max Kramer. Visually guided and controlled by signals transmitted across two fine steel wires paid out from bobbins in the wing tips, this missile had the advantage of being proof against electronic counter measures. It became the forebear of all future wire-guided anti-tank missiles and torpedoes.
Centre: Schmetterling surface-to-air missile.

The First Long-Range Semi-Guided Weapons: V 1 And V 2

Germany's attempt to off-set the failure of her long-range bomber programme with rocket weapons found its main expression in the V 1

Warhead

Automatic pilot

A pitch gyro (approx 2/3 scale)

Pick off points to control servos at tail

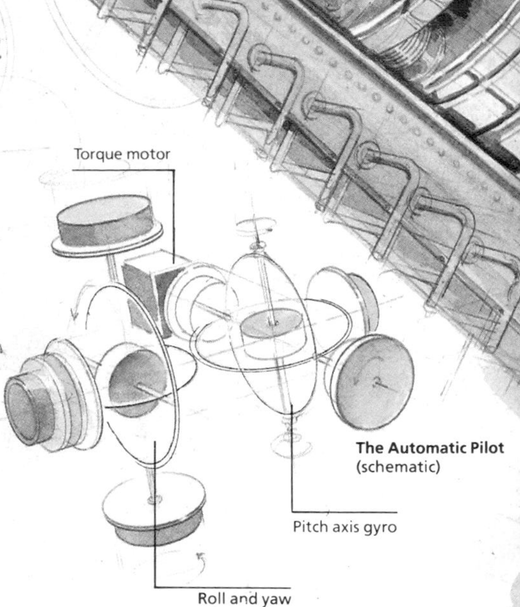

The Automatic Pilot (schematic)

V 2's autopilot gyroscopic guidance system was a wonder of technology, the electric torque motor to drive the gyros being the smallest of its kind yet made. Inaccuracies in launch and shut-off of motor were the principal causes of failure to close on the target. These arose from frequency variations in the current supply from the motor, poor pick-up from the pitch gyros to the servo motor which controlled the graphite vanes and aerodynamic tabs, and errors in estimates of burning time.

Launching vehicle raising V 2 to vertical position in readiness for launching from pad

V 1 was an inaccurate weapon which operated without dependence on radio guidance and was therefore immune to electronic interference. It was cheap but vulnerable in flight to fighter aircraft and AA guns, particularly the latter firing radio proximity fused shells.
V 2 was much more expensive than V 1, but shared similar disadvantages, such as inaccuracy, while also being immune to electronic counter measures. It was far less susceptible to attack, however, and only vulnerable when in transit from factory to launching pad.

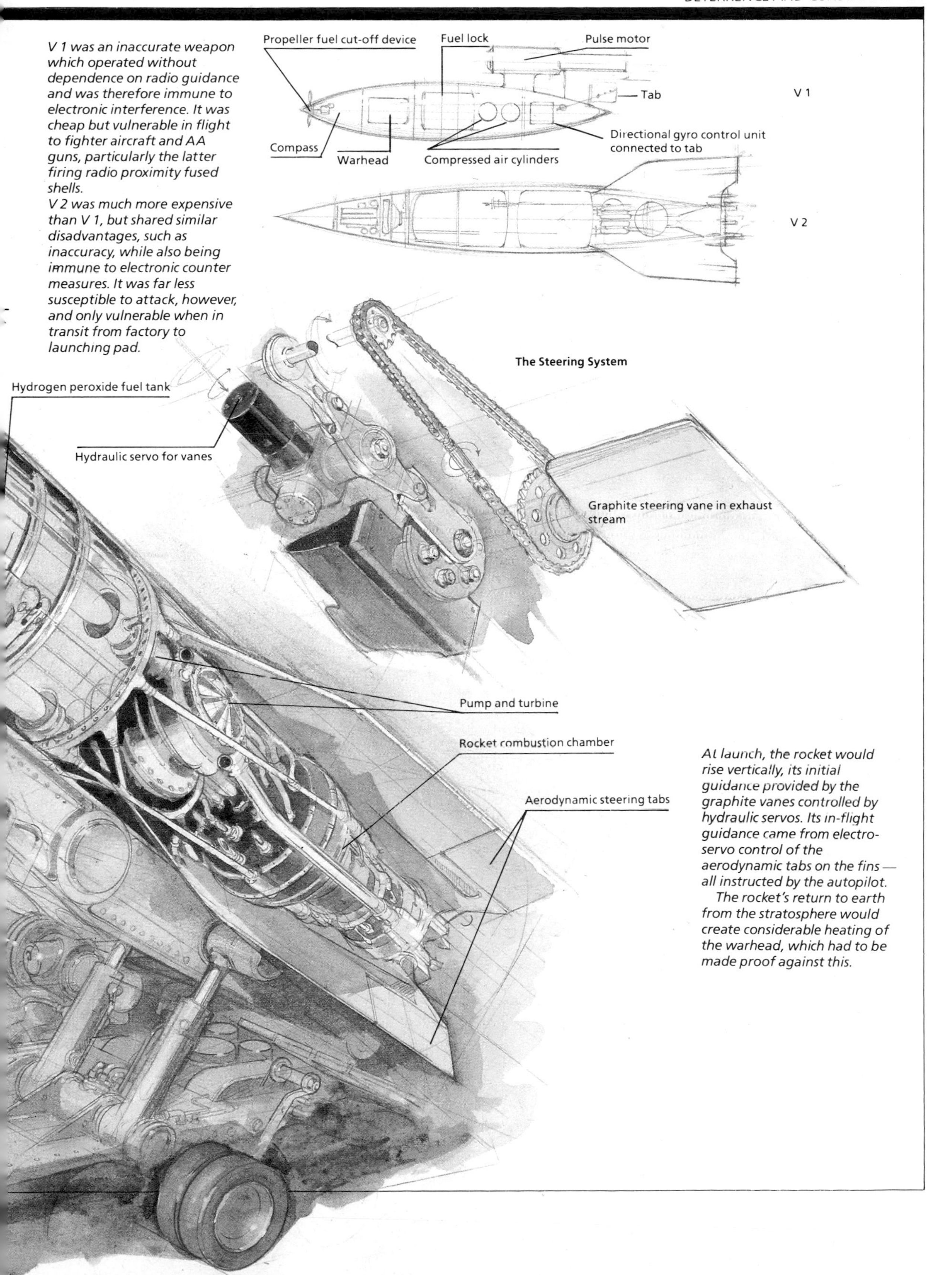

At launch, the rocket would rise vertically, its initial guidance provided by the graphite vanes controlled by hydraulic servos. Its in-flight guidance came from electro-servo control of the aerodynamic tabs on the fins — all instructed by the autopilot.
The rocket's return to earth from the stratosphere would create considerable heating of the warhead, which had to be made proof against this.

which gave 25 knots submerged; and information concerning the firing of a solid-fuel rocket from a submerged U-boat. There was a tank with a 150-mm gun, armour 240-mm thick and weighing 190 tons; and several examples of night vision devices which enabled an observer, a gun layer or a sniper to illuminate an object with infra-red radiations and see the target through a special receiver sight, without being seen in return unless the enemy happened to have his own infra-red detector. Quite apart from novel metallurgical discoveries intended to make better use of scarce materials for alloys while maintaining standards for armour and armour-piercing shot, there were 'squeeze' guns to increase shot velocity; high-velocity, fin-stabilized projectiles fired from smooth-bore guns as an improvement on the 'spun' kind from a rifled barrel; and lightweight, recoilless guns which directed their discharge backwards and so dispensed with the need for heavy recoil mechanisms.

These were products of sustained investigative vision and enthusiasm by the Wa Prüf group, which had been active since 1912, plus the air force's own research department. The sheer imagination, let alone volume, of revolutionary air weapons was staggering. There were aircraft in various stages of development or mock-up which could overcome compressibility beyond the speed of sound, including delta-winged machines such as the four-engine Junkers EF 130 with its speed of 625 mph and range of 3700 miles; a fighter with swivelling engines for vertical take off and landing; a helicopter; and a variable geometry fighter with swing wings. And to improve crew survival from machines which flew so fast that escape by ordinary parachute was almost impossible, there was a ribbon-canopy parachute with aneroid-controlled release set for 13 000 ft and an explosive ejector seat which saved some 50 fighter pilots in the closing months of the war.

Then there were rocket missiles, of which those designed to destroy aircraft above 30 000 ft were far more important than V 1s and V 2s, it being deemed no longer feasible, with reasonable economy, to produce a gun with the desired accuracy at the higher altitudes. Surface-to-air missiles were essential. They included von Braun's beam-guided Wasserfall with its speed of

The Casualty Bill And Medical Science

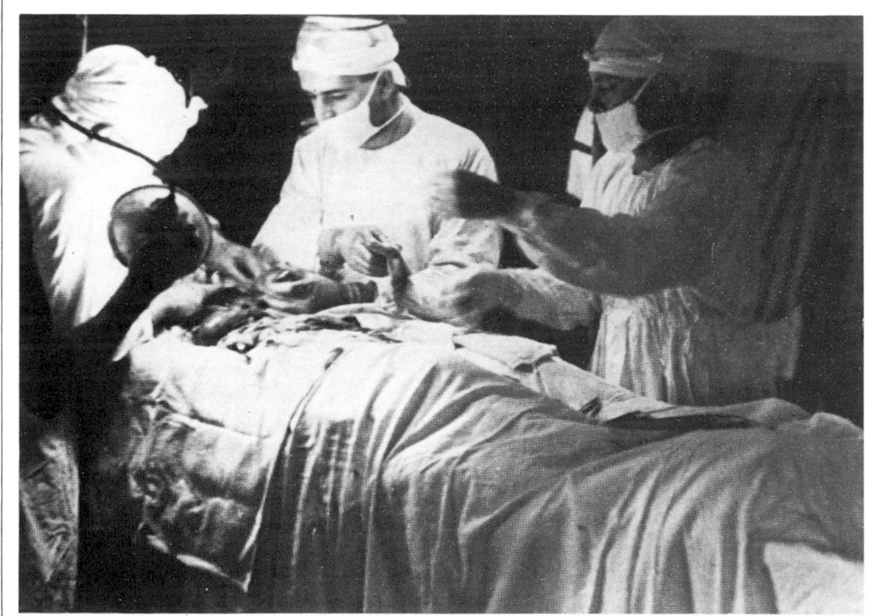

Left: A field surgical team operates on a wounded man in a tented hospital. Advances in technique and the invention of antibiotics made possible a vast saving in casualties compared with the First World War.

Above: Alexander Fleming, whose discovery and development of penicillin enormously assisted surgeons in saving lives by controlling infection.

Medical science advanced to a far greater extent in the Second than in the First World War. The death toll from wounds would have been far higher had not Oswald Robertson demonstrated the practicability of storing blood (around 1917) and had not the storage of blood by separation of its plasma from the red corpuscles been made possible in the 1940s. War, as usual, acted as the catalyst of progress. It stimulated, for example, the practice of healing and, by skin grafting, repairing damage from the burns which were so much more terrible in mechanized warfare. But the greatest life-savers of all were antibiotics for control of infection. Resulting from the discovery of penicillin by Alexander Fleming prior to 1939, these were made in quantity in America after 1941 once the method for so doing had been devised in Britain.

1700 mph and range of 16.5 miles, and Dr Max Kramer's Ruhrstahl X-4, a 520-mph air-to-air missile with a range of 3.4 miles, which was controlled to its target from an aircraft through two fine wires paid out in flight from a bobbin on the missile's wing tips. X-4 had a distinct advantage over Wasserfall; it could not be jammed and was line-of-sight, making it ideal for hitting targets which were vulnerable only to direct hits. It was thus the immediate progenitor of the X-7 anti-tank guided missile, which never progressed beyond the planning stage. The technique of scoring hits, of course, lay at the heart of the matter and demanded ever more automatic systems as ranges of engagement lengthened. 'Homing' devices of four basic modes were investigated. The first was *radar*, an 'active' kind reacting to reflections from an illuminated target while a 'passive' kind reacted to enemy radar radiations. *Infra-red* devices responded to heat radiations from the target (such as exhaust systems). *Radio* homed onto radio transmissions, and *television cameras* in the missile nose enabled the controller to watch the target until the moment of impact. Despite difficulties, the Germans had proved the feasibility of all these systems without putting any into service, and associated them with acoustic and infra-red proximity fuses — a potentially deadly combination.

Bizarre and fruitless attempts had been tried to make high-frequency, long-range vibrators designed to cause physiological effects on people. Sinister and in production were the liquid nerve gases Tabun, Sarin and Soman which inhibited the body enzymes and caused over-stimulation of the parasympathetic nerve system leading, within 15 minutes of contact with the skin or by inhalation, to sickness, convulsions and death. Because they were relatively easy to dispense by spray, were persistent and effective even in low concentrations, as well as unknown to the Allies, these gases posed a dreadful threat which, had they been used by surprise in large quantities, could have induced devastating effects on unprotected people. Fortunately, production was delayed and Germany was denied a weapon which might have been decisive in its own right. For in the search for a nuclear explosive, the one material which did have the potential to end the war, her research was still at the exploratory 'pure' stage and, by Speer's estimate, ten years behind that of her enemies.

The revelation of Germany's 'Pandora's Box' was both startling and corruptive beyond all constraint. As the lid came off, all things evil within the Box, as well as the scientists and technologists who had created them, fell into the hands of the Russians, Americans, British and French to be dispersed to all corners of the world, examined at leisure and, over the next two decades, developed into weapons of unheard-of power.

The Destruction Of Japan

The Japanese possessed nothing approaching Germany's technical treasure-house and, once their initial strength was dissipated, were therefore totally outclassed by the Allies' — above all by the Americans' — preponderance of material and technology. The reduction of the Japanese Empire is shown overleaf. Here are highlighted the principal features of the technical process which brought it about, in face of the skill and outstanding bravery of the Japanese fighting men.

Compromise of Japan's machine codes was, of course, fundamental to defeat. Nothing secret was secure. All major plans were read by the enemy. Radar was denied her as a German *Würzburg* sent by U-boat (one of several remarkable long, blockade-running voyages) arrived too late for copying. In parts of the world where malaria was endemic, the absence of suppressive drugs such as mepacrine (in Allied use) proved fatal to thousands. For example, when the last Japanese land offensive collapsed in Burma in 1944, the pursuing and immunized British and Indian troops found quite as many Japanese dead from malaria and other diseases as from wounds. Blockade had a crippling effect on production of existing weapons, let alone those of a next generation. The raw materials were interdicted by air but, most of all, by submarines which sank 1153 merchant ships, amounting to 57% of the 8 500 000 tons lost by the Japanese throughout the war — and this despite US torpedoes of which, for part of the war, 37% were defective.

The prosecution of war by the USA symbolized a nation which tended to put directness of approach before subtly; and, whenever possible, overwhelming application of material in preference to sacrifice of her men's lives. It was experienced most dauntingly by the Japanese in Burma, where their tactics of surrounding an opponent were baffled when hundreds of planes parachuted in ample supplies to American troops besieged in jungle terrain: and in June 1944, in what became known as the 'Marianas Turkey Shoot', when, for the loss of only 14 fighters and one bomber out of about 1000, the Americans sank three aircraft carriers and shot down 298 out of 406 Japanese aircraft in what was the fifth and last great carrier action of the war. It occurred yet again in October 1944, during the last great naval battle of the war off the Philippines, when a

American Sea, Land And Air Power In The Pacific

Above right: *American amphibious power on a beach at Okinawa in April 1945, as LSTs, LCTs and smaller craft unload stores onto a beach-head. In the background, transports wait to unload, while to the fore, dumps of supplies are picked up by trucks, with bulldozers improving the beach approaches for subsequent shipments. Compare this with the illustration of the Crimea on page 16.*
Top: *The initial assault across a Pacific beach. US Marines drag an anti-tank gun from a landing craft assault.*
Above: *The American P 47 Thunderbolt fighter. Its presence made logistic scenes like those above safer to accomplish.*
Right: *American infantry at the front in communication by walkie-talkie, the troops' presence made possible by logistics and air power.*
Centre: *The Boeing B29* **Enola Gay,** *a pressurized Super Fortress that dropped the first atom bomb 'Big Boy' on August 6th 1945. Previously such aircraft had been used on devastating conventional raids on Tokyo, killing 83 000 people on one night in March 1945.*

REDUCTION OF THE JAPANESE EMPIRE

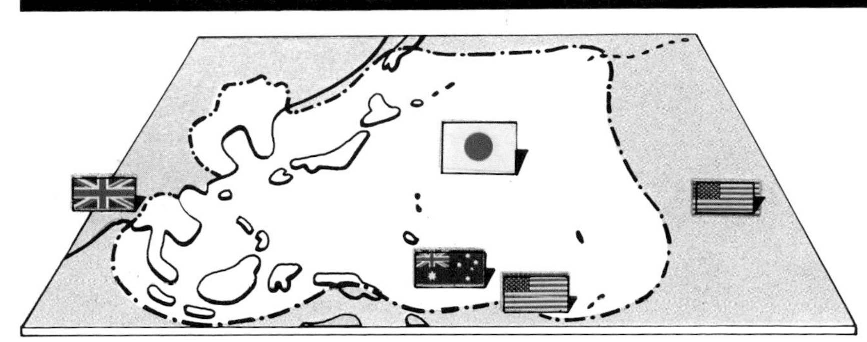

July 1942 In the aftermath of the battles of the Coral Sea and Midway, with the destruction of the main Japanese aircraft carrier force, the Americans launched a risky, under-strength invasion of Guadalcanal in the Solomon Islands. The subsequent fighting by sea, land and air became attritional until January 1943, by which time the Japanese forces had been fatally worn down. Meanwhile the Americans used their vast reinforcements to build up bases in Australia and the islands for an immediate expansion of their offensive.

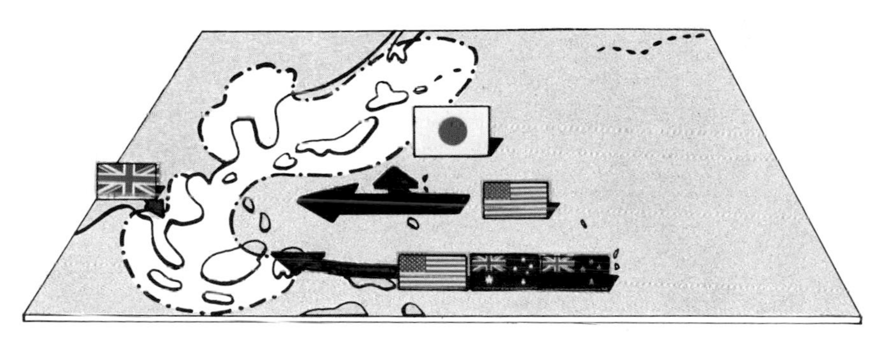

1943-44 Three Allied axes of attack broke the Japanese defensive ring intended to prevent assault upon the homeland. First came the advance from Guadalcanal to the Philippines via the Dutch East Indies. Then followed island-hopping strokes, based on Hawaii in the direction of the Marianas, from whence long-range air attacks could be launched against Japan. Finally, in 1944, after a Japanese attempt to invade India, came the advance through Burma aimed at Rangoon and Malaya.

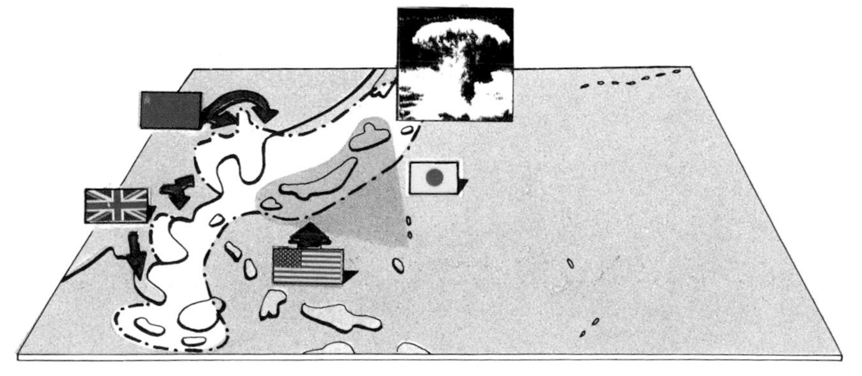

August 1945 By now the Allies had control of the south-west Pacific, large areas of the Dutch East Indies, the Philippines and the island of Okinawa on Japan's doorstep. On the mainland, Burma had been retaken, the invasion of Malaya was imminent, the Chinese were advancing towards Shanghai and the Russians were about to invade Manchuria. At any moment atom bombs would be dropped on Hiroshima and Nagasaki, to convince the Japanese that resistance to invasion was pointless.

balanced force of 218 fighting ships overwhelmed a mere 64 Japanese ships which entered the battle largely deprived of intelligence about their opponent. The battle of Leyte Gulf was indeed an Armageddon, with battleships fighting their last major action in a struggle dominated by underwater craft and aeroplanes. It included, too, a notable substitution of human courage in lieu of technical inadequacy — from the dedicated suicide pilots who flew their bomb-laden Kamikaze aircraft into enemy ships. Yet this sacrifice had a certain logic in perspective to reasonable survival chances. Any mission which could evade a fleet's early-warning radar screen, slip past combat air patrols, avoid being shot down by intense radar-directed gunfire with proximity fuses, hit its target and *then* escape, had a touch of the miraculous. Air attacks against a fleet were suicide missions under almost any circumstances. But the Kamikaze attacks did teach basic lessons. By accounting for 80% of ship casualties from all kinds of enemy action and by sinking 34 vessels (including three aircraft carriers) and damaging 288 from 1900 missions, they exposed the vulnerability of ships to guided missiles and a need for all-round armour as demonstrated by the armoured flight decks of British aircraft carriers compared to the wooden American kind. While the latter proved extremely vulnerable, the former could be struck, cushion the blow and soon be back in action. That made a lot of difference to pilots aloft who were running short of fuel and wondering where to land.

Maybe it was in logistics that the American contribution to warfare against the Japanese was most pronounced — and certainly vital. The search for means to carry out amphibious operations which had been forced upon the British in 1940 had led to such inventions as landing ships and craft which could operate on beaches (adjusting their draught by flooding internal tanks to suit gradients); wheeled and tracked amphibians; floating piers; and, mightiest project of all, the artificial Mulberry harbours built to supply the invasion force in Normandy across open beaches. The ability to supply major forces with immense tonnages (ammunition, fuel and vehicles predominating) without the need to capture ports was tactically as well as strategically crucial. Because German fortifications were centred upon ports, in the misplaced assumption that they were essential, the main plank of their defences was eliminated by Mulberry. And because the Japanese could not conceive a 'fleet train', capable of replenishing an armada at sea and thus enabling extremely long-range landing operations to be made, they were faced with the hopeless task of attempting to defend everywhere instead of simply those key islands within shorter range of Allied main bases. By applying technology to this end, the Americans added a new dimension to mobility. The setting ashore of armies was aimed mainly at seizing key localities, such as ports and airfields, in a stepping-stone advance under Admiral Chester Nimitz towards the enemy homeland, bypassing the rest of the defended islands. Heavy though the fighting was, the numbers of combat soldiers, except in Burma and China, were relatively small compared with those which had been deployed in Europe. And they were vastly fewer than those employed with engineering plant to build airfields and base facilities and to man the complex communication systems actuating the logistic machine. Even to the Japanese who clung despairingly to power, it was apparent, as the latest pressurized heavy Boeing B 29 bombers rained fire and high-explosive bombs from 30 000 ft upon their highly inflammable cities, that the game was up. Yet as final persuasion that the time to surrender had arrived and, in an ironic way, to reduce the possible casualty bill, the Americans took the opportunity to demonstrate what seemed, at last, to be the ultimate weapon.

Nuclear Science And The 'Clash Of Giants'

It was, of course, a grand irony that means to heal were intimately associated with means to kill. The chain of discoveries leading to the manufacture of the atomic bomb, which killed 70 000 Japanese on 6th August 1945 at Hiroshima, was begun in the 1890s with the discovery of radioactivity by Henri Becquerel; of radium by Pierre and Marie Curie; X-rays by Wilhelm Röntgen; and coherent knowledge of penetrative alpha and beta rays as published in 1900 by the New Zealander Ernest Rutherford. They thought only of peaceful applications of their knowledge. But three decades of research by many brilliant mathematicians and physicists, leading to a demonstration of the release of energy from nuclear reactions, was to wreck altruism — and crucially in 1938 when it was noted that a sustained chain reaction continued to 'burn'. Then the possibility of 'fission' to induce a new and awesome explosive was recognized. Then, too, its sinister threat in the hands of a ruthless individual or nation — Adolf Hitler's Germany, for example — was

*Right: **The nuclear age dawns with the bursting of the atomic bomb over the city of Hiroshima, its deadly mushroom cloud rising to 20 000 ft.***

Above: ***Nuclear devastation at Nagasaki.***

quoted, particularly since his scientists were known to be working on the subject. Yet the alienation of many scientists by Germany's and Italy's racial policies not only denied the Axis several great brains but also prompted, under American sponsorship, the formation of a brilliant group of physicists.

Beginning in 1943 under the leadership of Robert Oppenheimer, Europeans and Americans worked together to produce nuclear explosives and a weapon which seemed to present a choice between the moral evils of wholesale destruction, on the one hand, and defeat of extreme political oppression, on the other. Against the background of past experience related to seemingly irresistible weapons, there was scarcely an alternative to making the bomb if freedom of choice was to be retained by individual nations. Be that as it may, by the summer of 1944, Oppenheimer's team had made an air-deliverable device some 10 ft long and weighing about 9000 lb in which a few pounds of uranium (or plutonium) could, by shooting one sub-critical mass into another, initiate fission. Over Hiroshima on 6th August and three days later over Nagasaki, explosions equivalent to 20 000 tons of TNT created a truly blinding flash, searing heat and a devastating shock wave — and released a force absent from all previous explosives, the phenomenon of radioactivity in the form of alpha, beta and gamma rays with their insidious destructive effects on living matter.

Ostensibly the atom bombs ended the war. Japan sued for peace on 10th August and brought to an end seven years of conflict which had cost something in the region of 45 million lives, including German and Russian genocide. Paradoxically, however, the atom bomb was among the root causes of the next round of conflict which began (though some might argue it had already begun) no sooner than the terms of Japan's surrender were signed on 2nd September. For during peace talks, mutual suspicion between America and Russia had already surfaced. The American President Harry Truman agreed with Winston Churchill that Russia's recent record of intransigence linked to territorial ambitions had to be checked. And Joseph Stalin, aggrieved at resistance to Russia's historic outward urge with its strategic destabilizing effect, took steps to rectify the technical imbalance.

Cold War, Limited Wars And Technical Deterrence

It required only an outline knowledge of history and insight into human behaviour to realize in 1945 that the makings of further global conflict, possibly leading to a Third World War, were present and that, sadly, the pause between one spasm of intensive weapon development and the next would be nothing like the ten year moratorium after 1918. As the war-impoverished European powers began a retreat from Empire, the peoples of the territories they governed strove for self-rule, often by armed force and frequently in a political vacuum. Into territories infected by nationalism and instability moved the two great expansionist powers — America and Russia — each promoting its own interests for diverse motivations of over two hundred years' vintage.

The sheer incompatibility of philosophy and outlook of the two super powers guaranteed a political clash that would stimulate technical rivalry and an arms race. Secretive and expansionist by nature, inculcated with the crusading fervour of the Communist ideology and cause since 1917, and committed to massive rearmament since the 1920s (initially as insurance against Capitalism), the Russians put armament production first, followed in order by heavy industry, agriculture and consumer goods — without the need to bother about democracy or the people's feelings since the most elaborate, oppressive police organization in the world took care of all levels of society. Inspired by the right of the individual to promote his own interest within a relatively free society, America and most Western nations were compelled by democractic principles to respect the wishes of the people — and the people tended to resent being denied a good standard of living and paying high taxes for defence. The West's priorities of production were therefore less clearly defined than those of the Russians, and would vary year by year, crisis by crisis, as apparent threats waxed and waned. Throughout a succession of political and propaganda confrontations (in what came to be known as the 'Cold War'), several outbreaks of fighting under the name of 'Limited Wars' (because nuclear weapons were not used), and the ever-present threat of a nuclear global war, the struggle intensified out of mutual fear. Overriding all was the pronounced ruthlessness of Russian policy (which did not baulk at genocide) and their flaunting of immense armed forces equipped with aircraft and tanks that technically were becoming a match for those of the West. Since the West was unwilling to commit a high proportion of its population or gross national product to the armed forces, it was compelled to concentrate on improving technology in order to compensate for an imbalance in numbers. Such a policy could be industrially stimulating if kept within the bounds of economic means; or corrosive if allowed, or driven, into getting out of hand. There evolved a situation in which both sides sought to acquire a strategic and technical edge over the other, with the corollary that it was desirable, perhaps by subterfuge, to provoke the opponent into over-expenditure on wasteful or useless projects in order to create a state of exhaustion. Nearly all the landmarks of the 'Cold War' and 'Limited Wars' of the next four decades were related to these conflicting, provocative methods.

Research and development were therefore not brought to a halt at the end of 1945. Each nation which had participated in the war set about consolidating its knowledge, disposing of obsolete equipment, improving the obsolescent and investigating revolutionary ideas such as had been initiated by the Germans. To the Americans and British this was relatively easy, since they were well organized and ahead. To the Russians it was a daunting problem because they had tended to concentrate on simple, rugged designs which could be produced in mass — the Ilyushin Il-2 Shturmovik, for example, of which 36 000 were built. This had proved a fine ground attack aircraft and, when armed with 37-mm cannon, effective against German tanks. But the advent of the jet engine made obsolete all such machines, compelling Russia to seize on German technology, copy the Jumo 004B engine, mount it in a Yak 3 piston-engine airframe and have it flying by April 1946. Turning also to captured

German designs, Russia soon had in production a swept-wing Mikoyan Mig 15 fighter with a speed of 683 mph, and by legal purchase was able to acquire British and American jet engines. But Russia had not captured, and could not openly purchase, the secrets of nuclear explosives. These had to be acquired through the espionage of Alan Nunn May and Klaus Fuchs, who by giving details of their work for the Americans and British perhaps halved the time it took the Russians to produce their own atom bomb. This they did by 1949, so ending the US nuclear monopoly.

The Berlin Blockade

Most attempts by the Russians to force a political issue brought forth a Western technological response of sufficient magnitude to deter predatory ambitions. When the Russians closed land routes to Berlin in 1948 (and left the West in no further doubt about Stalin's true intent), the blockade was broken by a massive airlift which in fifteen months carried in 2 343 313 tons of supplies by 277 264 transport aircraft sorties. Quite apart from providing a splendid illustration of the reliability of modern aircraft and the intensive flight schedules then possible by day and night in all kinds of weather, by use of radar and radio control, this was a superb example of how an economic embargo could produce counter-economic difficulties. For if Stalin's intent had been to exhaust the Western Allies, as well as take possession of Berlin, all he managed instead was to strengthen their resolve, hasten the creation of the North Atlantic Treaty Organization (NATO), stimulate the growth of many great air transport companies (along with machines of greater capacity) and, ironically, induce a counter-blockade which brought the East German economy so close to ruin that the lifting of the Berlin blockade was of necessity enforced.

Airpower In The Korean War

Right: The helicopter proved useful for a variety of tasks, including reconnaissance and logistic support. It was principally valued in Korea, however, for hastening evacuation of the wounded from the front line to treatment centres. Here the wounded could be made ready for movement by road or air to the base hospital. Speed of evacuation thus saved an enormous number of lives.
Top right: The Russian Mig 15 swept-wing trans-sonic fighter, designed by Artem Mikoyan, achieved some superiority over American fighters above North Korea. It made extensive use of German and British technology to produce in 1950 a fighter to outmatch all but the US F8-86 sabre swept-wing fighter, and certainly outclassed the more conventional obsolescent Shooting Stars (bottom right), which could exceed a speed of Mach 1.

The Korean War

Between 1945 and 1977 there were, on average, 11 armed conflicts taking place somewhere in the world every day. Most were of the minor, guerrilla kind linked to the Cold War, nationalism or some local revolt, and were rarely technically innovative. It was the larger kind of conflict which served to introduce new weapons or experiments. When the Russian- and Chinese-backed North Korean Army sent its spearheads into South Korea in June 1945, its onrush was checked to some extent by United Nations' aircraft dropping new, air-bursting fragmentation bombs and flaming napalm (jellied petroleum also used as flame-thrower fuel) on their AFVs, infantry and logistic columns (these area weapons proving more damaging than inaccurate non-ballistic rockets). But for the most part the land weapons used in this three-year war were those of 1945, perhaps the most important new arrival being the British Centurion III tank with an 83.4-mm high-velocity gun mounted in a fully gyroscopically stabilized turret. This retained the gun at the selected elevation and azimuth, thus assisting a gun layer, for the first time, to fire with moderate accuracy while on the move.

It was in the air that the revolution flourished. At the lower altitudes, for example, helicopters were used in close support of land forces. Since the early days of flight, attempts had been made to invent a vertical take-off and landing machine, the autogiro of Juan de la Cierva coming quite close to the ideal in 1923 with a machine which had unpowered rotor blades. Not until 1936 did Heinrich Focke's FW 61 twin-rotor helicopter fly and later reach an altitude of 11 000 ft and a speed of 76 mph; but its military development was curtailed and the lead passed to America where Igor Sikorsky's single-rotor machine flew in 1939. Yet helicopters were only experimented with during the

The Attack Helicopter

The French Alouette II helicopter, built in the 1950s and powered by a gas turbine engine, had a speed of 110 mph and could carry five armed men. Its ability, like that of all helicopters, to take off and land in small spaces offered enormous advantages in military operations over land and sea. By carrying all manner of weapons, the helicopter also vastly extended its versatility, as for example, a submarine hunter or, as here, a tank hunter.

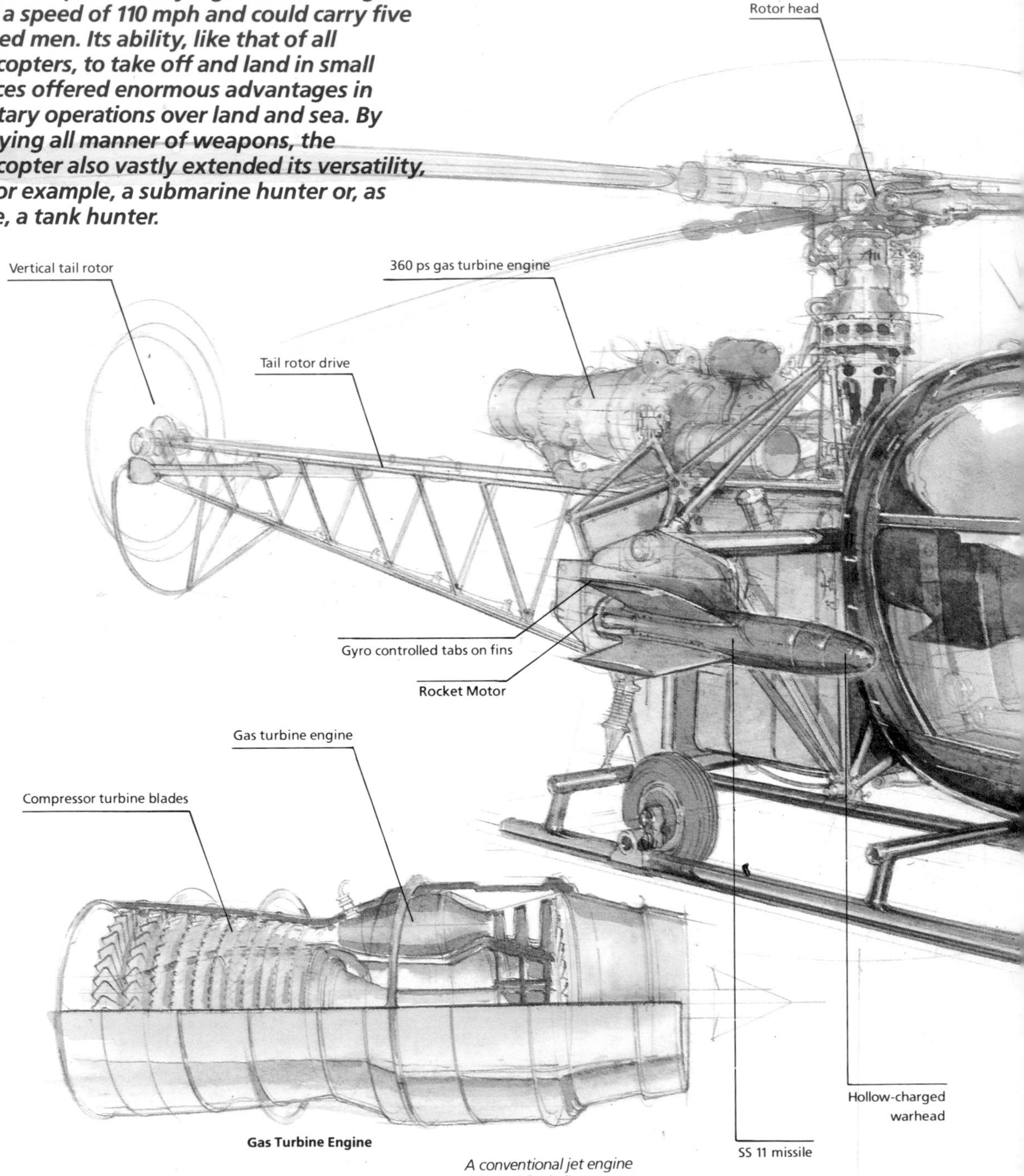

Rotor head

Vertical tail rotor

360 ps gas turbine engine

Tail rotor drive

Gyro controlled tabs on fins

Rocket Motor

Gas turbine engine

Compressor turbine blades

Hollow-charged warhead

SS 11 missile

Gas Turbine Engine

A conventional jet engine adapted for helicopter use — one of the instances of its application.

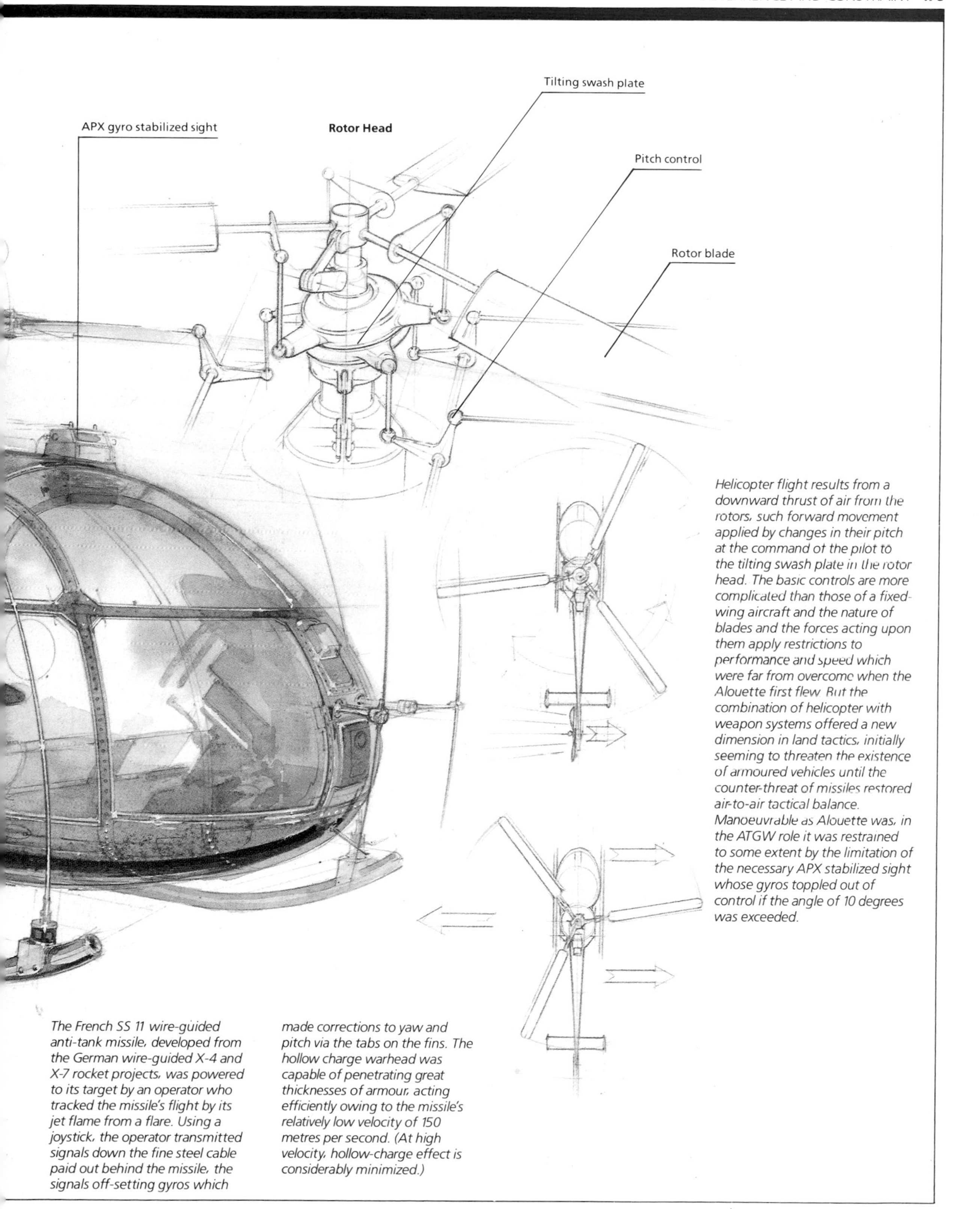

Helicopter flight results from a downward thrust of air from the rotors, such forward movement applied by changes in their pitch at the command of the pilot to the tilting swash plate in the rotor head. The basic controls are more complicated than those of a fixed-wing aircraft and the nature of blades and the forces acting upon them apply restrictions to performance and speed which were far from overcome when the Alouette first flew. But the combination of helicopter with weapon systems offered a new dimension in land tactics, initially seeming to threaten the existence of armoured vehicles until the counter-threat of missiles restored air-to-air tactical balance. Manoeuvrable as Alouette was, in the ATGW role it was restrained to some extent by the limitation of the necessary APX stabilized sight whose gyros toppled out of control if the angle of 10 degrees was exceeded.

The French SS 11 wire-guided anti-tank missile, developed from the German wire-guided X-4 and X-7 rocket projects, was powered to its target by an operator who tracked the missile's flight by its jet flame from a flare. Using a joystick, the operator transmitted signals down the fine steel cable paid out behind the missile, the signals off-setting gyros which made corrections to yaw and pitch via the tabs on the fins. The hollow charge warhead was capable of penetrating great thicknesses of armour, acting efficiently owing to the missile's relatively low velocity of 150 metres per second. (At high velocity, hollow-charge effect is considerably minimized.)

1939–45 war, the armed forces preferring more reliable light monoplanes such as the German Fieseler Storch and the American Taylorcraft (known in Britain as the Auster) for flying in and out of small fields. Not until 1946 did the American Bell 47 achieve operational proficiency and become the first of a family of helicopters. Soon, in Korea, these machines lifted the wounded from the front line straight to hospitals, flew in men and supplies to inaccessible places and enabled army observers or commanders to exercise surveillance and control over frontages as wide as 25 miles, with the added capability of landing among the troops to give verbal orders plus inspiration.

But it was at high altitude that the greatest change took place in air combat, when US jet fighters imposed their superiority upon Soviet-built, piston-engine aircraft. When the defeated North Korean army fled pell-mell and the UN stood poised to occupy the whole of their country, the Chinese army struck southwards from Manchuria in November 1950 and the Russians supplied Mig 15 jets to challenge successfully the older American-made F-80 jet fighters. A few months later when the line settled amid dense fortifications roughly aligned to the 38th Parallel and peace moves began, fighting between jets frequently occurred in the north. The Mig 15s were based in Manchuria and rarely flew far beyond the frontier. So combat took place over home territory with minimum loss of North Korean pilots (many of whom were East Europeans) and maximum risk to UN pilots who, if forced down, could then call for helicopter rescue operations behind the North Korean lines.

From the start, the Mig 15 with its three cannon could be matched in combat only by the swept-wing North American F-86 with six heavy machine-guns. Their encounters took place mostly at 30 000 ft or more at speeds, in a dive, above Mach 1, and to American advantage as the US pilots were better trained than their opponents, who were rotated frequently in order to gain experience. Soon it was found that, at such heights and speeds, it was practical only to fight in small groups of four, guided to contact by radar and radio, and that the chances of hitting fleeting targets was extremely low without radar-assisted gun sights. Not only was the day of larger fighter formations over; so, too, was that of the gun as a primary weapon. Already the substitute was to hand, the 'homing' solid fuel rocket (modelled on the German X-4) which found its victim by seeking the infra-red radiations from its engines.

When first used in action, the Sidewinder missile was decisive. Eight Chinese Nationalist F-86s took on 20 Chinese Communist Mig 15s over the Formosa Straits in 1958 and claimed four out of the ten Migs shot down to the credit of the missile. At once ECM (electronic counter measures) became necessary to detect and deflect (by decoy flares) this missile, with its range of two miles. Tactics, too, were altered because the missile had to be launched dead astern and could be shaken off by its quarry's violent manoeuvres.

Nevertheless, the general conduct of air warfare remained unchanged. The struggle for supremacy decided the percentage of effort which could be allocated in support of surface forces and impinged upon diversions of effort to protect strategic targets in depth. The replacement of long-range anti-aircraft guns by beam-riding or radar-guided missiles (copied from the German Wasserfall) which homed onto their targets, merely introduced a new threat of dubious efficiency in face of feasible counter-measures. Although it was sensational when in 1960 a Russian missile shot down an American Lockheed U-2 powered glider at about 90 000 ft while it was on a reconnaissance flight across the Soviet Union, the event did not accord dominance to the missile. The main and crucial change centred, of course, upon the undeniable capability of air power to deliver a knock-out blow with the nuclear weapon. Technology still could not prevent the bomber getting through. It was necessary for only a few, carrying atom bombs, to evade the defences and cause awful devastation — especially after 1952 when the first thermo-nuclear 'fusion' device was detonated by the Americans.

The Ultimate Weapon

The creation of the fusion bomb (provided by the isotopes H_2 and H_3 of hydrogen) was as inevitable as all previous dominant weapons of various kinds. But its ability to generate the equivalent of 14 megatons of TNT and vast quantities of radioactive debris ('fall-out') automatically outlawed its use by those responsible — providing they were sane, in command of the situation, and that their potential opponents had a like weapon to act as a balancing deterrent. Soon, indeed, the Russians and British did have an H-bomb to match the temporary American advantage. Henceforward attention tended to centre on the potency of delivery systems and the availability of enough weapons to achieve sufficient, guaranteed levels of destruction to preserve the credibility of the deterrent. But whereas, with milder weapons, debate on deterrence might easily, and without wholly disastrous consequences, give way to force of arms in order to profit from a transient technical or numerical

Above: ***Explosion of the fusion, so-called H (Hydrogen) Bomb.***

advantage, to do so with the ultimate weapon was prohibitively risky — or so it was hoped and calculated in a plethora of erudite discussions and documents in what might be termed a war of wordy threats related to evolving power-politics and diplomacy.

Until size, weight and yield of atomic bombs were substantially reduced, heavy bombers had to be used to carry both strategic and tactical weapons. The first demonstration of a small-yield, fission warhead took place in the USA in 1951, and was followed in 1953 by the firing of a nuclear shell from a 280-mm gun to indicate that optimum miniaturization had been achieved with yields reduced below 2 kilotons for close support on the battlefield. Nevertheless, aircraft remained the most practical carriers, with rockets gradually assuming a larger role. During the 1950s, Britain, America and, to a lesser extent, Russia built fleets of fast, long-range bombers whose main task it was to carry nuclear weapons to the enemy heartland and act as the principal arbiters of the force of nuclear deterrence. At immense cost, machines such as the 645-mph four-engine Avro Vulcan and the 630-mph eight-engine Boeing B-52 came into service along with intricate electronic equipment and, in due course, guided stand-off bombs respectively to help them find and hit their targets. For although it was initially supposed that, for example, an unarmed Vulcan flying at 65 000 ft should be able to reach its target unmolested, the advent of solid-fuel, surface-to-air guided missiles (SAM), let alone improved high-flying missile armed fighters, soon out-moded that notion. The Russian two-stage SA 2, for example, could reach a ceiling of 80 000 ft at a range from launching of about 27 miles. Guidance systems, warheads and detonation devices reflecting the full range of ideas produced during the Second World War were acquired.

The cost of defence was as great if not greater than that of offence. The mere hint of some new bomber in service, such as the relatively innocuous turbo-prop Russian Tupolev Tu 95 (Bear) in 1955, could prod NATO into vast expenditure on elaborate systems of radar coverage, fighter forces and missiles to meet a threat which was probably a partial bluff. The radar and cypher chess games which were played between contending sides were of a complexity and erudition to put events of the Second World War into the shade. Of necessity, long before the bomber became totally vulnerable as a carrier, substitutes were in hand — surface-to-surface guided weapons of considerably longer range, improved reliability, accuracy and immunity to counter-measures.

It was to be expected that versions of the liquid-fuelled V 1 and V 2 with their gyroscopic guidance systems would progressively be improved and merged with American and British solid-fuel types and radio and radar navigation methods. The American Honest John of 1950, a 23-mile-range free-flight solid-fuel rocket with nuclear capability, was a prime example of a relatively simple and reliable missile for tactical missions. And the US Atlas, $1\frac{1}{2}$ stage Inter-Continental Ballistic Missile (ICBM) of 1958, with a range around 5000 miles, was typical of rocket weapons to replace the bomber. But none would have been feasible had it not been for miniaturization of innumerable components, above all of on-board control and guidance systems and, in particular, the 'inertial' kind which were virtually proof against counter-measures. Three tiny gyroscopes similar to those in V 2 were integrated with components which could automatically measure acceleration in relation to time, and steer the missile on a preselected course from launch-pad to target. Gravity, of course, imposed its influence and had to be compensated for by measurement of deviations through a fast computer, which also had to be miniaturized and which was impossible so long as bulky thermionic valves were the best available amplifiers.

Miniaturization And The Transistor

Although large and heavy inertial navigation equipment was practical in 1954 for position-fixing in earthbound vehicles and ships, it was incompatible with space vehicles until a much smaller and lighter amplifier than the thermionic valve appeared. That proved to be the transistor. Small so-called 'semi-conductors' had been in existence since 1904 and used as 'cat-whiskers' rectifiers for radio, but their properties as amplifiers were only revealed in 1949 by three American physicists who worked for Bell Telephone — John Bardeen, Walter Brattain and William Shockley. All three had, to some extent, been associated with military projects and were keenly aware of the valve's shortcomings. Each foresaw the almost unlimited impact on all walks of life of small devices made of selenium, copper oxide or silicon carbide. Not only would these affect vital savings in weight, power consumption and cost, but would also enhance reliability by a factor of 1000 — besides

The True Submarine Threat

The appearance in 1955 of the first nuclear submarine, USS* Nautilus*, initiated a new era in naval warfare through the exploitation of this boat's almost limitless capability for sustained submerged operations.

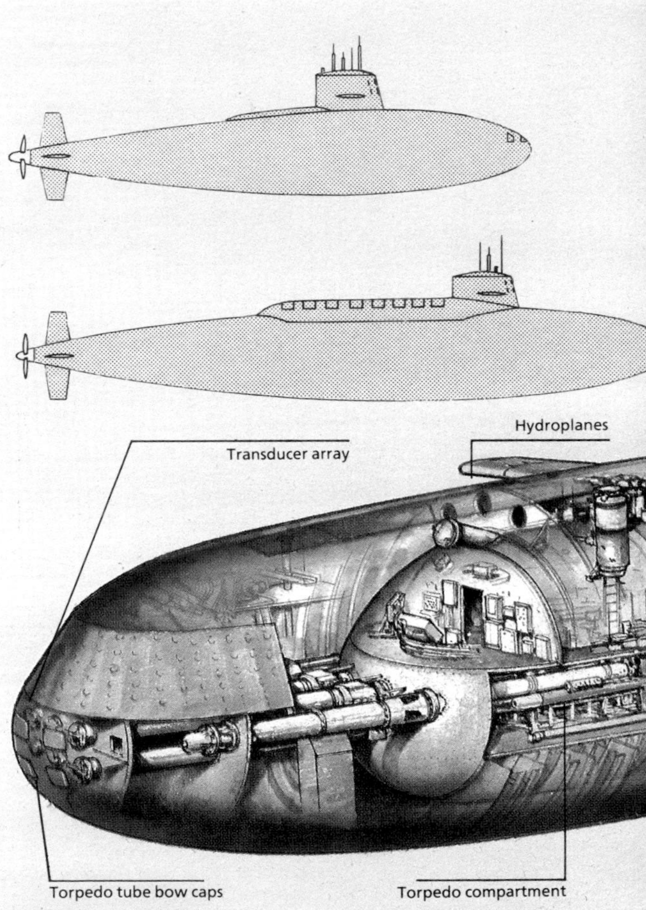

Provided with facilities to recirculate air; constructed to dive to unprecedented depths to avoid detection; and given a speed well in excess of 20 knots, craft of this kind not only possessed a conventional tactical ability far in advance of earlier vessels, but posed a strategic threat of unimagined potential.

By 'stretching' boats to carry nuclear-armed rocket ballistic missiles of the Polaris kind, it became possible with the aid of inertial (and later satellite-assisted) navigation to position the submarine so exactly that an undetected underwater attack could be made with precision on strategic targets in the enemy interior. Defence against this kind of threat was immensely difficult, calling for diversion of enormous enemy effort to find and track nuclear missile boats hiding in the vast ocean expanses.

__Below:__ Representation of a __Resolution__ class Polaris-armed nuclear submarine showing the full array of periscopes, aerials and snort which, when submerged, would be retracted.

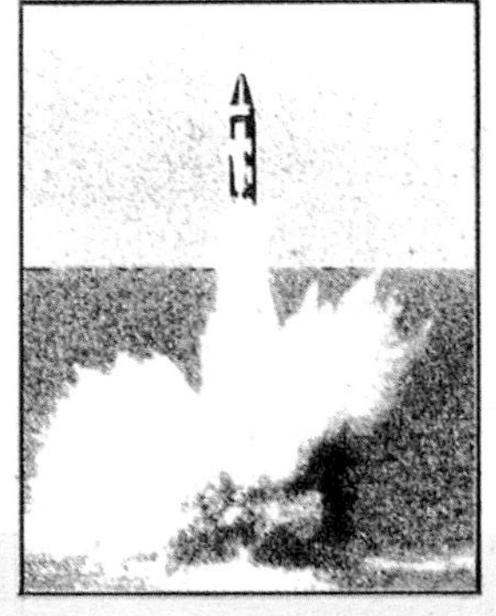

The Polaris missile, developed in the 1950s, was propelled by a solid-fuel rocket and launched from a vertical tube in the submarine. Guided by a miniaturized inertial system, it could deliver both thermo-nuclear or nuclear warheads to their targets at ranges which in the 1960s were progressively increased from 1380 miles to 2880 miles, mainly by the adoption of higher energy propellants.

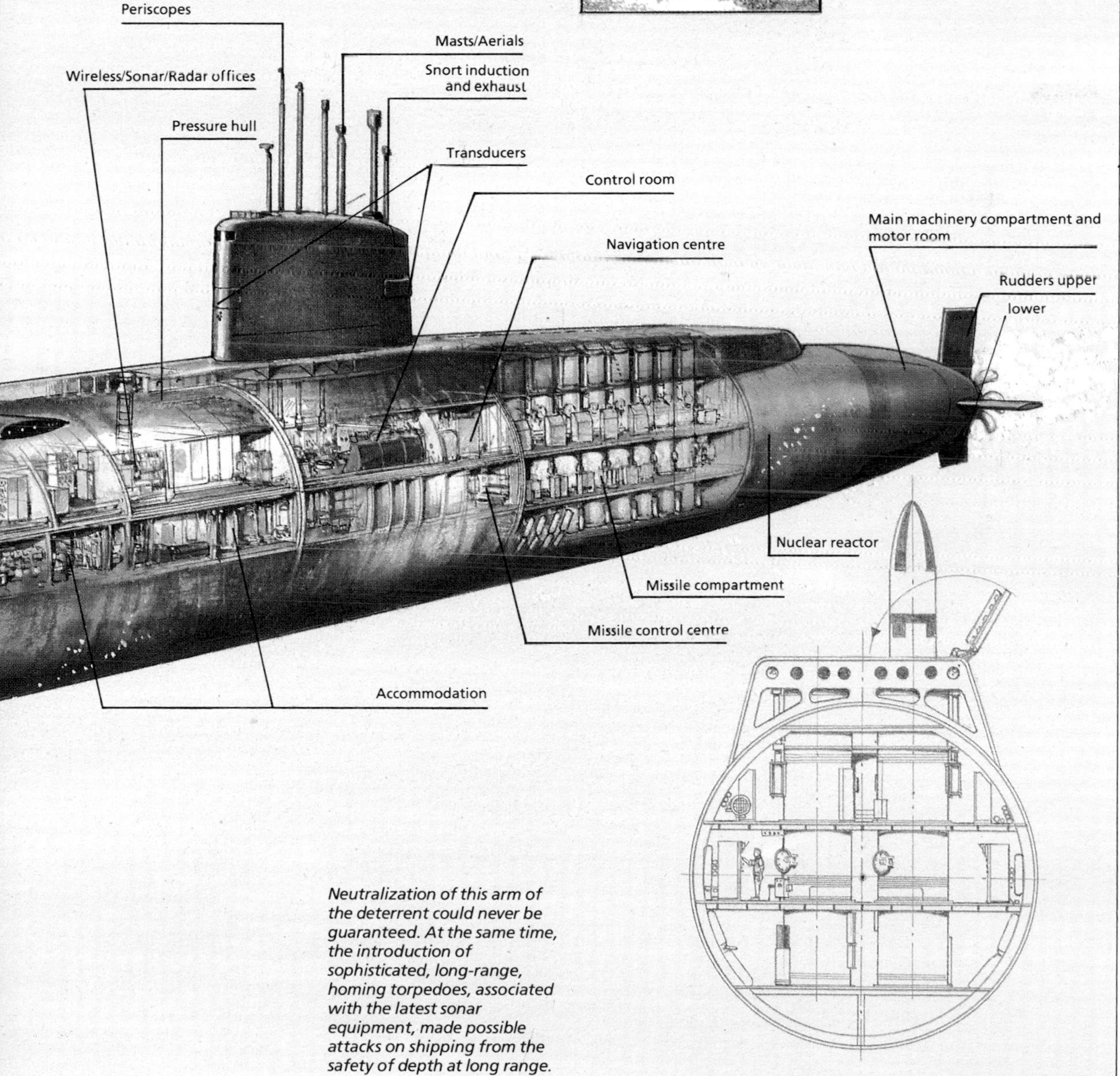

Neutralization of this arm of the deterrent could never be guaranteed. At the same time, the introduction of sophisticated, long-range, homing torpedoes, associated with the latest sonar equipment, made possible attacks on shipping from the safety of depth at long range.

opening up almost unlimited fields for development, among them small computers to resolve discrepancies with inertial navigators. Soon the 2400-valve Colossus of 1944 was left behind by faster electronic, digital and analogue computers built in British and American universities.

These computers included enlarged memories, mechanical programming and semi-permanent wiring. Commercial usage followed in 1953 for implementation of finance and stock control and the solution of engineering and production problems, but with machines which demanded great room, had limited memory storage and were relatively slow. The production of the transistor on a large scale towards the end of the 1950s offered enormously improved computer efficiency without at once enlarging stored memory, which still depended largely on punched paper and magnetic tapes. But already the frenetic drive of entrepreneurs and, above all, the military, was focusing attention upon combined analogue and digital computers with miniaturized blocks of transistors which, in due course, were themselves to be further miniaturized to form pin-head, solid-state memory cells, known in the 1960s as micro-chips.

In line with the appreciation by the West that the best way to match the numerically strong Russian and Chinese was through advanced technology, there emerged a formidable stream of demands for weapon systems which in 1945 had been deemed inconceivable due, in no small part, to lack of adequate miniaturized communication equipment. At the dawn of the nuclear electronic revolution, the scientific élite told their sponsors that possibly nothing was impossible — at a price. Within the USA (and soon copied by the Russians) stood governments prepared to pay a very high price indeed for wonders just round the corner. Outer space beckoned as a vital battleground, but primarily in the early 1950s it was the nuclear-powered, missile-carrying submarine which seemed to offer the greatest potential in combined invulnerability and striking power.

Nuclear energy as a source of industrial power and propulsion was inviting from the start and seemed, to most of its inventors, far more desirable than its use as an explosive. By 1954, so-called power reactors of the fission kind were operating in the USA, Britain and Russia. The following year, to satisfy every submariner's craving for a true submarine, came the USA's *Nautilus* powered by a nuclear reactor. In revolutionary terms, *Nautilus* was far more significant than had been the iron-clad and the dreadnought. She not only had the ability to stay at sea to the limits of her food supply and crew's endurance, but the specially strengthened, streamlined hull also allowed her to dive to well over 1000 ft, travel at well over 20 knots and, by recirculation of air through carbon-dioxide scrubbers, remain submerged indefinitely. Hence she was extremely difficult to detect by sonar in the lower depths beneath successive temperature layers of water; she was invisible to radar; and she was fast enough to elude most hunters above. Moreover, not only did the latest homing torpedoes with proximity fuses enable an attack to be launched from the depths at long range, but the advent in 1957 of a 1250-mile-range, solid-fuel, two-stage rocket with a compact nuclear warhead (the Polaris) made it possible to employ the submarine strategically as a deterrent element, capable of firing a ballistic missile. Solid fuel enabled Polaris to be stored aboard without the complexities of volatile, liquid propellants. It could be launched from beneath the surface, from a position exactly fixed by inertial navigation and programmed to its target by small computers in the submarine and in the missile. Only by devising the most elaborate defensive systems based on still more new and expensive technology could an enemy hope to find and track the missile-carrying submarine. Furthermore, it was by no means certain that every enemy detected could be sunk, even by the latest rocket-propelled, homing anti-submarine torpedoes.

Entry Into Space

Remarkable and deadly as the nuclear-powered, deep diving submarine was, its significance was soon overshadowed by developments at the other end of the environment — in space. The V rockets which had first entered the stratosphere and demonstrated their capability, through shielding, of returning to earth without 'burning up' in the atmosphere were the mothers of successive generations of vehicles despatched ever deeper into space. Space vehicles became symbols of competition between America and Russia because they alone had the financial resources and political urge initially to build them. The expensive race into space, dictated as it was by the classic military desire to take possession of commanding heights, was also valued as a prestige activity in the war between ideologies — as a propaganda gimmick advertising whose ways of technology and life were most impressive and likely to prevail.

Russia scored a series of early victories (and demonstrated an ability to originate as well as copy) by being the first in 1957 to launch an 184-lb sphere (Sputnik I) to orbit the earth at a height of between 142 and 588 miles. Sputnik I transmitted information on atmospheric conditions until gravity prevailed and it

Above: ***The Russian Vostok I satellite vehicle could be used either to take a cosmonaut into space (he being ejected on reaching the earth's atmosphere and parachuting to safety) or a reconnaissance camera for surveillance on ground targets. It was the forerunner of thousands much more sophisticated such vehicles to come.***

burnt up in the earth's atmosphere after 1400, 96-minute circuits. A month later a dog was placed in space by the Russians, followed shortly by the USA's Explorer I satellite to discover the inner Van Allen radiation belt. Ambitions soared as technology, which depended on nearly 100% reliability, improved. Radio communication deep into space became commonplace along with the transmission of photographs of the earth's surface and weather conditions from satellites. In 1960, the first recovery of a space capsule after its journey through the atmosphere led to the Russian triumph in 1961 of launching into orbit and safely returning a man.

By then the military potential of satellites was defined. Unmanned, they could provide reconnaissance pictures of the entire earth's surface using all manner of cameras, including infra-red and television; they could act as relays for radio and television signals from earth to space and back again; they could work as dispensers of missiles by remote control against targets in space or on earth; and they could be used to help surface vehicles navigate with an unheard-of accuracy, particularly once surveys from satellites remapped the earth and corrected a plethora of previous errors. Manned, they could establish a base and perform as fighting craft — although the time when the latter would be possible was far in the future.

To whatever purpose they were to be put, space vehicles, including those dedicated to future lunar landings and exploration of deep space, generated a vast increase in the scope and intensity of development of light-weight materials and structures, radar and communication equipment, cameras and sensors and automated machinery. They also necessitated the building of larger and ever more sophisticated spacecraft along with their enormous, multi-stage rockets and the guidance and control equipment needed to steer them into space, monitor and adjust their progress and bring them safely home again. An entirely new industry was in the making, with peaceful derivations motivated mostly by military urgency, and one from which mankind benefited to an extent that might otherwise have taken far longer had unsubsidized commercial motivations alone appertained. It is impossible and unnecessary to list here the host of inventions linked to space programmes; but interesting to point out that they were accelerated by a traditional evolutionary process, fear of political revolution and war and that, although combat in space above 70 000 ft was for several decades non-existent, space technology at once had a profound effect upon war in the troposphere and on the surface itself — and nowhere more so than Vietnam.

Vietnam And Mid-20th-Century Technology

Technically and tactically there was little original about the war which broke out in Indo-China in 1946 between the French colonial power and the nationalist/communist Vietnamese forces supported by Russia and latterly China. The weapons available were those of the Second World War and none too plentiful at first. The combat was initially of the guerrilla type, in terrain ranging from forested highlands to swamp and lowland paddy in which centres of population were tenuously connected by poor roads and railways. By 1953, however, it was plain that the Vietnamese, inspired by their cause and the command of General Vo Giap, had formed themselves into a coherent army with adequate signal communications and logistics to fight a prolonged campaign. Indeed, when the French, with American material aid, chose to challenge the Vietnamese Army to a major trial of strength by dropping an air-supplied garrison at Dien Bien Phu, within Vietnamese territory, it was the French who were defeated and compelled to sue for peace. In the event, France withdrew and Indo-China was split into the predatory communist-led North Vietnam and the non-communist South Vietnam. Then, as pressure from the North grew, the South turned to the USA for economic and military assistance.

While it is true that the initial pressure on the South was by Viet Cong terrorists, it remained a long-held misconception that the ongoing struggle throughout the 1960s was a guerrilla war when it was actually a conventional campaign by a well-organized army. The political ramifications of the escalating war will be set aside here. Suffice it to say that American aid (supplemented by that of Australians and other Pacific nations in due course) was extended in 1961 to allow military advisers to take part in combat; and that participation became direct when helicopter squadrons were deployed to enhance the mobility of anti-Viet Cong ground forces. In response, Russia and China stepped up their material support of North Vietnam leading to the moment when, in response to attacks by US carrier aircraft on targets in North Vietnam, the North Vietnamese Army entered the fray in force in 1964.

There were wide discrepancies between the two sides. The North was dedicated to persevering with long-service soldiers of sound skills but lacking the technical backing and firepower of their enemies. They would prosecute the war as a highly mobile swarm, well served by espionage, merging with the population and diligently supplied by material from sources which, beyond North Vietnam, were inviolate. The South was also supplied from inviolate sources but dominated by the traditional American methods of seeking a quick result by striking with massive firepower and technology at the roots and branches of enemy strength. Unfortunately for them, they repeatedly fell short of the ideal. Political constraints vetoed the sheer totality required for fear of upsetting a world and home opinion which tended, in the glare of television and adverse press coverage, to be sceptical of the entire business. Short-service conscripts were no match for professionals. Although fighting between men on the ground was intense, cutting off the North Vietnamese Army from its base was never fully attempted and the forces of the South suffered.

Blockade — The Sea And Air War

The laying of mines by the North in coastal seaways in 1961 was one factor which drew the USA into the war, compelled as they were to start sweeping operations. Opposed only by light forces, the US Navy, along with the South's coastal craft and armed junks, were able to seal off deltas, countless inlets and beaches from Viet Cong supplies. But it was a costly business, involving all manner of craft, including in 1968 a new kind to patrol the Plain of Reeds — the aircushion craft. Pioneered by the British inventor Christopher Cockerell in the 1950s, it had an advantage in that it

The Hovercraft Comes Into Its Own

Above: *A British-designed air cushion vehicle (hovercraft) of the US Navy, used in Vietnam after 1966 to penetrate coastal regions and marshes which otherwise would have remained safe refuges for Viet Cong forces. Very noisy, they were also vulnerable in a fire fight. But as a means of searching for mines at sea they were extremely useful since their characteristics only marginally influenced a mine's detonators.*
Left: *The more conventional way of enforcing blockade in inland waters — a vital function in crippling the Viet Cong's clever logistic system.*

could carry men, weapons and supplies at speeds up to 70 mph from waterways to reeds or across paddy without stopping and without being susceptible to mines. Hence it could also be used for mine-sweeping and anti-submarine work.

The blockade leaked until targets in inviolate territory, such as neighbouring Laos and Cambodia and the North Vietnamese port of Haiphong, were violated. The majority of stores for the Viet Cong entered these sanctuaries by sea, and thence by a myriad routes into South Vietnam. Laos and Cambodia could only be hit by air and land, but it was not until 1970 that this was authorised — and not until 1972 that the closure of Haiphong was permitted with thousands of air-delivered magnetic, acoustic and pressure mines. This at once caused a supply crisis in the North which was closely related to a massive air offensive and ground fighting taking place at the culmination of the war.

Air warfare rose to unheard-of peaks of effectiveness and flexibility, made possible by the versatility of aircraft performing tasks for which they were not designed. Setting aside for the moment the part that air power played in the land battle, the overriding task of the South between 1965 and 1972 was to prevent interference with that effort by hitting the North's Mig fighters at their bases and in the air while endeavouring to suppress the North's ground defences. At the same time they aimed to strike at strategic targets and lines of communication to make the North sue for peace. The burden had to be carried by fighter bombers from aircraft carriers in the Gulf of Tonkin and from airfields in South Vietnam, Thailand and Guam. For although the giant B-52 bombers were available to drop 15-ton loads of high explosive on well-defined military targets, their use close to politically sensitive centres of population was often vetoed and their vulnerability to Russian SAMs uncomfortably demonstrated.

The problem therefore amounted to delivering fewer explosives with great accuracy against small, dispersed targets and to give tactical aircraft a long-range strategic capability. Air refuelling had been practised in Britain and America since the 1930s and in the 1950s air tankers had been introduced to enable reinforcement of threatened regions for strategic purposes. Boeing KC 135 tankers (military versions of the 707 airliner) could carry 30 000 gallons of fuel and so enable reconnaissance and strike aircraft to operate deep into enemy territory, their route and protection against enemy Mig 17s and supersonic Mig 21s being carefully co-ordinated. Frequently a tanker was the saviour of an aircraft leaking fuel from enemy action; one actually towed a damaged fighter home attached to its fuelling boom.

The tactics of air-to-air combat and low-level bombing were conditioned by the missile (ranges for the latest kind having already been extended beyond six miles), electronic counter-measures, the weather and air refuelling (which did not apply to North Vietnamese pilots, who never used this technique). The need to conserve fuel by a high approach, followed by a dive to the target and a climb to rendezvous with the tanker, gave enemy radar warning of raids and time to position their fighters. The threat from North Vietnamese air and ground defences unsettled pilots who frequently failed to score hits even with missiles (which sometimes malfunctioned, were decoyed or 'ducked'). The North lost only a few more aircraft than it destroyed and the US Navy and Air Force had to conclude that automated weapons did not compensate for pilot skill and sound tactics. Once they tightened up training, results improved.

The problem of hitting targets with bombs and missiles remained difficult even though target-location with comprehensive electronic aids was almost guaranteed. Bridges were, as ever, difficult to hit, particularly when low-flying aircraft were met by a hail of well-aimed, radar-directed fire from automatic guns and heat-seeking missiles. Losses in these attacks could be heavy even though special suppressive fire support and ECM missions were flown against anti-aircraft sites. The use in 1967 of the 1000-lb Walleye gliding bomb, incorporating a television camera in its nose, enabled the parent aircraft to stand off and allow a controller who was relatively unhindered, providing visibility was good, to pick out and strike any target with precise economy.

Far more deadly, however, was the next generation of so-called 'smart' bombs, the 2000-lb Electro-Optical Guided Bomb (EOGB) and the 2000- or 3000-lb Laser Guided Bomb (LGB). EOGB was an improvement on Walleye, enabling its controller to locate the target on his TV screen, designate it to the bomb and then leave the bomb, after release, to find its own way to the target. LGB capitalized on an entirely new form of energy radiation to achieve a similar fire-and-forget guidance. Conceived in 1958 and reported by T H Maiman in 1960, the laser in various forms (ruby, gallium arsenide, etc.) could amplify or create coherent light waves which could be in the visible or infra-red portions of the spectrum. It could be focused, if required, to a fine point of great intensity and be adapted for cutting metal, cloth or, during surgery, human tissue. Its military uses were also numerous: communications; range-finding; radar tracking; and,

Above: ***Air refuelling over Vietnam of Phantom fighters by a KC 135 tanker aircraft. Alongside the long-range tanks beneath the wings of the Phantoms (F-4) may be seen laser guided bombs which shortly will be fired against enemy bridges.***

in the case of the LGB, as a source of energy which could be aimed optically from its parent aircraft to the target, leaving a laser sensor in the bomb's nose to detect the reflected 'illuminations' from the target and automatically guide its bomb home. Fire-and-forget bombs almost guaranteed knocking down bridges and hitting vital pin-point targets such as electricity generating stations. For example, it required only eight such bombs to shatter the robust steel Thanh Hoa Bridge, plus another 24 finally to detach it from its concrete abutments. In three months in 1972, no less than 106 bridges were destroyed in North Vietnam with logistically disastrous consequences.

Somewhat obscured by the vociferous anti-war and anti-American feeling generated by propaganda throughout the world, between 1966 and 1973 US forces several times brought North Vietnam to its knees. The saving factor hinged on political unwillingness, on the apparent verge of victory, to finish the job. Air power's direct part in the land campaign will be described below, but its indirect strategic role in strangling the North's Tet Offensive of 1968 was every bit as important — and even more so when, in 1972, a round-the-clock, no-targets-restricted offensive by all types of aircraft (including B-52s) brought such

THE VIETNAM WAR ZONE 1965-72

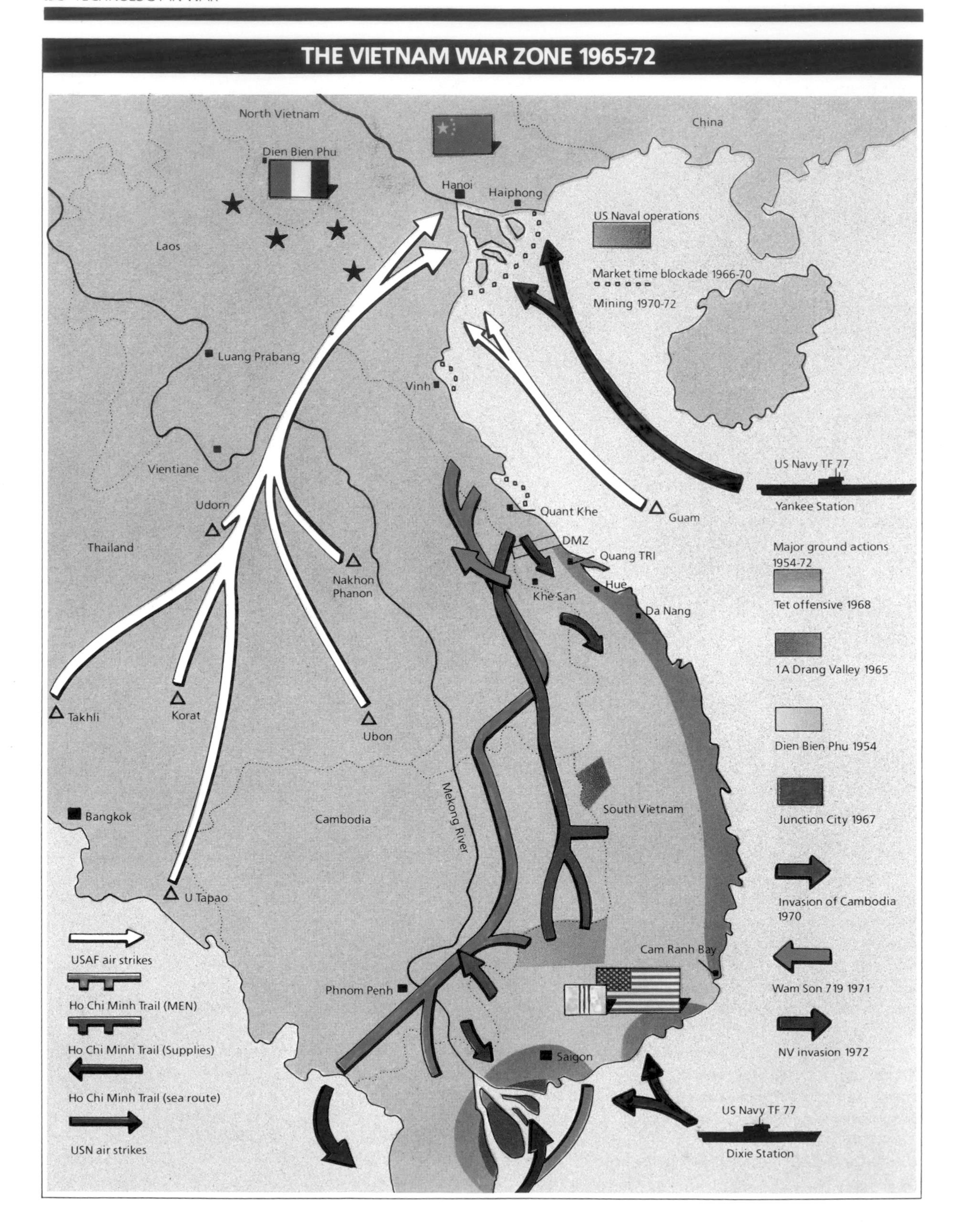

devastation to routes, fuel stocks, base installations, airfields and missile sites that the North Vietnamese were brought to the peace table. Setting aside the wiles of the subsequent diplomatic solution, to the North's advantage, it was shown at a cost of 92 aircraft that Soviet Russian supersonic fighters and SAMs could be overwhelmed by skilful attacks supported by ECM. A moment arrived at which the defences of Hanoi and Haiphong collapsed due to the blotting out of their radar control system, a shortage of missiles (which usually missed) and the loss of 193 aircraft because US supersonic fighters dominated their opponents in combat. All 18 B-52s lost fell only to SAMs because, when flying straight and level at 37 000 ft to bomb, they were at their most vulnerable.

Firepower Versus Fieldcraft In South Vietnam

As already stated, ground combat was only partially of a guerrilla nature, even if the Viet Cong (VC) and North Vietnamese Army (NVA) did follow the dictates of Mao Tse Tung (who merely urged mobility like others before him) when he demanded withdrawal when the enemy attacked, harassment when he defended and attack when he was exhausted. Uncluttered by heavy sophisticated equipment, the VC and NVA were able to move at great pace under officers who were experienced in the tactics required. Methods which had been too much for the French and would have defeated the ill-trained South Vietnamese and 'green' American infantry were mainly countered by technology linked to air and firepower. Only gradually were armoured troops promoted by the Americans, a consequence of the initial belief that this was only a counter-insurgency campaign and that the nature of the terrain precluded vehicular movement off the frail roads and tracks. Not until 1968 would both misconceptions be dispelled — one as the result of a study which showed that 46% of the terrain could be traversed all year round by AFVs; the other by the Tet Offensive in which the NVA, abandoning Mao's concepts, came into the open aiming to overwhelm the South in pitched battles. Coincident with Tet (the

Left: The long war in Vietnam and surrounding lands, illustrating the widespread characteristics of the various campaigns and the manner in which, curtailed or supported by air attacks and blockade, the respective forces managed to infiltrate and infest each side's heartlands.

Above: REMBRASS, a remotely monitored battlefield sensor system for Infantry to detect the speed, direction of movement, number and classification of intruder activity.

Chinese New Year), strong NVA forces attacked a variety of key fortified communications located at Khe Sanh, through the port of Hue down to Saigon, the capital city, and across the Mekong Delta in the south.

Furious NVA infantry assaults enjoyed some success because they came as a surprise in magnitude and technique. But they were met by stiffening resistance from infantry whose conventional wired and mined perimeter positions were improved by a variety of new weapons, night vision and alarm systems. The use by both sides of gas-operated, high-velocity, semi-automatic rifles and rapid-fire grenade dispensers immensely increased both infantry firepower and ammunition expenditure — and the tendency to spray fire rather than aim. Severity of wounds also increased, as did reluctance to move about in the open, both by night and day. To match the enemy's preference for night action, infra-red searchlights, soon to be followed by passive image-intensifiers (which utilized the ambient light, electronically amplified, as illumination) were brought into use along with infra-red 'fences', doppler radar and seismic sensors to detect objects on the move — be they men, vehicles, or quite frequently, some inoffensive farm animal. Lacking AFVs and facing the American preponderance of artillery and mortar fire, the NVA had little chance of overrunning entrenched positions so long as

their enemy's reinforcements and supplies continued to flow. And flow they did through a classic combination of armour and air action.

On the ground, fast reaction by tanks and infantry in armoured personnel carriers frequently broke up NVA concentrations and carried men and supplies by road to besieged garrisons. Helicopters in great numbers supplemented the armoured effort and also carried supplies to places inaccessible to ground vehicles. But the helicopters often had to fight their way through and were very vulnerable (about 3000 were lost between 1962 and 1973). They fitted their own machine-guns, but depended mainly upon fire support from helicopter gunships armed with the latest multi-barrel, Gatling-style guns which could pump out 7000 rounds per minute; and upon multi-rocket launchers designed to suppress light anti-aircraft gun positions. North Vietnamese troops caught in the open suffered terribly and were held. Their most telling rebuff came at Khe Sanh (which was held by the best of US troops, the Marines) which they looked upon as a prestige objective akin to Dien Bien Phu in 1953. For five months the NVA, for the first time using a few tanks, encircled it but failed to penetrate the main defences as supplies continued to be flown in and the wounded out by fixed-wing aircraft and helicopters. Far from being cut off, indeed, it was the defenders of Khe Sanh who began to strangle the NVA by the interdiction of supply routes, by ground raids and a storm of fire from the air upon their depots, much of it delivered by B-52s. In places, too, rockets and bombs from aircraft were accurately dropped within a few yards of forward ground troops, among virtually defenceless assailants. The North's air force took no part in these battles, being confined by the air attacks upon the homeland (already described above). Close support missions of ground forces, however, revealed the need for special aircraft. Cheaper, simpler, subsonic machines of high payload were far more suitable than the sophisticated supersonic kind. For example, the venerable Douglas Dakota of the 1940s, when fitted with side-pointing, rapid-fire

Above: **Three 123 Fairchild transport aircraft fitted with internal tanks, spray defoliants and herbicides in an effort to destroy hideaways under the jungle canopy and help enforce blockade in Vietnam by destroying crops. Considerable destruction was caused without conclusive results.**

weapons, proved a powerful weapon when strafing enemy positions and supply routes by night. And low-speed, training-type aircraft were found to be extremely useful in seeking out the hidden enemy. When the NVA abandoned its attempt to take Khe Sanh it conceded that overall failure was due to being wrecked by intense firepower and technology. It would need four years to rebuild its offensive capability.

Those four years were spent in simultaneous attempts to build up the South Vietnamese Army and eliminate the VC and NVA — the latter an almost hopeless task until supplies were cut off. Fighting took place along the borders and then in 1970 erupted into Cambodia as the South's mechanized forces crossed the border to raid the North's supply depots, destroying vast quantities of stores and crippling its logistical organization for more than a year. Once more it had been proved that armoured forces can be viable in the trickiest country providing they practise mobility and all-arms co-operation.

Naturally this conventional form of attack attracted far less comment than the chemical warfare campaign waged by the South, with air-sprayed herbicides and defoliants against enemy crops and the forests which hid him. Tried on a small scale by the British in Malaya in the 1950s, with marginal results, it cannot be proved that the spraying of 300 000 acres of crops out of 2 million cultivated acres in Vietnam was decisive, or that the defoliation of 8 million acres of trees was catastrophic — militarily in the short term or ecologically in the long. But the entire operation was politically counter-productive since it raised the spectre of widespread biological warfare and presented the propagandists with cogent arguments.

In 1973 the USA withdrew from South Vietnam, leaving the indigenous people to fight alone against an NVA which was resurgent once air attacks upon its bases stopped. When the South collapsed in 1975, one reason was the appearance of strong NVA armoured forces with a prowess much improved since their first memorable appearance in mass at Binh Long in 1972. For on that occasion they lost 80 out of 100 AFVs, a few being destroyed by helicopter-launched, wire-guided missiles. A new era of anti-armour operations had begun.

Like so much else, the anti-tank guided missile (ATGW) had its origin in German technology, specifically the X-7 missile which was unproven in 1945 but developed by both the French and the British in the 1950s. Infra-red guidance systems having proved susceptible to battlefield conflagrations (let alone counter-measures), main development centred on the transmission of signals by an operator along a wire trailed by the missile, which he tracked by the rocket motor's glow or a flare. Because speed was sub-sonic, chemical warheads were mandatory; the more so since research into hollow-charge warheads progressively raised penetration from a multiple of about three times the cone diameter in the 1950s, to seven in the 1970s — and the promise of a factor of 10 in the 1980s, always supposing the hole was big enough. Guidance systems were also improved from the manual kind of the 1950s to semi-automatic computer control (such as was used in the TOW missiles fired by the Americans against the NVA in 1972). But far more telling in the debate which inevitably surrounded the advent of ATGW, was the outcome of the intensive, if short-lived, battles which periodically broke out in the Middle East.

The Yom Kippur War, October 1973

The campaign which began on 6th October 1973 was in complete contrast to the one taking place in Vietnam, even though the contenders were, in certain respects, similarly equipped — the Arab nations mainly with Soviet weapons and the Israelis with American, French and British.

The state of Israel had been born in war in 1948. Ever since it had been at odds with its Arab neighbours, on terrain which was mostly desert and, in certain upland sectors, rugged. Although nearly all the combatant nations bordered the sea, they depended chiefly upon land and air forces for defence and attack, and the Arabs, to begin with, held the initiative while the Israelis built up their strength under the incentive to survive. Since their inception armoured forces have dominated desert campaigns and been heavily dependent upon air forces. It was upon these two arms of decision that the Israelis concentrated once the battles of 1956 had again shown them to be better investments than a mechanized infantry/artillery force. When, in another period of tension, the Israelis had made a pre-emptive strike against Egypt in June 1967 (in what came to be known as the Six Day War), it was on classic lines similar to those demonstrated by the Germans against the Russians on 22nd June 1941 — but marked by even greater speed and violence. Aircraft struck without warning against Arab airfields, crippling their opponents at low cost in an attack of unprecedented accuracy with success measured by the loss of 20 Israeli aircraft compared to 308 destroyed. Armoured forces overwhelmed opposing armies with concentric strokes which destroyed all their equipment and threw the Egyptians out of Sinai, the Jordanians out of Old Jerusalem and the Syrians from the Golan Heights.

At the heart of the Israeli victory lay superb planning executed by highly-trained pilots, flying mainly French-built fighters; and by armoured forces, of which the British Centurion tank with its 105-mm gun provided the core. Victory in the air ensured victory on land within six days. The architect of the success on land was Major General I Tal, who formed and led the tank forces and who placed his faith in relatively simple gunnery techniques and tactics to provide his men with a pronounced advantage in fire-power over the Russian-built T-54 and T-55 tanks of their opponents. So pleased was Tal with the execution wrought by high-velocity guns on enemy armour, that he dismissed the recently introduced ATGW (of which the Arabs had a few) as unimportant. He indoctrinated his successors with a philosophy which elevated the tank above all else as the dominant weapon — to the detriment of artillery and other battlefield weapon systems.

So when the Egyptians and Syrians, after a prolonged build-up and following the so-called War of Attrition, decided to take the offensive in 1973, they tackled an opponent who was not only relaxing on a Holy Day, and militarily off-balance, but was sublimely over-confident behind fortifications along the banks of the Suez Canal and in the Golan Heights. Moreover, the Egyptians were aware of Israeli armoured counter-attack plans if a crossing of the Canal took place, and could thus lay an ambush in the assurance of predicted enemy moves! They would infiltrate some 8000 troops armed with 2000-metre-range Sagger ATGW and short-range RPG 7 anti-tank

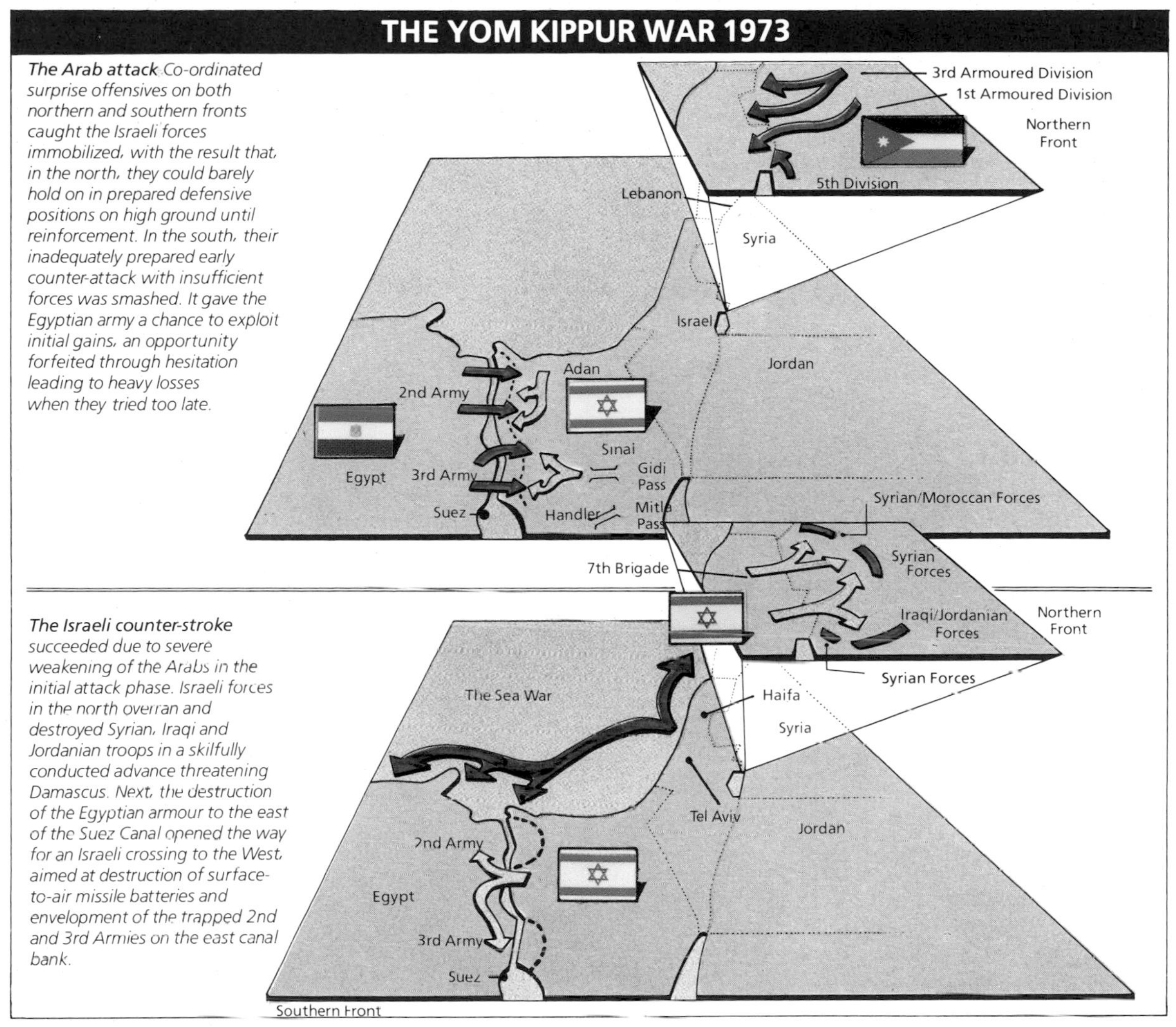

THE YOM KIPPUR WAR 1973

The Arab attack Co-ordinated surprise offensives on both northern and southern fronts caught the Israeli forces immobilized, with the result that, in the north, they could barely hold on in prepared defensive positions on high ground until reinforcement. In the south, their inadequately prepared early counter-attack with insufficient forces was smashed. It gave the Egyptian army a chance to exploit initial gains, an opportunity forfeited through hesitation leading to heavy losses when they tried too late.

The Israeli counter-stroke succeeded due to severe weakening of the Arabs in the initial attack phase. Israeli forces in the north overran and destroyed Syrian, Iraqi and Jordanian troops in a skilfully conducted advance threatening Damascus. Next, the destruction of the Egyptian armour to the east of the Suez Canal opened the way for an Israeli crossing to the West, aimed at destruction of surface-to-air missile batteries and envelopment of the trapped 2nd and 3rd Armies on the east canal bank.

launchers, installing them two miles inside enemy-held territory. Yet, apart from this massed use of ATGW in an offensive-defensive role, there was little innovative about the Egyptian daylight crossing of the Canal, which was covered by the fire of more than 2000 guns and mortars — except, perhaps, for the use of high-pressure water hoses to breach the 80-ft embankment of the Canal side to expedite construction of crossing sites through which heavy equipment could be moved.

It was meticulous planning and execution by well-rehearsed troops that enabled 1000 rubber boats carrying the ATGW squads to reach their objectives on schedule; for key Israeli strong points in their Bar Lev line to be captured by infantry and engineers; and for 60 'cuts' to be made in the banks in time to build bridges and put the armour across at night. And it was careless, arrogantly-launched Israeli tank attacks, insufficiently supported by artillery and infantry, which brought about their downfall in the ATGW ambush. But far worse was to occur two days later when the Israelis, already fully extended by the heavy Syrian offensive in the north, attempted a major counter-offensive with numerically inferior forces against the Egyptian penetrations. Unco-ordinated, due to command confusion; ill-founded, due to lack of information about the enemy; and conditioned by a belief that the Egyptians (being Arabs) would collapse once pressed, it also lacked adequate support since the Israeli Air Force, too, had been ambushed by the Egyptians.

When Israeli jets attacked they unexpectedly found themselves the target for the latest heat-seeking SA 6 and 7 missiles, as well as the older SA 2s and 3s. Pilots

Israeli Armour In The Yom Kippur War

This desert war, dominated by armour and air power, gave indications of the potential and weaknesses of high technology and missiles in opposition.

Right: *A US-built M60 tank supporting mechanized infantry in Second World War Whyte half-tracks.*

Above: *A British-built Centurion tank advancing special defensive devices into a minefield gap.*

Top right: *A captured Russian T-55 tank equipped with its 100 mm gun.*

who had learnt during the War of Attrition to dodge a single missile found three or four at once a handful. They also discovered that when they tried to knock out the SAM sites from low level, the fire from the protecting ZSU 23-4s was highly lethal. Shaken and suffering heavy losses, they not only frequently failed to hit targets but accidentally attacked their own ground forces. It was a technical as well as a tactical rebuff.

When the main Israeli counter-attack opened on 8th October it ran into an unshaken opponent who declined to bow to shock action. Most Israeli attacks tended to culminate in an unsupported tank charge which invariably was met by long-range picking off by Sagger ATGWs, backed up by tanks in sniping positions, artillery and rocket concentrations, and by fusillades of RPG 7s against those vehicles which came within close range. The spread of ranges at which effective fire was opened was in itself remarkable — anything from 2500 metres down to point blank, the tanks manoeuvring in clouds of dust which hindered one side as much as the other. By mid-afternoon the Israelis had been held and an Egyptian counter-attack was starting against shattered survivors who now had the setting sun in their eyes. The Egyptians had won and stood poised, according to plan, to seize the key passes through the hills to the east as the Syrians moved down from the north.

Simultaneously with the attack on the Canal, a strong Syrian armoured force had struck at the Golan Heights where, from good natural defensive positions held initially by meagre forces, the main clash of armour would erupt as the Israelis, committed to crushing the Syrians before the Egyptians, threw in strong reinforcements. From prepared, hull-down positions on ground of their own choosing, the thin but ever-deepening line of Israeli tanks was able to destroy at something like a ratio of 10 tanks to 1 against Syrians endeavouring by sheer weight of numbers to rush the vital crossings over the River Jordan in their enemy's rear. Only in the air did the battle on this front assume a similarity to that in the south. SAMs took their toll of Israeli aircraft, losses of which on all fronts during the first week of this war would amount to approximately 80 machines, nearly all downed by ground-based missiles and guns.

The initial Syrian onrush was held short of its objectives in daylight, and the night action did not go their way either. For in this, the first-ever major battle between armoured forces in which both sides were using infra-red night vision equipment, the natural advantage with the defenders was even more pronounced than by day. AFV crews endeavouring to pick their way across unfamiliar terrain by squinting at the

Above: ***Russian T-54 tank advancing with infantry and wheeled armoured personnel carriers. This tank, a considerable improvement with its 100 mm gun on the earlier T-34/85, has an infra-red searchlight mounted above the gun mantlet to aid the crew in night fighting.***

flickering green images on their screens or in their infra-red binoculars would acquire only a hazy picture of the ground and would soon tire because scanning could be maintained only for about 30 minutes without rest to relieve eye strain. So the ideal of an uninterrupted spell of night fighting was found unattainable — and unrealistic if attempted after a hard day's combat. Within 36 hours, in fact, both sides felt naturally bound to call a halt and go to sleep.

The Syrians were doomed when they failed quickly to reach the Jordan, thus allowing the Israelis possession of the vital ground. Their confidence was further sapped as they discovered that even the latest Russian tank, the T-62 with its 115-mm gun, was no match for the proven British Centurions and American M-60s of their enemy. The cramped fighting compartment of the T-62 was incompatible with prolonged combat and the vehicle's tendency to catch fire when hit was, to say the least, discouraging. Within three days the Israeli deployment of their reserves was complete and the Syrians were in shattered retreat before superior forces which exacted a toll of AFVs amounting to more than 1000 destroyed against 150 losses — a ratio augmented when an Iraqi tank brigade, bravely trying to stop the rot, was crushed by the high-velocity guns of the skilfully-handled Israeli Centurions and M-60s.

On the Syrian front, ATGW played only a minor role, not only because the Israelis fought from cover but also because, when they did move in the open, they either screened the ATGW operators with smoke or bothered them with neutralizing fire. ATGWs take up to 30 seconds to fly to their target and a nervous operator's aim can be upset by a shaky finger.

The Egyptians, who at last (and half-heartedly)

AIR DEFENCE IN THE 1970s

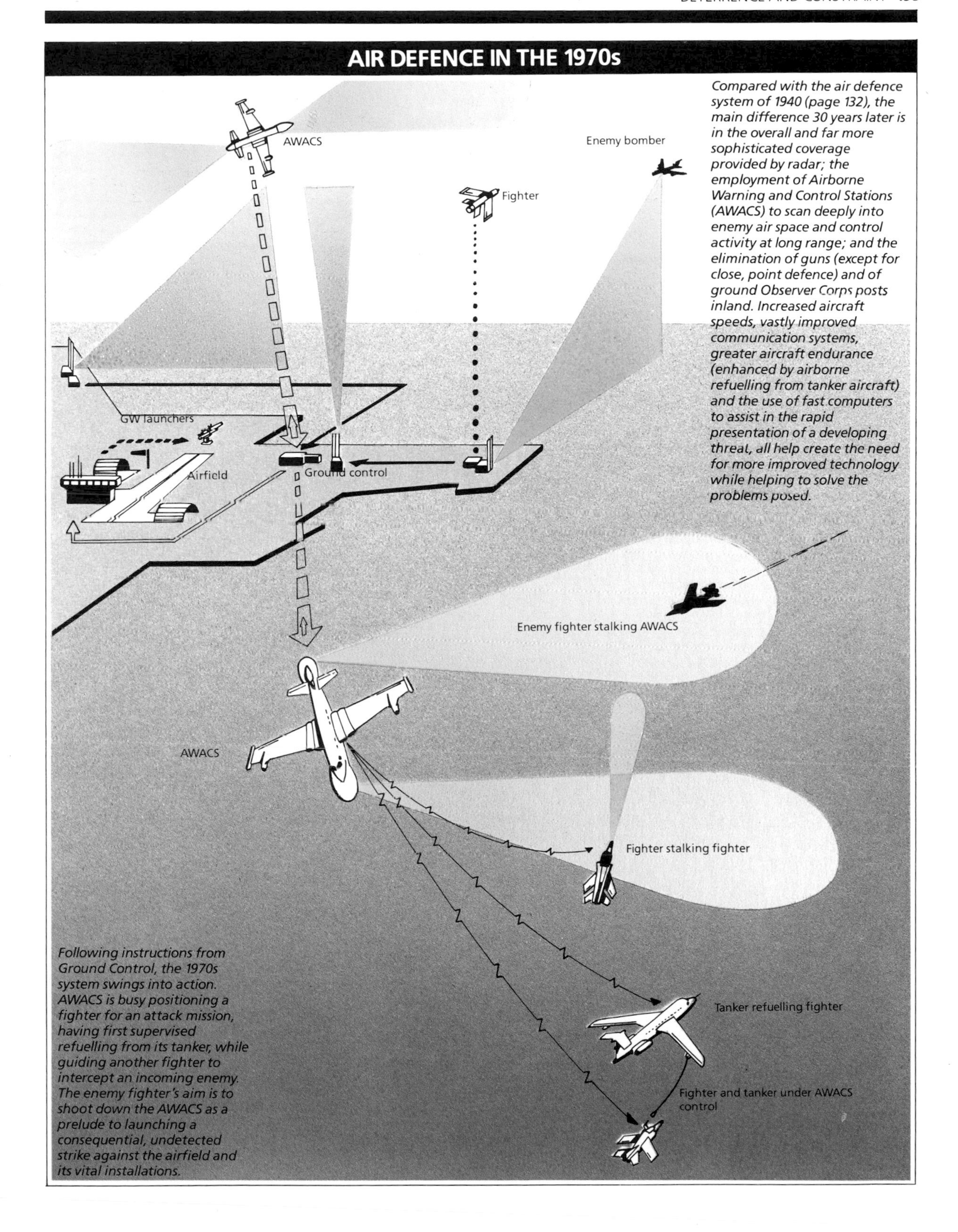

Compared with the air defence system of 1940 (page 132), the main difference 30 years later is in the overall and far more sophisticated coverage provided by radar; the employment of Airborne Warning and Control Stations (AWACS) to scan deeply into enemy air space and control activity at long range; and the elimination of guns (except for close, point defence) and of ground Observer Corps posts inland. Increased aircraft speeds, vastly improved communication systems, greater aircraft endurance (enhanced by airborne refuelling from tanker aircraft) and the use of fast computers to assist in the rapid presentation of a developing threat, all help create the need for more improved technology while helping to solve the problems posed.

Following instructions from Ground Control, the 1970s system swings into action. AWACS is busy positioning a fighter for an attack mission, having first supervised refuelling from its tanker, while guiding another fighter to intercept an incoming enemy. The enemy fighter's aim is to shoot down the AWACS as a prelude to launching a consequential, undetected strike against the airfield and its vital installations.

attempted to advance on the passes in support of their hard-pressed allies to the north, now found themselves at the disadvantage of most tank forces when approaching high-velocity guns in the open. When they attempted to insert tank-hunting teams among the Israelis by helicopter, these extremely vulnerable aircraft were annihilated. And when their tanks repeatedly tried to get to grips, they were assailed by a hurricane of gunfire from tanks which no longer made the mistake of charging. Moreover, the Israeli Air Force came back into its own once the Egyptians moved beyond the umbrella cover of their SAMs. Air combat went almost entirely the way of the Israeli pilots who brought down the bulk of 440 Arab machines destroyed — only some 25 being credited to surface-to-air missiles. Air power and US satellites had already provided the Israeli command with vital information about the Egyptian columns approaching the passes. Now they plotted the location of the main positions guarding the Canal.

When the Israelis took the offensive with the aim of crossing the Canal to knock out the SAM batteries and cut off the main enemy force from its base, their plans of deception and main punch were precisely directed by information from the latest cameras, sensors and communications. The operation was expedited because prefabricated bridges were nearby, ready instantly to carry strong armoured forces to the west bank and complete the defeat of a weakened opponent by a classic penetration operation.

New technology, gun and missile also contributed significantly to skirmishing at sea between 14 Israeli power boats, armed with 76-mm guns and the French Gabriel radar-guided missile (with a range of 12 miles), and Egyptian and Syrian torpedo boats armed with the Russian Styx missile (range 25 miles). One of the latter had scored a 'first' during the Six Day War when, launched from the shelter of Alexandria harbour, it had sunk a destroyer at sea. Rarely caught napping twice, the Israelis held back until enemy boats on a bombardment mission on the first night of the war came close inshore. Then speed and agility baffled the Styx-armed boats which were routed without sinking a single Israeli vessel. But the threat of sea-skimming, radar and infra-red guided missiles was very real, as the Israelis showed when they took advantage of the Gabriel's long-range accuracy effectively to engage Syrian boats sheltering in harbour. It was bad luck on the three neutral merchant ships which were also destroyed by the missiles, but homing devices such as these had yet to be given the ability to discriminate between friend and foe — let alone unwilling spectators.

In the aftermath of the Yom Kippur War a wide-ranging reappraisal of waging war with new technology was set in motion — and not simply because so much new equipment was used and shown to be effective or wanting. The unbridled and often uneducated interest of the media, notably with exaggerations of the effect of missiles on land, at sea and in the air, created false impressions and delusions. Pernicious reports which favoured the rejection of some viable systems in favour of the acquisition of devices of dubious quality and application had to be resisted. Programmes to acquire, for example, much-needed and operationally sound aircraft and tanks were threatened by competing agencies seeking advantage on behalf of diplomatic, political, industrial, financial and personal interest. Sometimes it is easier to influence a busy decision-maker with popular distortions in capsulated form than it is to convince him of grim essentials by an unavoidably lengthy and scientific dissertation. It is always central to the technological struggle that acquisitions are essential and within budgets. It is also part of modern confrontation that an opponent should be deluded into unnecessary, wasteful and bankrupting purchases.

Regardless of the efforts of those who sought disarmament or arms limitation, the proliferation of weapons continued unabated in the decades following the Vietnam and Yom Kippur wars. Nothing in history promised otherwise. Mankind remained true to his inherited assertive and self-protective motivations, closely related to the pursuit of causes in a period of worsening political and economic instability. Worldwide in 1978 there were no fewer than 1000 terrorist incidents (compared with 100 in 1968) in addition to some six small wars — virtually all politically inspired and the majority intertwined with the contest between the USA and Russia. By 1985 annual expenditure on weapons was running at £630 billion without hint of reduction, but with every danger of worsening the financial situation of all, including major contenders, in the arms race.

The Falklands War Of 1982

The power and diversity of the latest technology was illuminated when Argentina by surprise made an old-fashioned amphibious conquest of the Falkland Islands in April 1982 and presented the British with the task of reconquest in an unexpected region. Being able to launch a strong amphibious task force at less than a week's notice and accomplish the mission in less than three months would have been impossible two

Weapons Proved In The Falklands Crisis

Above: ***Atlantic Conveyor*** *holding stores and containers for the campaign. Large container ships are easily modified to operate vertical VSTOL machines and open up endless possibilities in combat and logistic application. Already aboard are two twin-rotor Chinook helicopters and landing is a vertical take-off Harrier fighter.*

Left: *Fitting the Sidewinder air-to-air missile to one of the Royal Navy Sea Harriers. This improved weapon gave the Harrier a distinct advantage over enemy aircraft.*

Tactical Missiles

The latest generation of ground-to-ground missile, the Pershing (above) on take-off. This weapon, with its simplicity of launching arrangements, can be teamed with the Tomahawk (right) to provide an effective strike capability hard to intercept since it does not depend on external control systems to find its way to the target.

Centre left: *Tomahawk cruise missile launched from mobile, land-based vehicle which considerably enhances its protective mobility.*
Left: *Approach of a Tomahawk cruise missile to impact on target. This particular missile was launched from a submerged submarine off the southern Californian coast. Here it is striking the target — a reinforced concrete structure the size of a warehouse.*

This warhead contains conventional explosive, but it could be nuclear material, or a special warhead designed to disrupt airfield surfaces.

decades previously. Impossible, that is, without a computerized logistics system which could rapidly control the assembly of vast and complex quantities of men and stores, subsequently moving them in the right order to meet every foreseeable contingency in rough waters and wild terrain 8000 miles from Britain.

The whole operation would have proved slow and vulnerable had not flight refuelling made it possible to fly short-range aircraft long distances and have them ready to attack the enemy as well as augment supplies without delay. The supreme example of this was the 7800-mile-round flight of a Vulcan bomber to drop 21 1000-lb bombs on the runway at Port Stanley. To do so required 17 flight refuellings, each demanding exact rendezvous in mid-ocean.

The task force would also have seen severely hampered if satellites had not only been aloft but also manoeuvrable enough to cover regions previously considered unlikely as combat areas. As it was, those at sea or on the ground in and around the Falklands were able to exchange messages with sufficient speed and volume to meet exacting calls for information and execution; and were able also to navigate accurately and, on many occasions, receive vital information about enemy intentions, particularly those of his ships. It was, for example, a feat of communications that a

The New High Ground

The introduction of reconnaissance satellites represents the latest stage in the age-old struggle for command of the heights in the endeavour to see what the enemy is doing on the other side of the hill. But this technology is now so enormously complex and expensive, and the results so revealing, that only the superpowers can afford such surveillance.

Above: *A photograph of Moscow taken from space on a very clear day shows the sort of detail that can be obtained. Excellent results can also be gained through cloud with infra-red photography.*
Right: *Illustration of surveillance of the earth's surface by various space vehicles, such as the shuttle, and orbiting satellites which, by one form or another, transmit their information to earth.*

British nuclear submarine could be accurately directed to intercept units of the Argentine fleet. For while submerged, it was able to hear long-wave radio instructions from London and, at decisive moments, momentarily raise its aerial to send a compressed 'burst' transmission (impossible to pin-point by DF) to report changes of enemy course. And it was a turning point in the campaign when, as the culmination of such communications, a submarine sank an Argentine cruiser and, by using its speed and depth characteristics, remained on hand to scare every enemy surface ship back to port. The Argentine's submarines were constantly kept at bay by British ships and aircraft fitted with the latest sonar and magnetic anomaly detectors; and by helicopters with dipping sonar, one of which managed to cripple a surfaced submarine with sea-skimming misssiles.

Indeed the consequences would have been disastrous had not British aircraft demonstrated their versatility. Constantly, helicopters moved supplies; repeatedly they saved men from burning ships, evacuated the wounded, engaged enemy targets with missiles, and acted as decoys against incoming enemy missiles — particularly the French-built Exocet sea-skimming, radar-homing missiles which sometimes hit and sank ships lacking the necessary ECM. But no counter invasion would have taken place had not Argentine air power been defeated. As it was, British missiles succeeded in their job of scoring enough 'kills' and providing distraction to pilots who were compelled to fly too low or evade too violently to destroy their targets (the failure rate of Argentine bombs being contributory). The vertical take-off Harrier fighter, with its capability to operate even from merchant ships, was able to outfly faster opponents and shoot them down with the latest Sidewinder AIM9L missile. This had an unprecedented 90% success rate due to an improved guidance system and agility that eliminated its predecessors' limitation of engaging only from astern. The result of the campaign would probably have been different if all the Argentine forces had been, like their airmen, highly trained, regular professionals, as indeed all the British were. The fine performance of the Argentine aircrew indicated, as so often, the vital power of a small élite compared to an ineffectual, poorly-trained conscript mass when confronted by an opponent whose technology was superior and knew how to make the best use of it. If nothing else, the Falklands underlined that the latest technology, handled by trained men, will frequently defeat an amorphous mass — always with the proviso that the minority can conserve its strength and has adequate reserves of similar quality to call upon.

The Space Age

President John Kennedy's determination for prestige and strategic reasons to put a man on the moon triumphed on 20th July 1969 and indicated that American space technology had overtaken Russia's. Although both nations in the next two decades sent up hundreds of orbiting satellites and several deep-space exploration vehicles, it was noticeable that the Russians seemed to concentrate more on the military sphere, with missiles and reconnaissance satellites of shorter life, while the Americans moved further ahead in all fields with considerable emphasis upon civil applications for mapping, navigation and communications — all of which also had military uses, of course. Benefiting from micro-technology and the utilization of an immense range of lightweight materials suited to extreme conditions, such as titanium alloys, plastics and carbon fibres (most of which also gave ordinary people a higher standard of living) it became possible from space to inspect rapidly and in detail with a plethora of cameras and sensors every inch of the earth's surface (and even examine the ocean's depths). The results could be sent to earth by radio link or capsule; an exact position in space or on the ground could be pin-pointed by the touch of a switch; clear pictures and messages could be re-broadcast with the greatest ease anywhere; and multiple weapons could be deployed and intercepted in space with assurance and accuracy. And while the scale and magnitude of this revolution was awe-inspiring, even if extensively cloaked in secrecy, its implementation would have been impossible without the latest computers to acquire, store, generate and speedily transmit for display enormous quantities of information in addition to playing an automated part in actuating and monitoring the decisions of those supposedly in charge.

Weapons themselves had also reached new peaks of power, adaptability, accuracy, invulnerability and reliability. Nowhere was this more pronounced than in missile surveillance and submarine technology, which assumed a dominant influence overall.

Missiles had largely overcome earlier unreliability in parallel with better performance, reduced size and enhanced mobility. They incorporated micro-

The Modern Main Battle Tank

The modern main battle tank (MBT) tends to be developed with a cross-country speed well above 20 mph; strong protection by well-shaped, composite Chobham-type armour against high-velocity armour-piercing shot and lower-velocity chemical energy warheads; a sealed environment against nuclear and chemical effects; and a high-velocity gun of about 120 mm which can fire armour-piercing shot in excess of 5000 fps, high explosive, smoke and illuminating rounds.

The heart of the tank is to be found in the fire control equipment under the commander and gunner. Pin-pointing of targets by day, by night and through mist or smoke is obtained, in addition to optical scanning, by a variety of complementary devices:
1. By active infra-red or white searchlight illumination.
2. By passive:
• Far infra-red thermal imaging equipment which can detect targets through light camouflage and smoke and at night.
• Electronic image-intensifiers using ambient light at night.
• Low-light TV at night.
3. By active radar (although this is rarely mounted on MBT).
4. By detection of enemy infra-red radiations through passive viewing devices.
Electronic images are displayed on a CRT to the commander and gunner. The former can exploit these for tactical purposes or help the gunner select the right target – either through vision screens or telescope. Laying on the target is controlled by computer once it has been fed information concerning atmospheric and wind conditions, range (by laser) and bearing of target, and ammunition to be fired. With this in his possession the gunner, or the commander, has only to aim at the target's mass to score a hit with the first round.
The turret and gun are electrogyro stabilized to keep the gun on a constant bearing in elevation and azimuth, making it possible to fire while on the move with a reasonable

electronic memories with overriding systems to make their guidance proof against decoys and jamming — or even independent of external active systems such as radio signals or infra-red homing. For example, the American sub-sonic Tomahawk 'cruise' missile (a descendant of V 1 but with a range of 1500 miles) could not only hug the terrain to reduce chances of detection, but linked its inertial guidance system to a pre-recorded radar map in order to find its own way to the target. So also did the Pershing Mach 8 two-stage rocket (a descendant of V 2 but with a range of 1000 miles).

No wonder the Russians applied intense political pressure in attempts to prevent the deployment of weapons which promised to be extremely difficult to detect and defeat, as well as exceedingly accurate in all conditions. For the time had arrived when it was also extremely difficult to conceal vital targets and to achieve surprise by deception. To supplement the existing familiar 'near' infra red (0.7 — 1.0 microns wavelength) 'active' night-vision instruments, there entered service 'far' infra red (1.0 — 100 microns) 'passive' devices which could display on a CRT a 'thermal image', even of objects behind camouflage nets or natural cover. Furthermore, the smallest target, such as an AFV, could be laser-designated and engaged by lightweight fire-and-forget missiles. No longer did the ATGW operator have to cope with complicated procedures and prolonged tracking of the target to the detriment of nerves and accuracy.

As for the latest submarines, the small, ultra-silent diesel kind, as well as the 25 000-ton Russian Typhoon class, with a reputed speed approaching 40 knots and the strength to dive between 2000 and 3000 ft, could not only baffle the latest passive, ship-borne and

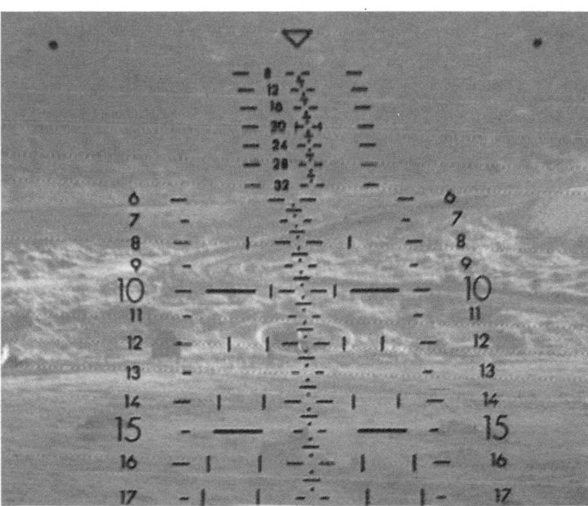

chance of hitting the target. Some MBTs are fitted with automatic loaders but most still employ a crewman who also acts as radio operator. Ammunition over 110 mm calibre is usually separated, its charges stowed in water-enclosed containers to minimize the risk of serious conflagrations.

***Centre left**: Abrams main battle tanks.*
***Left**: Laser sight graticule and ballistic aiming mark.*

***Below left**: A detail of the Challenger MBT showing the external components for the fire control system. These are principally the Thermal Imaging Sensor Head (TISH) in the barbette and the input/output window of the Barr and Stroud laser rangefinder/optical sight.*

seabed detectors, but made necessary a new generation of complex weapons which could plunge to extreme depths without being crushed by water pressure. High explosive depth charges were almost out-moded; torpedoes did not guarantee a kill; small nuclear depth charges were perhaps the only sure antidote to the deepest diving boat which, equipped with an assortment of missiles and torpedoes, could tackle any target at sea and a great many far inland.

No weapon system was invulnerable — not even Mach 8 missiles in space, against which anti-missile missiles, controlled by the most complex and expensive electronic warning, detection and guidance systems, were installed to give sufficient protection for vital centres. And as for the latest Mach 2 +, variable geometry multi-purpose aircraft, such as the European Tornado, these were anything but safe. Even with air-to-air radar over a range of 100 miles, with supersonic missiles capable of taking on multiple targets out to 25 miles, computer control to enable the crew to 'fly-by-wire' (derived from space technology), extensive ECM to baffle enemy missiles, and armed with all manner of weapons to home in on enemy radar, missile sites and airfield runways (quite apart from all other kinds of target), still such an aircraft was vulnerable. Its own airfields might be put out of action before it could take off, or the enemy might pierce its ECM defences. Or the fortunes of war might simply be adverse and allow destruction by old-fashioned cannon fire from an opponent who, by guile, came closer than desirable. Tactics and the electronic chess game were as potent as ever. The human factor remained omnipotent, even though the dawn of the computer which could think and originate for itself was nigh.

It was the same on land. AFVs protected by the latest Chobham-type armour, which could defeat both high-velocity shot and the latest chemical energy projectiles, could still be penetrated by clumsier, enlarged missiles; by mine blast through the thinner belly plates; or by bomblets showered upon the thinner top armour. True, the latest tanks containing 'far' infra-red surveillance and laser ranging sights, were equipped to hit targets first time before coming under fire. But that was no guarantee against ambush, nor was armour proof against a nearby nuclear burst.

Below: ***US F 15 fighter firing a fire-and-forget missile of the sort which has been used to home onto a satellite in space.***

Chapter 7

Towards Armageddon? 1987-2000

Until 1984 the concept of deterrence by the fusion bomb remained unchallenged, and there resided a pious hope that nobody could be so stupid as to start using even a tiny proportion of the 15 000 megatons reckoned to be stored away in various arsenals, missile sites, aircraft and submarines at sea. Then computer projections and studies based on examination of the after-effects of gigantic volcanic eruptions suggested that, in addition to the catastrophic damage caused by fire, blast and radiation following a nuclear attack, the massive quantities of debris and smoke which would be thrown up into the stratosphere might well blot out sunlight for several years. This would result in lower temperatures on earth (to −30°C) and bring about a condition of so-called Eternal Winter to kill livestock, bring desolation to the environment and produce total famine. Unproven and challenged as this theory had to remain, the dangers could not be ignored. Certainly, doubts were injected into the minds of those contemplating the use of fusion and very large fission explosives. Regardless of how safe an élite leadership might selfishly feel from nuclear attack in the deepest, air-conditioned shelter, the nagging fear of doom from a self-induced destruction of Nature could not be glossed over. In a manner of speaking, a deterrent to the deterrent had been promulgated. The fusion weapon — along with bacteriological weapons of indefinable harm — had to be outlawed. But the less devastating fission and chemical weapons somehow remained within concepts of tolerance.

In consequence, the scenario of Global War had to be rewritten for the later 1900s with American and Russian leaders, most of all, forced to reconsider the viable options and accept much stricter limitations to their contingency plans than previously.

Hereafter everything is conjecture, with history giving way to selective imagination in what follows. Certain options are put forward as possible courses of action — always on the strict understanding that nothing is impossible and that events may bear little resemblance to what is pictured here. What follows, indeed, is not a synopsis of a Global War history but pointers to the effects of the latest technology upon war in space, at sea and on land.

On the assumption that all-out and widespread attacks by the larger nuclear weapons are excluded, the options presented to a major aggressor in 1995 could be as follows.

1. Attempts to seize key localities, worldwide,

through insurrection, subversion or invasion (or combinations of all three) but using minimum force. In other words, an orchestrated projection of the process long practised by Russia and, to a lesser extent, by America, to achieve a dominant strategic position with the aim of imposing a political solution, short of major war.

2. An attempt to win a strategic advantage by limited nuclear war at sea, in barren areas (notably the Arctic region) and in space, in order to dictate a political solution without recourse to massive nuclear strikes and the attendant risk of Eternal Winter.

3. An attempt by Russia, without warning, to seize Western Europe by invasion after NATO has been further weakened by disagreements within the alliance and rejection of the latest most potent weapon systems, including small nuclear devices. In fact, an old-fashioned war of military conquest by the masses, shorn of subtlety!

Option 1 is obviously the most attractive to Russia, and could be successful, but might forfeit its attraction if, for example, Russia's ruling hierarchy found itself in difficulties (maybe because of famine and internal dissent). *Option 3* might be impossible to achieve by surprise and could lead to stalemate and/or an all-out nuclear exchange which none desired. Therefore, it might only be undertaken if *Option 2* had achieved marked initial success in the train of undermining by *Option 1*. For argument's sake, therefore, *Options 1* and *2* will be discussed as being more probable, with *Option 3* finally included as a possible event if one side or the other finds itself in desperation.

Option 1 — The Struggle For Key Points

Sophisticated late 20th-century society, with its dependence upon technological centralization, offers rich dividends to predators who strike at its muscles through brain and stomach. Vital targets for a strategy of disruption are signal communication centres, major computers controlling national defence and economic organizations, sources of fuel and power (above all electric power) and means of transportation. Deny or destroy these on a large enough scale through political intervention, civil disturbances, strikes or, as a last resort, by destruction, and a nation as great and complex as the USA or Russia could be weakened or stunned for a sufficient period to produce spasms of collapse. At the heart of a worldwide campaign of co-ordinated disruption, and of the counters to it, would also be comprehensive and secure communication systems to permit the aggressor contact with a host of agents and cells pre-positioned within the selected target areas.

Left: ***MX, the latest orbiting space missile system produced by the USA. It has the capability of releasing a number of accurately guided missiles to targets on land.***

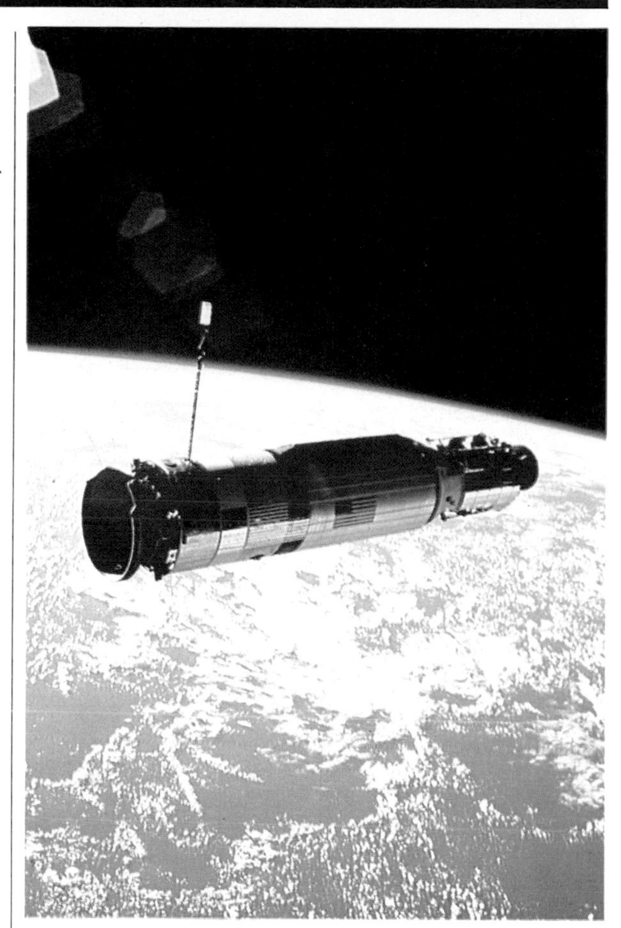

Above: ***Agena target docking vehicle, used for experiments to show the feasibility of homing onto an enemy satellite in orbit.***

Politically and geographically, by 1995 the advantage will lie with Russia. It will be far easier for her to infiltrate the fabric and centres of democratic Western nations, which thrive upon a relative freedom of social intercourse and cross-frontier movement, than it would be for America and her allies to penetrate Russian territory. With Russia's opponents held in suspense in nearby territories such as Western Europe, friendly Middle East and Far East countries, she could foster aggressive elements within the USA while enjoying, too, the benefits of proximity to Central America and the Caribbean Sea as well as the Northern Pacific area.

The true causes of a Third World War may never be known, if only because diplomacy, Russia's above all, is inherently secretive. But the signs of its coming would be impossible to conceal no matter how strenuous the attempts to do so. Modern organizations which depend upon massive communication networks and high technology are bound to expose many secrets to a technologically strong opponent. The test comes when those who synthesize the available intelligence have to translate it into reports upon which action is based. Patterns of aggression provide warnings which trigger counter-actions. As *coups d'états*, industrial strikes, sabotage and civil dissent proliferate, those under threat will have to take counter-measures which will absorb effort but will provide increased readiness against emerging threats. As one aggressive squeeze, with its attendant counter-measure, follows another, undercover operations will lead to overt moves in the historic Soviet manner. Most noticeable of all would be associated maritime deployments — evidence of an increase in the number of surface and undersea units at sea, extra activity in dockyards, changed patterns of commercial shipping movements close to sensitive regions, and redoubled surveillance by surface craft, aircraft and satellites seeking to track potentially hostile elements and discover their intentions. Indeed, the measures taken by both sides to intercept and search merchant ships suspected of carrying raiding troops with helicopters, weapons, and even nuclear charges into port areas, would probably initiate the next stage of hostilities.

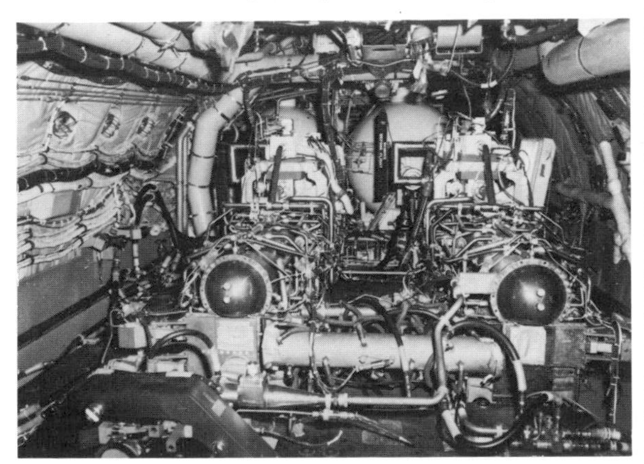

Above: ***An experimental, airborne laser weapon system set up to discover the feasibility of destroying targets in space or perhaps even on land. Primitive though it is within a KC 135 tanker, it does indicate the complexity of such equipment with its fuel tanks and various devices to make it function.***

Option 2 — Limited Nuclear War

It could well be attempts to interfere with space surveillance systems which would trigger *Option 2*. A process of satellite destruction would quickly get into top gear as both sides seek to blind the other. Russia would probably have the edge because, true to history, she tends to invest in quantity before quality. She has invariably had four to five times more satellites in orbit, because they were short-lived, for every one of the more durable kind built by the USA. As a result, the USSR would have fewer targets to attack and a greater capacity to replace her own losses as they occurred, having, at the same time, reached near-parity with the USA in the development of attack systems. Not that the Space War would be conducted solely by satellites. The handful of 50-ton, re-usable manned space vehicles (the American 'shuttles') and the smaller reusable space planes and tugs which the Russians are developing would come to grips. Energy beams would also be employed for the first time, but with minimal effect. For although earthbound particle-beam projectors would be available, they are likely to be still in the development stage, and the more advanced chemical laser beams, while capable of crippling satellites, will require considerable power and be by no means easy to aim and operate. It would be the much more cost-effective aircraft- and ground-launched missiles which would reduce the satellite population and also hit the few and expensive space vehicles.

At the same time the missile-versus-missile war would be in progress, with attacks upon static launching sites and depots in the less populated hinterlands of Russia and the USA, and by anti-missile missiles attempting to intercept incoming projectiles. Mostly, at first, these would be armed with high-explosive warheads. It would only be later, as the conflict became more general, that low-yield fission weapons would come to be used as the contenders either grew more desperate in their endeavours to eliminate their enemy's surveillance capability or felt confident enough to risk reprisals.

There would doubtless be a few surveillance satellites which would go unrecognized among the many communication and navigation types which would tend to survive unmolested. Both sides probably would prefer to preserve communication and

Development Of The Shuttle Programme

Above: *The space shuttle. The concept of a reusable space vehicle was a revolution in space travel and warfare.*
Left: *The Martin Marietta X - 24A lifting body first flew in 1969. It was a test vehicle in a series of experiments into winged re-entry from space. This research was to be vital to the development of the space shuttle, yet had its origins in the early 1960s. The Boeing Aircraft Company, in response to demands for a manned orbital weapons system, developed the X-20 Dyna-Soar (dynamic-soaring) and had produced plans for a series of space bombers before the project was cancelled in 1963.*

Manned Orbital Weapons Systems

This impression of a space shuttle under development by both the USA and USSR illustrates essential components of the revolutionary recoverable vehicle. In space it behaves like a satellite, but in the earth's atmosphere can be controlled like a conventional aircraft during approach and landing after a mission.

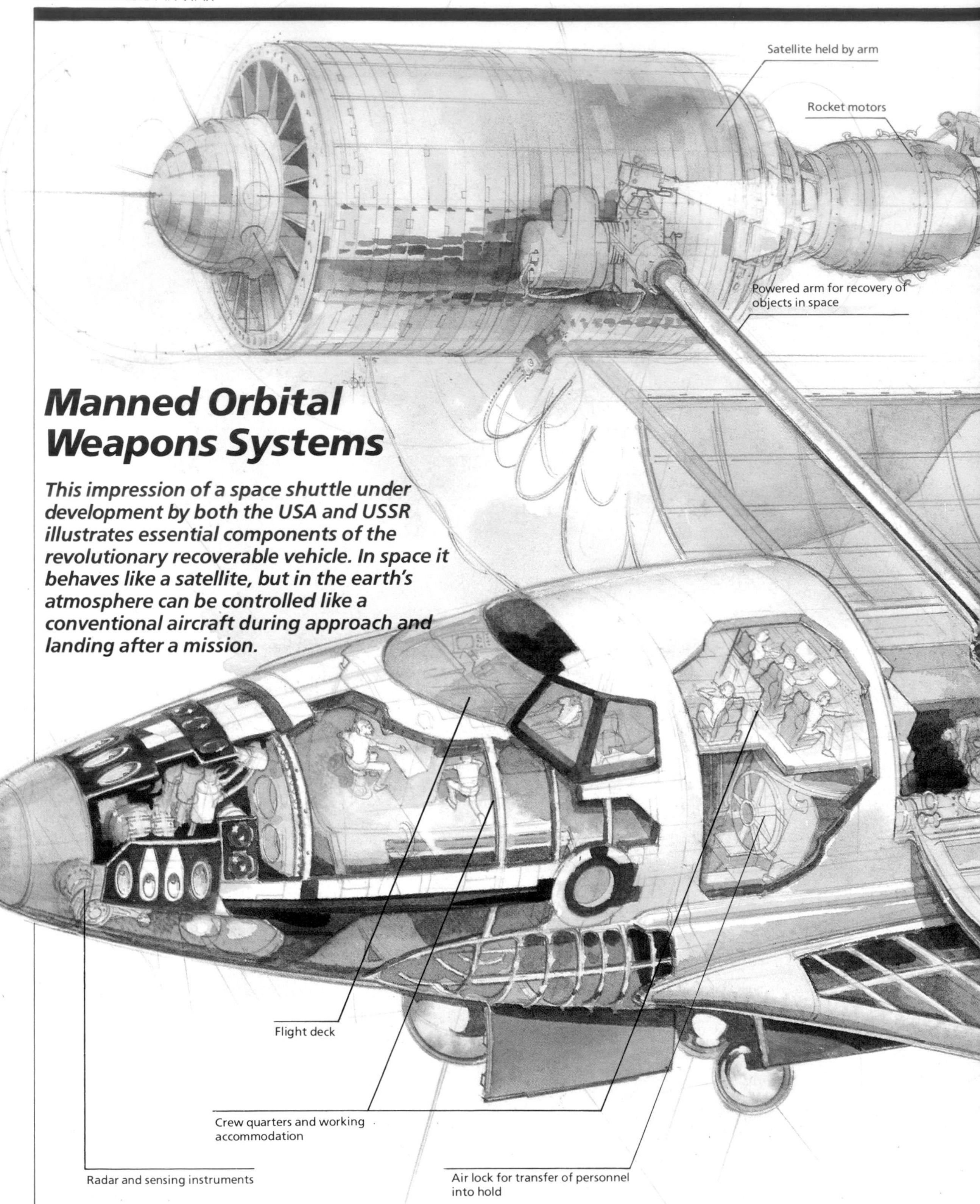

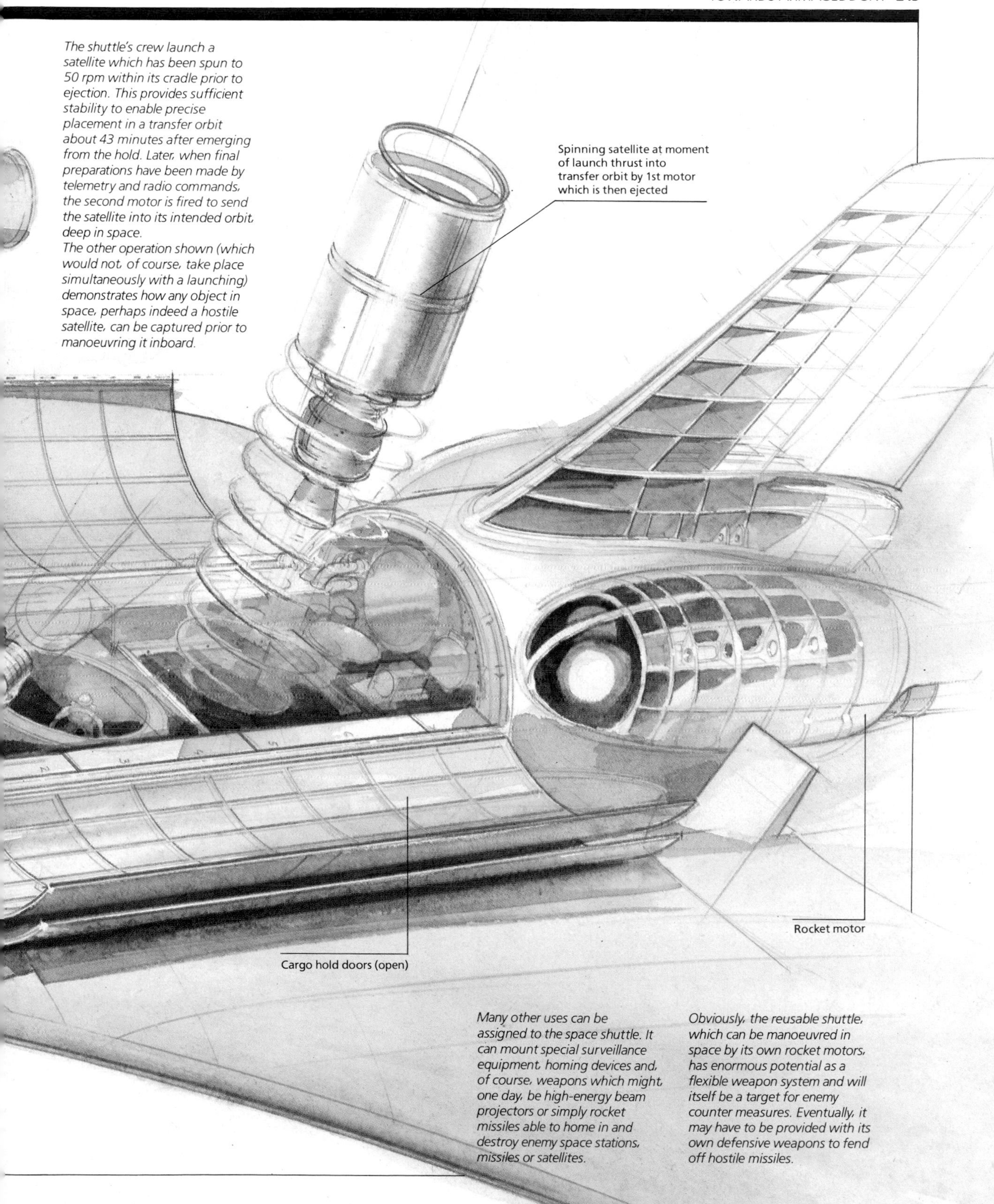

The shuttle's crew launch a satellite which has been spun to 50 rpm within its cradle prior to ejection. This provides sufficient stability to enable precise placement in a transfer orbit about 43 minutes after emerging from the hold. Later, when final preparations have been made by telemetry and radio commands, the second motor is fired to send the satellite into its intended orbit, deep in space.
The other operation shown (which would not, of course, take place simultaneously with a launching) demonstrates how any object in space, perhaps indeed a hostile satellite, can be captured prior to manoeuvring it inboard.

Many other uses can be assigned to the space shuttle. It can mount special surveillance equipment, homing devices and, of course, weapons which might, one day, be high-energy beam projectors or simply rocket missiles able to home in and destroy enemy space stations, missiles or satellites.

Obviously, the reusable shuttle, which can be manoeuvred in space by its own rocket motors, has enormous potential as a flexible weapon system and will itself be a target for enemy counter measures. Eventually, it may have to be provided with its own defensive weapons to fend off hostile missiles.

*Above: **The US carrier** Carl Vinson **has on deck a vast array of aircraft of different types for different tasks. Powered by nuclear energy, the ship can stay at sea for a considerable time provided it is replenished. It represents a powerful weapon system but is also a prime target in the event of all-out war. Note the aircraft ready for catapult launch.***

navigation facilities for their own purposes, which would seem more beneficial than denying them to the enemy. The detonation of nuclear weapons in space, with consequent disruption of all radio communication through disturbance of the ionized layers, would be avoided since to retain maximum communications for command and control of global operations would be deemed vital. Selective jamming, co-ordinated with specific operations, is generally rated more profitable since it can be switched on and off as required. But anti-jamming measures (including 'random frequency hopping' radio sets which could avoid jamming altogether) would enjoy considerable success in keeping radio channels working. Moreover, it is probable that dense land-line networks, many comprised of fibre optic cables capable of carrying 2000 simultaneous channels at rates as high as 140 pulses per second, would adequately cater for most signal communication needs. Furthermore, an enemy broadcasting system which could be monitored is a source of intelligence not to be lightly forfeited.

Important as the space campaign would be in denying information to the contenders and rupturing anti-missile defences, it is the sea war and its overspill, in terms of destruction to ports, upon which most attention would be focused. Within hours of the first actions involving suspect shipping and the related outbreak of anti-satellite operations, major naval units could be expected to clash. Both sides would move to protect their merchant or pseudo-merchant shipping — it would be impossible to tell one from the other. Hostile submarines shadowing aircraft carriers and other key vessels would themselves come under surprise fire aimed at forestalling their intended attacks. Within minutes of intelligence of these activities being broadcast, a great convulsion would spread across the oceans and in the narrow waters (such as those off Western Europe, the Icelandic Channel, the Caribbean, in Japanese waters and in the Red Sea and the Persian Gulf.) Located submarines would pay a terrible price for any improvidence or technical inability to avoid detection. Lurking submarines which managed to evade (and there would be many), would take a toll by long-range missile and torpedo attacks against aircraft carriers, cruisers, escort vessels and any other ships unlucky enough to get in the way. Submarines, once detected, would come under attack, both sides paying studious attention to the big deep-diving types, reserving for them the relatively limited quantity of nuclear depth charges. In the approaches to known naval bases, mines (laid to plan by air and sea) would catch several boats as they were putting to sea. Air attacks, with bomb and rocket, would strike harbours as well as supply, fuel and ammunition depots — the known instability of Soviet munitions encouraging hopes of major destruction through sympathetic detonations.

Because land warfare would start more tardily than had for decades been expected, war in the air would tend to concentrate at first upon support of naval operations in the classic manner. The struggle for surveillance would predominate, with each side attempting to use fixed-wing machines and helicopters to seek out and track enemy craft while, simultaneously, endeavouring to prevent the enemy doing the same. Conjointly with this game of hide and seek and aerial combat, strike aircraft would range far and wide, hitting enemy radar installations and airfields on land, sinking ships at sea, laying mines — trying, above all, to establish a superiority which would enable their own surface and undersea craft to operate freely and keep supplies and reinforcements moving. The planners would be aiming to preserve, unnoticed, the SSBN (nuclear warhead submarines) as the final arbiters of political pressure in the event of no diplomatic solution to the war being found.

In this struggle it would be the Americans and their allies who would have the edge, based as their sea power is on generations of naval experience, superior techniques and more reliable vessels operating from numerous, widespread ports against a Russian navy which, impressive in size though it may be, lacks ingrained expertise and is hampered by relatively limited access to the oceans. Severe as the losses would be to both sides, it is the Russians who would become the hunted, unable easily to resupply vessels at sea or replace them through narrow channels from ports which would be severely damaged. Logistics would be as crucial as ever they had been. Remain indefinitely at sea as a nuclear vessel can, it has to replenish its weapons and, as the British found during the Falklands War in 1982, even against a minor foe, expenditure of missiles, ammunition and torpedoes can be colossal. A condition of atrophy would not be felt at once. Nevertheless, the wastage of naval assets, the severe damage to radar early warning systems and the destruction of many valuable aircraft over the sea and the Arctic, with the West probably having the better of the exchanges, would present the Russians with a dilemma which suggests they would either call off their challenge (and make the best of a bad job before irreparable damage was done to their prospects) or stake everything on an escalation of the conflict by activating *Option 3* — invasion of Western Europe.

Option 3 — Conquest Of Europe

Elements of *Option 3* would have been operative from the start. The mere threat of naval warfare adjacent to such sensitive places as Northern Norway, Iceland and Greenland; in the entrance to the Baltic as well as to the Mediterranean and Black Sea; in the Red Sea and Persian Gulf; in South East Asia and the Sea of Japan; and adjacent to the Panama Canal, would make necessary the stationing of forces to seize or retain control of vital areas and air space. Troops which might otherwise have been sent to Central Western Europe would find themselves on guard or in action elsewhere, contributing to a dilution of effort on the anticipated battleground of Germany, and causing a further wearing down of men and equipment on peripheral fronts.

It is impossible to disentangle maritime from land operations. They interact upon each other. For example, a Russian attempt to seize North Norway to secure access for her naval and air forces into the Atlantic from Murmansk, would trigger the planned NATO response of sending forces as flank protection to Denmark and Germany. And the skirmishing in the Arctic by a few highly-trained units (whose technology would be as much geared to survival against the extreme cold as against the enemy) would be related to the maritime war (bearing in mind that many submarines would be lurking beneath the thick ice), to the air war (in counter-radar operations) and to the land war (as yet another distraction from the central battle arenas). But wherever land forces clashed, long-established principles would govern and human beings would be the most important factor in taking the fullest advantage of what technology offered.

At all levels, from Army HQ down to the individual soldier, firepower would be used massively to impose crippling destruction. At the same time every effort geared to survival by protection would be practised. The air war above and among the ground forces would be closer tied to these aims than ever before, because co-operation between the elements concerned has been made so much more sophisticated by elaborate surveillance and communication facilities that contribute to slick command and control. Yet the initial air battles would tend to leave the ground forces unmolested while airmen concentrated upon winning air superiority in a battle which would be related, of course, to the parallel struggle in support of navies. Central to this struggle would be the assault on radar installations and airfields. Both sides would employ élite ground raiding forces (some inserted by helicopter or parachute, some indigenous) to sabotage vital installations, cripple equipment, destroy fuel and ammunition supplies and crater runways. Desperate guerrilla-type battles with varying degrees of success would be fought behind the main battle fronts, but in close co-ordination with the aims of the higher commands which would themselves be constantly under threat of attack and in need of protection. Essential to retaining air superiority would be the AWACs aircraft (Airborne Warning and Control stations) which, from altitude, would scan deeply into enemy territory and monitor even the landing and taking off of aircraft. First used in the Indo-Pakistan war of 1971, AWACs would themselves be prime targets for attack and invariably the central element in air battles fought at maximum range — each AWACs' information being used to vector fighters to its own protection as well as for operations in general.

Battle honours would perhaps be fairly even. The Russians would benefit from their central position, tighter security within their own territory and larger numbers. The West would hold the advantage in radar, electronics and training, allied to an imaginative aiming of cruise missiles and homing missiles effectively to crater airfields and destroy bridges with special non-nuclear warheads. Inevitably, as the attenuation of early warning systems and base installations progressed, the efficiency and numbers of high-performance aircraft would decline, and not only from losses in combat. This would give greater scope to lower-performance fixed-wing machines and helicopters working closely with ground forces and would enable, too, the continued flying in of reinforcements to threatened spots — above all from the USA to Europe as the armies locked in battle.

Whenever land forces clashed, artillery, missiles and AFVs would hold the stage and frequently cancel each other out by the sheer accuracy with which they could tackle each other. It is probable that losses inflicted upon vehicles would be extremely high but, to confound those who persist in claiming that AFVs could not survive in the face of the latest high-velocity fin-stabilized shot, fire-and-forget missiles, air-delivered bomblets and mines, enough would continue in action. Wherever a well-armoured tank, or even a light APC or a self-propelled gun held its ground, infantry would have to pause and call for help from AFVs and missiles. Only because AFVs exist in enormous numbers could the generals afford to sacrifice them. And soon it would be noticeable that, as among some aircrew, if losses were extravagantly high, shaken tank crews would become over-cautious. The anticipated onrush of armoured hordes would slow to

Combat Tiers Of The 1990s

Diagrammatic representation of the tiers in which future combat could take place, bearing in mind that nothing is arbitrary, that conditions will vary and that one tier will often overlap another.

Above: *The Boeing E-3 AWACS now in service with NATO.*

Above all else in strategic and tactical priority is the struggle for control of dominant heights. Hence the need to establish and defend reconnaissance satellites against predatory shuttles and killer satellites sent to destroy them in the upper stratosphere; or to prevent the shooting down of surveillance aircraft, such as U 2, in the lower stratosphere. Here (as also on page 199) AWACS control the entire troposphere battle while, between 1000 ft and 3000 ft, a slower control aircraft watches over engagements at ground level. Upon the victory of the fighters between 3000 ft and 30 000 ft depends the survival of aircraft engaged in attacking airfields, upon logistic systems, radar and vehicles at the forward edge of the battle area. Land forces operate under threat of attack from MCRA and helicopters armed with fire-and-forget SMART missiles; while they themselves, with radar-assisted SA missiles, pose a threat to slower, low flying attackers.
In this hostile, ground-level environment, reconnaissance is of traditional importance and carried out by every possible optical or electronic means. Remotely controlled drones have an important part to play here, to avoid sacrificing far more costly, manned aircraft on suicide missions.
While it is true that air power does not absolutely dominate land operations, there is no doubt that modern airborne surveillance equipment makes the achieving of surprise extremely difficult. It therefore poses the need for extensive electronic counter measures and 'spoofing' in an ether flooded by radio transmissions on all frequencies.

1. Killer satellite
2. Reconnaissance satellite
3. Shuttle with high energy weapons, or missile, stalking shuttle
4. Shuttle
5. U2 reconnaissance at 65 000 ft
6. F15 launching anti-satellite missile
7. AWACS
8. Top cover fighter combat
9. Controlling aircraft
10. Multi-role combat aircraft aim for ground targets
11. Anti-radar aircraft
12. Attack helicopter
13. A10 ground attack aircraft
14. Remotely controlled reconnaissance drone
15. Armoured combat team protected by radar-controlled missiles and guns

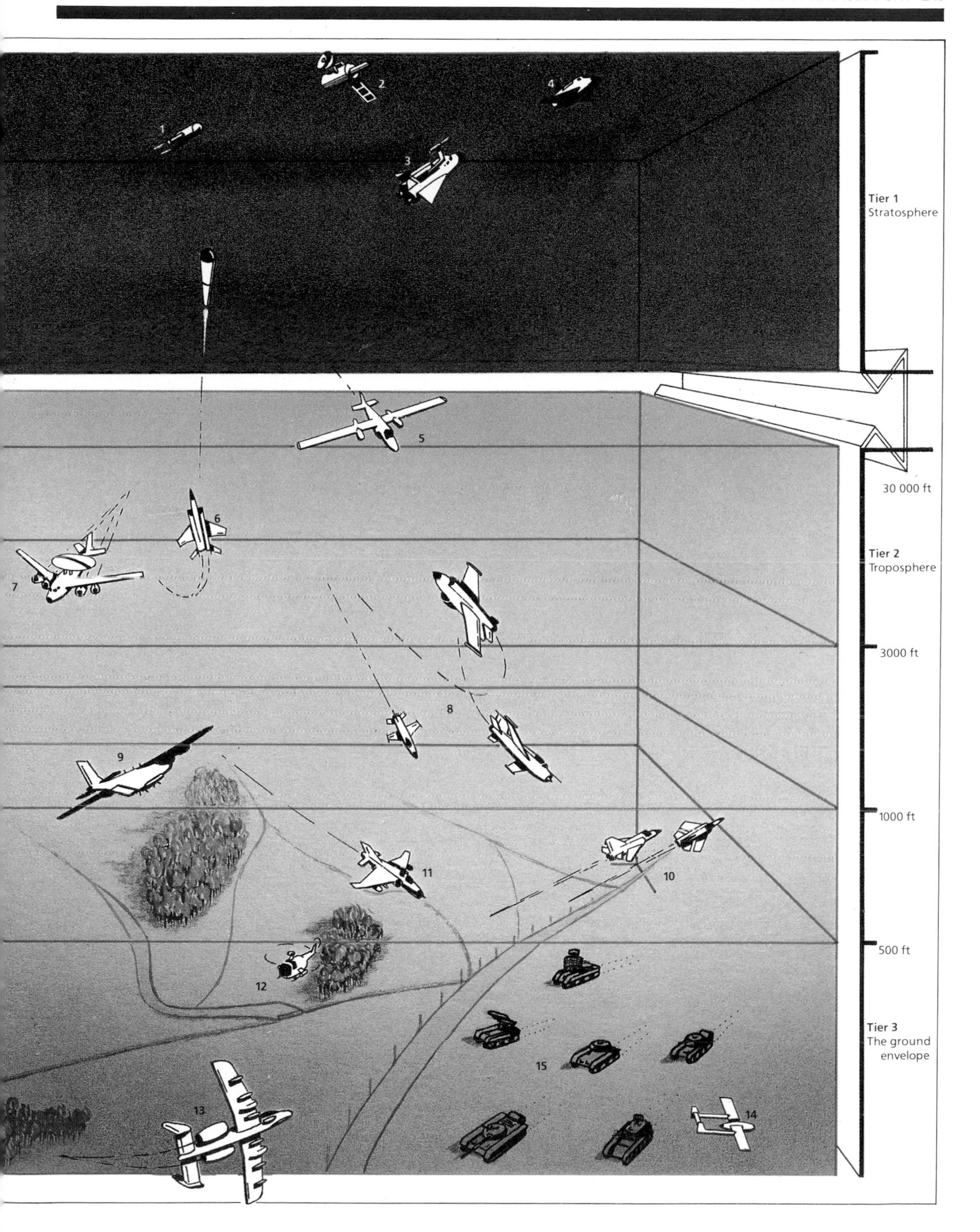
1
2
3
4
5
6
7
8
9
10
11
12
13
14
15
Tier 1
Stratosphere
30 000 ft
Tier 2
Troposphere
3000 ft
1000 ft
500 ft
Tier 3
The ground
envelope

Above: ***US Air Force A10 Thunderbolt, a ground attack aircraft. This highly manoeuvrable machine, with considerable armour, is capable of carrying a vast variety of stores for ground attack against all targets, including tanks.***
Right: ***US Army's Bradley infantry fighting with TOW guided missile. The turret mounts a 25-mm gun and full night fighting equipment.***

something more approaching the nature of a shuffle.

Nowhere would this be more evident than in Germany, where the collisions of armies would only happen gradually. Contrary to long-held expectations, the Russians may not indulge in a headlong drive to the west. Instead, by relentless pressure against sensitive points they may well try to lure the NATO armies into a Verdun-like attritional contest which, assuming equal losses, should logically go the way of their more numerous, though less efficient, AFVs. They may reason that the advantages of defence against their massed armoured forces will not far outweigh those of attack; that vehicles moving about in the open would be far easier prey than those of NATO, which choose to fight from cover within a carefully considered mobile defensive scheme. It may also be realized that the enormous spread of urbanization has converted terrain which had previously been ideal for offensive mobility into quite the opposite. In breadth and depth the defenders could establish interlocking bastions based on rivers, forests and townships which would narrow down the open terrain to be guarded and make it almost impossible for an attacker to launch a long-range penetration untrammelled by fear of hidden forces in his rear which would cut his vulnerable lines of communication.

High-velocity guns and missiles would dominate. Ground attack aircraft of all kinds would be compelled to hit and run, rarely daring to loiter in combat areas and often at peril within their own territory from helicopter-infiltrated hunting parties. In fact, aircraft might win only marginal success at high cost against hard point targets in the battle zones. They would probably achieve their best results, as so often in the past, by attacks on command centres and logistic targets, thus justifying those airmen who have always declared it wasteful to endanger valuable aircraft against targets which could be more economically and effectively engaged by ground units. Supply services would be hard hit and hard pressed to function. Air reconnaissance, particularly by helicopters and remote-control drones, fitted with an assortment of sensing and surveillance equipment, would provide most of the vital information upon which ground force commanders would depend for their decisions. This too would give an edge to the West, whose better trained and equipped forces would benefit most from these facilities. When mechanized or helicopter-borne troops massed or made a dash into the open, they would rarely escape retribution unless the means of surveillance had previously been eliminated — and this would be difficult. And even if small-scale manoeuvres were attempted, they would often be detected electronically, both by day and night, to draw upon themselves a deluge of fire which would make aggressors reconsider as fears of escalation of the conflict multiplied. For as frustration supervened, recourse to chemical weapons would be made on some fronts (to the far greater harm of civilians than well-clad soldiers, some of whom would be safe within sealed air-conditioned vehicles and buildings) and low-yield nuclear strikes would be launched against major missile-launching bases, located in under-populated hinterlands, to forestall a massive launching of missiles into space. These warning shots could not be ignored.

The Fulcrum

If neither side could win a clear-cut superiority within a relatively short period by the use of the three *Options*, those in charge would either have to call off the war or commit the mass of stagnant nuclear weapons in the hope of breaking the impasse and cracking enemy resolve — but at the risk of Eternal Winter. With this the projection closes, leaving the reader with the hopeful suggestion, amid the horror, that some sort of progress in the relationship between peoples must be gained if Eternal Winter is to be averted.

Dismissing as most unlikely the theories of those who suggest that, in the aftermath of defeat, one or other of the monolithic contenders would collapse (if only because the apparatus of technologically controlled political systems is not easily disrupted) it is suggested that a modification of the existing balance of power, alloyed with a movement towards a realistic merging of interests, might take place. If it does, the chances of peace with disarmament are improved. If not, mankind will pursue its natural combative technological bent, ever seeking all-powerful weapons.

Rarely, prior to the First or the Second World Wars, were accurate predictions made of the course of conflict, not even by those in possession of the majority of the data. For example, the underestimation of land weapon defences before 1914 and the overestimation of the offensive power of air weapons before 1939 led to serious miscalculations, unfounded assumptions and mistaken ambitions. Certainly no claim is made here as to the absolute validity of the above notions, which are founded only upon information made public. No doubt revolutionary weapons of staggering impact are under consideration and development in secret. Maybe among them are one or two which can be produced rapidly and relatively cheaply (as was radar) to change radically the outlook overnight.

It is naturally important that war should be avoided, or at least contained. Technology itself is becoming too strongly over-persuasive to be denied except by technology. There is a danger that, with increased automation, 'thinking' computers which originate and plan by the combination of an almost infinite number of random impulses (in the manner of the human brain) will take over affairs with which only human judgement can reasonably be trusted. That way we could begin to lose control of our destiny as the initiation of war for purely technical and logical reasons becomes more likely.

Index

Picture Credits

Page	Credit
6	Mary Evans Picture Library
9	TRH/US Navy
10	MEPL
11	MEPL
14T	MEPL
14B	MEPL
16T	BBC Hulton Picture Library
16B	MEPL
20	MEPL
21	National Army Museum/TRH
24	Sjofarts Museet–Gothenburg
26	US National Archives/Grub Street
30	TRH
31	MEPL
34	Royal Navy Museum/TRH
35	Grub Street
36	Grub Street
36/7	Grub Street
39T	MEPL
39B	Grub Street
41	MEPL
44L	Marconi/TRH
44R	BBC Hulton Library
46	Imp War Museum
47	MEPL
49T	RAF Museum/TRH
49B	RAC Tank Museum/TRH
51	TRH/Smithsonian
53	TRH/US Navy
52/3	TRH
55	MEPL
56	TRH/US Navy
58/9	National Maritime Museum/TRH
61	TRH/Belgian Air Museum
65T	MEPL
65BL	IWM/TRH
65BR	IWM
70	TRH/Belgian Air Museum
75	National Army Museum/TRH
76	IWM
77	IWM
79	MEPL
84	BBC Hulton
86	MEPL
89	BBC Hulton Library
90	Grub St/IWM
95	Grub St/National Archives
98	RAC Tank Museum/TRH
99	RAC Tank Museum/TRH
100	BBC Hulton Library
100/1	BBC Hulton Library
102	BBC Hulton Library
103	IWM/TRH
105T	Associated Press
105BL	Science Museum
105BR	Vickers
106	Vickers/TRH
110	Robert Hunt/IWM
113	TRH/IWM
114	Robert Hunt
116/7T	Robert Hunt
117	IWM/TRH
116/7B	TRH/US Air Force
119T	BBC Hulton Library
119B	IWM/TRH
120/1T	TRH/RAF Museum
120/1B	TRH/IWM
121	J C Scutts
122	TRH/IWM
126	BBC Hulton Library
127L	Grub St
127R	Grub St
129	TRH
130T	Robert Hunt
130B	French Railways CNET
133	RAF Museum/TRH
134/5	K Mackey/USAF
136/7	US National Archives/MARS
137T	MARS/IWM
137M	David Tytler/MARS
137B	MARS
138/	TRH/RAC Tank Museum
141	Novosti
142/3	TRH/US Navy
144	Yasuholzawa
148/9	US Navy/TRH
150/1	IWM/TRH
152/3	Grub St
154/5	TRH/IWM
156T	TRH/IWM
157M	John Foreman
157B	John Foreman
159	Chaz Bowyer
162	Grub St/Public Records Office
164	Quadrant Picture Library
165	Grub St
165B	Grub St
168L	Grub St
168R	Grub St
170TL	Fairchild (Col. R F Toliver)
170BL	US Marine Corps/Carina Dvorak
170/1T	Carina Dvorak/National Archives
170/1B	MARS
173	TRH
174	Carina Dvorak/Dept of Defense
176/7	BBC Hulton Library
177T	Malcom Passingham
177B	US Air Force/TRH
181	TRH/Dept of Defense
185	Ken Gatland
187T	TRH/US Navy
187B	TRH/US Navy
189	TRH/US Air Force
191	TRH/RCA missing
192/3	TRH/US Air Force
196	K Mackey
196/7T	K Mackey
196/7B	K Mackey
198	Tass
201T	Min of Defence via API
201B	Press Association
202	TRH/US Air Force
202/3	Dept of Defense/TRH
203T	Dept of Defense/TRH
203M	Dept of Defense/TRH
203B	Dept of Defense/TRH
204	TRH/NASA
204/5	TRH/US Air Force
208	TRH/McDonnell Douglas
210	TRH/USAF
211	TRH/NASA
212	TRH/US Air Force
213T	TRH/NASA
213B	Grub St/NASA
216	TRH/US Navy
220T	US Air Force?
220B	FMC/TRH

TRH–The Research House
MEPL–Mary Evans Picture Library